MICROBIOLOGY OF TROPICAL SOILS AND PLANT PRODUCTIVITY

Developments in Plant and Soil Sciences

Volume 5

Series ISBN 90-247-2405-8

Microbiology of Tropical Soils and Plant Productivity

edited by
Y. R. DOMMERGUES *and* H. G. DIEM

1982

MARTINUS NIJHOFF / DR W. JUNK PUBLISHERS
THE HAGUE/BOSTON/LONDON

Distributors:

for the United States and Canada
Kluwer Boston, Inc.
190 Old Derby Street
Hingham, MA 02043
USA

for all other countries
Kluwer Academic Publishers Group
Distribution Center
P.O. Box 322
3300 AH Dordrecht
The Netherlands

Library of Congress Cataloging in Publication Data
Main entry under title:

Microbiology of tropical soils and plant
    productivity.

    (Developments in plant and soil sciences ;
v. 5)
    Includes index.
    1. Soil microbiology--Tropics. 2. Micro-
organisms, Nitrogen-fixing--Tropics. 3. Crops
and soils--Tropics. 4. Soils--Tropics. 5. Optimal
soil fertility--Tropics. I. Dommergues,
Y. R. (Yvon R.) II. Diem, H. G. III. Series.
QR111.M4   631.4'6   81-22536
                    AACR2

ISBN-13:978-94-009-7531-6   e-ISBN-13:978-94-009-7529-3
DOI: 10.1007/978-94-009-7529-3

Foreword

It is an established fact that we must continually increase and improve agricultural production if we are to meet even the minimum requirements of a growing population for food, shelter, and fuel. In recent years, the introduction of new plant varieties and the extensive use of fertilizers have effectively increased crop yields, but intensifying agricultural methods has often led to depleting soil fertility. Two examples of the harmful consequences of intensive farming practices are the loss of up to 2.5 cm of topsoil every 15 years in the United States through erosion and the alarming rise in environmental pollution through widespread use of pesticides. Countless other processes affecting the activity of soil microflora and the interactions between microorganisms and plants may pose an equal danger to soil equilibrium, but their potential hazards are often overlooked because of an insufficient understanding of soil microbiology on the part of scientists.

In the first published study of its kind, the authors of this book have attempted to address major aspects of the microbial activity of soil in the tropics. Tropical conditions serve as an ideal context for a discussion of soil microbiology, since biological processes in the soil are particularly active in tropical environments in comparison to other settings and in relation to physical and chemical processes.

The book's essays represent contribution from a variety of schools of thought and encompass the major concepts in the field of soil microbiology. Our purpose will have been fulfilled if these discussions help to stimulate and quicken the development of farming techniques designed to regulate the interactions between soil microorganisms and plants in order to increase plant productivity while maintaining or increasing potential soil fertility.

Y.R. DOMMERGUES
H.G. DIEM

Contents

List of contributors

Baldensperger, J., ORSTOM, 24 rue Bayard, 75008 Paris, France.

Becking, J.H., Institute for Atomic Sciences in Agriculture, 6 Keyenbergseweg, Postbus 48, Wageningen, The Netherlands.

Diem, H.G., Laboratoire de Biologie des Sols, CNRS/ORSTOM, B.P. 1386. Dakar, Senegal.

Dommergues, Y.R., Laboratoire de Biologie des Sols, CNRS/ORSTOM, B.P. 1386, Dakar, Senegal.

Dreyfus, B.L., Laboratoire de Biologie des Sols, ORSTOM, B.P. 1386, Dakar, Senegal.

Freney, J., Division of Plant Industry, CSIRO, P.O. Box 1600, Canberra City, A.C.T. 2601, Australia.

Garcia, J.L., ORSTOM, 24 rue Bayard, 75008 Paris, France.

Ganry, F., Institut de Recherches Agronomiques Tropicales, ISRA (IRAT), Bambey, Senegal.

Gianinazzi-Pearson, V., Station d'Amélioration des Plantes, INRA, BV 1540, 21034 Dijon, France.

Gibson, A.H., Division of Plant Industry, CSIRO, P.O. Box 1600, Canberra City, A.C.T. 2601, Australia.

Jacq, V., Laboratoire de Biologie des Sols, ORSTOM, B.P. 1386, Dakar, Senegal.

Redhead, J.F., Faculty of Agriculture and Forestry, University of Dar Es Salam, P.O. Box 643, Morogoro, Tanzania.

Reynaud, P.A., Laboratoire de Biologie des Sols, ORSTOM, B.P. 1386, Dakar, Senegal.

Rinaudo, G., Laboratoire de Biologie des Sols, ORSTOM, B.P. 1386, Dakar, Senegal.

Roger, P.A., Laboratoire de Biologie des Sols, ORSTOM, B.P. 1386, Dakar, Senegal.

Tiedje, J.M., Departments of Crop and Soil Sciences, and of Microbiology, Michigan State University, East Lansing, MI 48824, USA.

Watanabe, I., The International Rice Research Institute, Los Baños, Laguna, Philippines.

Wetselaar, R., Division of Land Use Research, CSIRO, P.O. Box 1666 Canberra City, A.C.T. 2601, Australia.

Yoshida, T., Institute of Applied Biochemistry, The University of Tsukuba, Ibaraki-ken, 300-31, Japan.

1. Nitrogen balance in tropical agrosystems

R. WETSELAAR and F. GANRY

1. Introduction

Nitrogen is an essential element for the growth of crops. In the last two decades, it has been considered a key element in attempts to increase the production of food crops to keep pace with the growth of the world's population. Problems associated with these attempts include the high energy demand incurred in the industrial fixation of fertilizer nitrogen, and the inefficiency of fixed nitrogen when used for the increase of crop production. This inefficiency is mainly due to nitrogen losses from the agrosystem to which it is applied, and these are, for the greater part, the direct or indirect result of microbiological processes in the soil. In tropical agrosystems the losses can be high because most microbiological processes are, to a certain limit, accelerated under high temperature conditions [99]. In addition, nitrogen loss due to microbiological reduction (denitrification) is enhanced under the extremely wet soil conditions associated with flooded rice fields, one of the most important agrosystems in the tropics.

In natural ecosystems, nitrogen losses (outputs) are generally fully compensated for by nitrogen gains (inputs). Interference by man, e.g. through the imposition of an agrosystem, disturbs this balance and may cause a decrease in soil nitrogen status. This reduces agricultural production unless it is compensated for by 'artificial' means such as nitrogen fertilizers. However, a decrease in soil nitrogen status is also invariably the reflection of a lowering in soil organic matter content, leading to lower rates of water infiltration and lower water holding capacity, and making the soil more vulnerable to erosion losses. A continuous decrease in soil nitrogen content over time will therefore ultimately result in agriculturally uneconomic land. Thus, changes in soil nitrogen content are sensitive indicators of the balance between inputs and outputs of nitrogen and of the stability or otherwise of an agrosystem. As a result, microbiologists, soil scientists, agronomists and ecologists are concerned with the assessment of nitrogen balances in different agrosystems. Such assessments are particularly appropriate in the tropics because of the instability of soils under annual dryland conditions.

In this chapter we provide an overview of the different inputs and outputs of the soil, plant, and animal pools of tropical agrosystems, and of the changes that can take place in the nitrogen contents of these pools.

Y.R. Dommergues and H.G. Diem (eds.), Microbiology of Tropical Soils and Plant Productivity. ISBN 978-94-009-7531-6.
© *1982 Martinus Nijhoff/Dr W. Junk Publishers, The Hague/Boston/London.*

2. The three main nitrogen pools

In any given agrosystem, the total N can be distributed over the three main pools: soil, plant, and animal (Fig. 1). The N in the soil is mainly in the organic form (between 95 and 100% of the total N). Its concentration is generally highest in the topsoil, reflecting the fact that nearly all N inputs enter the soil through the surface. The total amount of organic N in this layer can be as low as $800\,kg\,ha^{-1}$ and as high as $10{,}000\,kg\,ha^{-1}$.

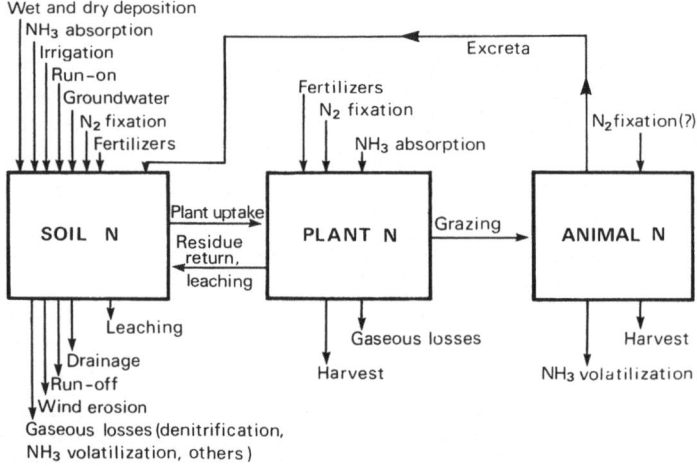

Fig. 1. The three main nitrogen pools and their inputs and outputs.

Mineralization of organic N, a microbiological process, results in temporary accumulation of inorganic N, mainly in the nitrate and partly in the ammonium form. The mineralization coefficient, i.e. the percentage soil organic N in the top-soil that is mineralized per growing season, is generally higher in the tropics (about 5%) than in temperate climates (1–3%) [87]. However, the size of this sub-pool fluctuates strongly over time because of crop uptake of this mineral N. If we assume for the tropics an organic N content of 0.07% in the top 15 cm of a soil with a bulk density of 1.43 and a mineralization coefficient of 5%, then the inorganic N sub-pool can be as high as $70\,kg\,N\,ha^{-1}$, assuming no residual accumulation of inorganic N from the previous season.

The plant N pool of an annual crop at sowing is equal to the amount of N contained in the seed. From then on, the pool size increases sigmoidally over time, and reaches a maximum at flowering (Fig. 2). In the case of a non-leguminous crop, the N in the plant pool can originate from mineralized soil organic N or from fertilizer N applied to the soil. In the case of a leguminous crop, a substantial percentage of the total N in the plant can originate from symbiotic fixation of N_2.

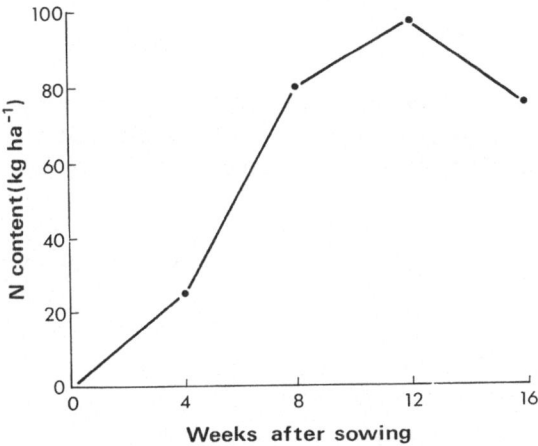

Fig. 2. Time course of plant N content (mean of 6 different fodder crops) in northwest Australia, for the wet season 1956–57, with 105 kg N ha^{-1} applied as ammonium sulphate [60].

Where grazing occurs, the size of the animal N pool will depend on the type of animal and the stocking rate. For instance, under the extensive beef cattle grazing conditions in northern Australia, the animal N pool size can be as low as 0.2 kg N ha^{-1}. On the other hand, using improved pastures, its size can be increased to about 10 kg N ha^{-1} at certain times of the year.

The net changes in the pool sizes (ΔN) can be determined directly or indirectly. In the direct method the amount of N in each pool is assessed at times 0 and t, and ΔN is calculated according to

$$\Delta N = N_t - N_0 \tag{1}$$

For the plant N pool, representative samples can be obtained easily at different times. For the animal N pool, $\Delta N/\Delta t$ is generally small and a representative sample is difficult to obtain owing to the heterogeneity of the materials that constitute this pool. Similarly, for the soil N pool the changes in ΔN per unit time are also small in relation to the pool size; a 5% change per year is high. This is complicated even further by the enormous spatial variation, often more than 15%, in the total N content of the plough layer. Only by increasing t to 3–5 years or more is it possible to get a reasonable assessment of $\Delta N/\Delta t$. An additional complication in the assessment of ΔN of the soil N pool is the possibility of change in the bulk density of the topsoil. The way to overcome this complication is not simple, but an elegant solution has been proposed [38], by (i) sampling to two soil depths, (ii) measuring bulk density and percentage N for all samples, (iii) estimating the weight of accumulated organic matter, and (iv) adjusting the results by a proposed iterative procedure.

Alternatively, the changes in the different N pools can be assessed according to the equation

$$\Delta N = \sum_0^t \text{inputs} - \sum_0^t \text{outputs} \qquad (2)$$

which assumes that all inputs and outputs of the pool concerned can be measured. This is very rarely the case, especially when effluxes of N_2 are involved. One way to determine such an efflux, or unknown output, would be by combining equations (1) and (2) thus:

$$\sum_0^t (\text{unknown outputs}) = N_0 - N_t + \sum_0^t (\text{inputs}) - \sum_0^t (\text{known outputs})$$

$$(3)$$

In this case, the size of the pool is determined at times 0 and t and all known inputs and outputs are monitored. Such an approach can be useful when using an ^{15}N-labelled input, if all but one output can be measured.

3. The soil N pool: inputs and outputs

3.1. Inputs

3.1.1. Wet and dry deposition. Nitrogen can be added to the soil from the atmosphere through rain, snow and hail (wet deposition) or dust and aerosols (dry deposition). Few if any estimates of dry deposition have been made in tropical areas, and most deposition estimates must be regarded as a combination of wet and dry ones (Table 1). In such cases, contamination of the deposition collector probably occurred and the data must be interpreted with caution. Furthermore, organic N is seldom included in rainwater analyses, resulting in gross underestimation of the contribution of wet deposition to the soil N pool.

Table 1. Nitrogen content in rainwater for different tropical countries (in many cases, the N contained in dry deposition is likely to have been included)

Country or Region	N (kg ha^{-1} yr^{-1})	Reference
Vindhyan plateau, India	23	Singh and Pandey [71]
Chiang Mai, Thailand	13[a]	Brown [9]
West Malaysia	3–10	Samy and Vamadevan [68]
West Malaysia	20–38	Pushparajah [66]
West Java, Indonesia	7.5	Brotonegoro et al. [8]
Northwest Australia	1[b]	Wetselaar and Hutton [93]

[a] Wet deposition only; includes organic N.
[b] Wet deposition only; does not include organic N.

On a regional scale, transfer of terrestrial N can occur from one ecosystem to another through wind erosion and subsequent deposition. The absence of reliable estimates of the deposited N can therefore affect the outcome of a nitrogen balance made for an ecosystem. On a global scale, however, the only true accession of N through deposition is the part contributed by electrical discharge, the amount being only in the order of 0.1 to 0.2 kg N ha^{-1} yr^{-1} [78, 90, 93].

3.1.2. Ammonia absorption. The concentration of ammonia N in the atmosphere immediately above the soil surface varies from 1 to 2 μg m^{-3} in unpolluted areas to 200 μg m^{-3} close to areas with industrial activity. Ambient ammonia concentrations are higher in the tropics than elsewhere [51]; in general the residence time of this ammonia, mostly only days, will depend on the rainfall frequency.

Absorption of atmospheric ammonia has been shown to occur for a wide range of soil types. Malo and Purvis [52] found this absorption to range between 25 and 100 g ha^{-1} d^{-1} under field conditions. Under controlled temperature and atmospheric ammonia concentrations, Hanawalt [35] found an average rate of absorption of 240 μg of NH$_3$-N 100 g^{-1} soil d^{-1} at 20 °C at an ammonia concentration of 38 μg N m^{-3} for six soils differing in clay content, organic matter and cation exchange capacity, and with a pH range of 5.2 to 6.7. His studies indicate that the sorption rate increases markedly with ammonia concentration and moderately with temperature. According to Malo and Purvis [52] atmospheric ammonia absorption might explain in part why maize yields were maintained in the absence of nitrogen fertilizers in long-term experiments in New Jersey, USA.

In conclusion, the phenomenon is likely to be of importance only for soils of low pH, in the neighbourhood of high industrial activity, and in a warm climate.

3.1.3. Irrigation, run-on and groundwater. Irrigation water may contain nitrogen, and this is of particular importance for flooded rice fields, where substantial amounts of water are added to the soil to maintain the required ponding conditions. Estimates of the N added in this way to the soil range from 6 to 16 kg ha^{-1} crop^{-1} [91].

Run-off from these fields either drains directly into a water way (creek, river, lake, or canal) or runs over less-elevated land. In the latter case, nitrogen in the run-on water might contribute to the soil N content, depending on the speed at which it moves over the land, its state (in solution or solid) and on particle size if solid. To our knowledge, such a contribution to the soil has not been documented.

Some paddy fields in northern Thailand [9] and some grey soils in Senegal [69] are underlain by relatively permeable soil layers, which may conduct large quantities of water. Such shallow aquifers are capable of transporting measurable quantities of dissolved nitrogen leached from the soil surface. This contribution to the soil N pool will largely depend on both past and present use of the land from which the nitrogen originated. For northern Thailand, Brown [9] estimates a maximum contribution of less than 0.1 kg N ha^{-1} yr^{-1}.

3.1.4. N_2 fixation. This input to the soil can take place through phototrophs, mainly cyanobacteria [74], and heterotrophs. The former may play an important role under flooded conditions such as prevail in a paddy field. Estimates [91] of their contribution to the soil vary from 1.5 to 50 kg N ha^{-1} crop^{-1}. However, when algae are in symbiosis with the water fern *Azolla*, their contribution may be much greater. Thus, 22 crops of the *Azolla-Anabaena* complex may accumulate up to 450 kg N ha^{-1} over 330 days [80]. In China and Vietnam, this complex is used as a green manure for rice [81].

For heterotrophs not associated with the rhizosphere, the main limiting factor is available carbohydrate in the soil. Consequently, the return of plant materials such as straw tends to increase the amount of N fixed by heterotrophs [53]. In a rice bay the amount of N fixed by heterotrophs ranges from 6 to 25 kg ha^{-1} depending on the amount of straw returned [91].

Evidence is accumulating to support the view that N$_2$ fixation in association with the roots of some Gramineae can contribute, though moderately, to the nitrogen economy of the host plant. In Australia, a fixation rate for *Sorghum plumosum* of 136 g N ha^{-1} d^{-1} was measured under laboratory conditions [82]. Nitrogenase activity in the rhizosphere of rice is about one-tenth of the activity of symbiotic systems [21, 22]. Assuming for the latter a rate of 100 kg N ha^{-1} season^{-1} [88], an associative N$_2$ fixation of about 10 kg N ha^{-1} season^{-1} can be expected. Based on integrated acetylene reduction values, Döbereiner and Boddey [19] estimated a maximum fixation rate for rice of 4.6 to 8.4 kg N ha^{-1} season^{-1}.

3.1.5. Fertilizers. In the tropics, organic fertilizers continue to play an important role in supplying the soil with nitrogen. An outstanding example of this occurs in China where two-thirds of the annual input of fertilizer N (ca 100 kg ha^{-1}) is applied in organic form, mainly as animal and human wastes and plant residues [73]. However, in the tropics as a whole there is a trend towards increased use of industrially-produced nitrogen fertilizers, partly due to the need to supplement the 'natural' organic fertilizers, and partly because some tropical countries, especially in southeast Asia, have found major sources of natural gas that can support NH$_3$ production (P.J. Stangel, personal communication).

The amount of fertilizer N applied per crop ranges from 4 kg ha^{-1} for rice in Burma to 300 kg ha^{-1} for sugarcane in tropical Australia. In some rice growing areas such as Java in Indonesia, where nitrogen-responsive varieties are used, the recommended rate is 90 kg N ha^{-1} crop^{-1}, but on average the farmers apply only about half of this. For a perennial crop such as rubber a total of 1500 kg ha^{-1} of fertilizer N is applied over the 30-year life of the tree in peninsular Malaysia, but the use of legumes reduces this to one-third [66].

In general, the amount of fertilizer N applied in the tropics is dependent on biological factors (e.g. land history, type of crop), physical factors (e.g. climate, access roads for fertilizer transport), and socio-economic factors (e.g. price ratio of grain to fertilizer [41], local customs).

3.1.6. Return of plant residues, plant leaching and animal waste products. These will be discussed in Sections 4.2.2. and 5.2.1.

3.2. Outputs

3.2.1. Run-off, leaching and drainage. When free water makes contact with the soil surface, certain fractions of the soil nitrogen may be dissolved. If the supply of water is greater than the rate of infiltration into the soil, this water will temporarily pond on the soil surface or run off. In the latter case, any soluble nitrogen in the water that originated from the soil is lost from the soil N pool. In addition, solid soil fractions can be removed as erosion products with the run-off water when the kinetic energy of the rain droplets at the time of impact on the soil surface is high enough to disperse soil aggregates. A dense plant cover over the soil surface absorbs much of the energy of the rain droplets and effectively reduces the loss of N through erosion (Table 2). The paucity of plant cover may lead to the removal of at least 5 t ha^{-1} during one season [11]. This is equivalent to a loss of 3.5 kg N ha^{-1}, assuming a soil N concentration of 0.07%.

Table 2. Estimate of nitrogen losses through erosion as a function of the plant cover, at the Séfa Agronomic Station, Senegal [11]

Plant cover	Erosion[a] (kg N ha^{-1})		
	Minimum	Maximum	Weighted mean
Forest	traces	< 1	< 1
Bush fallow	2	10	5
Crops	2	19	7
Bare soil	6	54	21

[a] A soil N content of 0.1% has been assumed.

Loss of N through run-off can be substantial in the wet tropics. Further, marginal land is continuously being opened for agriculture because of increasing population pressure, and such land is often steeply sloping. On land permanently under flooded rice conditions, overflow of excess rainfall or irrigation water can lead to a maximum loss of 1 kg N ha^{-1} $crop^{-1}$ in the absence of nitrogen fertilizers [91]. When a soluble fertilizer is broadcast into the floodwater, nearly all the applied N will be dissolved in the water in the first few days [98]. Substantial nitrogen losses can occur in that period should excess water lead to overflow of the paddy field.

The movement of water from one paddy field to another is a planned feature in many rice-growing areas, and under such conditions the loss of fertilizer N from one field due to overflow is the gain of this N to the adjacent area at a lower level. The total overflow loss of fertilizer N from the irrigation scheme as a whole will depend mainly on the flow rate of the water in relation to the total length of the slope, the type of fertilizer, and edaphic characteristics such as cation exchange capacity and pH.

Once water starts to infiltrate into the soil, only negatively charged soluble nitrogen fractions (principally nitrate and nitrite) will move down the soil profile by convection (leaching), due to the mass flow of the soil solution and, to a much lesser extent, by diffusion due to concentration gradients. The movement of positively charged ions (principally ammonium) is restricted by the cation exchange capacity of the soil [86]. Leaching of the gaseous nitrogen fraction nitrous oxide has recently been shown to be of importance [23]. This fraction is a product of denitrification, a process of particular importance under wetland rice conditions.

In general, the rate of leaching is highly dependent on rainfall amount [85], and to some extent its intensity; therefore this loss pathway is of particular importance in the humid tropics. Once a leached nitrogen fraction has been moved permanently beyond the reach of plant roots, it may be regarded as lost. This is likely to occur sooner in a shallow soil profile than in a deep one. In the latter case, the nitrogen can be taken up at a later stage by deep-rooting crops such as pearl millet [95]. In general, soil depth can have a pronounced effect on the prevention of N losses.

Where artificial drainage systems have been installed at a shallow depth, any nitrogen contained in the drainage water will be a loss to the soil, i.e. a drainage loss is comparable to a leaching loss from a shallow soil profile. Under the artificially induced leaching conditions, zero or positive soil water potential is required for actual drainage to occur, i.e. anaerobic conditions conducive to denitrification may prevail under such circumstances. Indeed, Dowdell *et al.* [23] have shown that drainage water may contain the denitrification product nitrous oxide at concentrations that are one or two orders of magnitude higher than in rainwater.

The actual amount of nitrate moved down the soil profile is largely a function of the total amount of organic and fertilizer N in the topsoil and the rate at which they are nitrified. Under high temperature conditions, as in the low-altitude tropics, this nitrification can be rapid when soil moisture is adequate [54]. In addition, the mineralization rate of organic N is increased when a period wet enough for nitrification is preceded by a long dry period [6]. This means that in the monsoonal and semi-arid regions of the tropics nitrate concentration in the topsoil can increase rapidly at the beginning of the wet season. When there are no plant roots to take up nitrate, it can be leached into the sub-soil. Subsequent rains may move this nitrate permanently below the root zone. In northern Australia, one quarter of the N applied as ammonium sulphate to a sorghum crop was found below the root zone in a clay loam at the end of the growing season, after an excessive wet season with 1170 mm of rainfall [86]. In the sandy soils of the Sudan-Sahelian zone with a rainfall of only 400 to 550 mm yr^{-1}, leaching losses have been estimated to vary between 15 and 50 kg N ha^{-1} yr^{-1} [63].

3.2.2. Ammonia volatilization. This process requires (i) presence of a source from which ammoniacal N can be formed, (ii) enhancement of the processes that lead to the formation and accumulation of ammoniacal N, and (iii) conditions for a low NH_4^+ to NH_3 ratio.

In the absence of nitrogen fertilizers or organic additives, the main N source in the soil is soil organic N. Under most dryland conditions, ammoniacal N formed from this source does not accumulate in the soil, because the rate of nitrification of the ammoniacal N is almost invariably higher than the rate of ammonification [54]. Under anaerobic conditions, however, ammoniacal N can accumulate over time owing to the anaerobic mineralization of soil organic N and the absence of nitrification. In a flooded rice field, where such conditions prevail, Wetselaar *et al.* [98] found an ammonia volatilization loss of only 1.5 kg N ha^{-1}, presumably because accumulation of ammoniacal N in the soil could not occur due to plant N uptake. In practice, significant accumulation only occurs with the application of high amounts of materials that contain or generate ammoniacal N, such as certain fertilizers, animal urine and large quantities of organic matter.

The dominant factor influencing the NH_4^+ to NH_3 ratio is the pH of the soil solution, because it affects the equilibrium represented by equation (4) [30]

$$NH_4^+ + OH^- \rightleftharpoons NH_3 + H_2O \qquad (4)$$

The higher the pH, the lower is the ratio, and therefore the greater the potential for ammonia to be lost [30]. For example, at pH 6, 7, 8 and 9 the approximate ratios are 10, 1, 0.1 and 0.02 respectively.

Other factors that influence the relative proportions of NH_4^+ and NH_3 are titratable alkalinity, buffer capacity of the soil and the temperature of the soil solution [30, 79].

In practice, the pH of the floodwater in a paddy field can rise substantially, sometimes up to pH 10, during the middle of the day owing to the high CO_2 assimilation rate of algae [81]. In addition, the high temperatures of the low-altitude tropics increase the ratio of ammonia in the gaseous form to that in solution.

When other conditions are favourable for a high NH_3 to NH_4^+ ratio in the soil solution or floodwater, the flux of ammonia into the atmosphere is positively related to windspeed [18, 31].

When ammonium sulphate is applied to a flooded rice field in the tropics, the losses vary from 3 to 9% of the amount of N applied, depending on the mode of placement of the fertilizer [98]. Under dryland conditions, urea N losses through NH_3 volatilization of up to 40% have been recorded when applied to a crop of millet in Bambey, Senegal [32].

3.2.3. Denitrification. Nitrate and nitrite may be subject to dissimilatory reduction (denitrification) to the gaseous products NO, N_2O and N_2 through the action of biochemically and taxonomically diverse types of bacteria [47]. Denitrification is a major loss pathway of nitrogen in an agrosystem and particularly in tropical agrosystems, for two reasons. First, the high and intensive rainfall induces temporary anaerobiosis in soils. Under such conditions, the strongly denitrification-inhibiting factor O_2 is removed, and when other conditions such as pH and supply of organic carbon are favourable, denitrification can take place. Second, the high temperature in the low-altitude tropics stimulates gaseous loss through denitrification, because the reaction is strongly temperature dependent (between 10 and 35 °C the reaction has a Q_{10} of ca 2.0 [72] and the rate increases with higher temperature until a maximum is reached at 60 to 75 °C [47]).

Denitrification in an extreme form occurs in wetland rice agrosystems. In flooded soils, the layer immediately below the floodwater generally contains sufficient oxygen, obtained by diffusion through the floodwater or from the activity of photosynthetic organisms, for nitrification to take place. The nitrate so formed may diffuse or leach to deeper, anaerobic, layers where it may be denitrified. Consequently, when ammonium-type fertilizers are placed on or in the top few mm of a paddy field soil, low recoveries of the applied N are found. Since, in general, such losses can only occur if the fertilizer N has been nitrified, they can be reduced by deep-placement of these fertilizers in the anaerobic zone. In addition, such placement will reduce losses by ammonia volatilization, due to ammonium adsorbtion within the soil.

In general, periodic flooding alternating with aerobic conditions will stimulate losses through denitrification. Thus, Ponnamperuma [64] estimates a loss of about 26 kg N ha^{-1} from wetland rice soils in the Philippines due to nitrate accumulation in the soil between harvesting and replanting of rice followed by flooding. On the other hand, when a soil is well draining, the absence of anaerobic macrosites can preserve nitrate in the soil profile for at least four years [86].

Since most denitrifying bacteria are heterotrophs [47], the presence of organic compounds in the soil is important for their growth. Denitrification is stimulated by organic carbon that can be extracted from soil water [7, 10], and by root exudates and plant residues. Ganry et al. [33], in lysimeter studies in Bambey, Senegal, observed increasing N losses from ^{15}N-labelled urea applied to pearl millet due to the ploughing in of straw. Those losses were attributed to denitrification. When 90 kg ha^{-1} of urea-N was applied, the losses ranged from 39% with no straw incorporated, to 55% with 30 t ha^{-1} of straw incorporated; a similar loss increase was obtained from 150 kg ha^{-1} of urea-N.

The main problem with the estimation of denitrification losses is the virtual impossibility of measuring the gaseous losses directly under representative field conditions. Denmead et al. [17] estimated that only a very small fraction (1.4%) of the nitrate N lost from a flooded soil was in the N_2O form, the rest presumably being lost as N_2. As the latter already constitutes about 80% of the atmosphere, the addition of a small amount of N_2 to this particular pool through denitrification is extremely difficult to estimate. Consequently, most assessments of denitrification losses are based on the difference method as given in equation (3). In this way, losses due to denitrification in an unfertilized paddy field are estimated to be between 3 and 34 kg N ha^{-1} [91].

When using ^{15}N-labelled material, equation (3) can be replaced by

$$D_e = T - \sum \text{(known losses)} \tag{5}$$

In this case, all known losses are determined using the ^{15}N method, and it is assumed that the difference between these losses and the total loss (T) is due to denitrification (D_e).

Under field conditions, such fertilizer balances, as applied to an annual crop, are mostly determined at the end of a growing season when the crop has reached maturity. It is usually assumed that any nitrogen taken up by the plants remains there until maturity. Recently, Wetselaar and Farquhar [92] have pointed out that this is not necessarily the case. A reduction in nitrogen content of 50 kg ha^{-1} crop^{-1} between flowering and maturity is not uncommon for annual crops such as rice, sorghum and wheat. The example given by Basinski and Airey [3] for rice (Fig. 3) suggests that the denitrification loss determined by the difference method (equation 5) could be much lower at anthesis than at maturity, i.e. any loss of N from plant tops could give an overestimate of denitrification loss. Further, losses from plant tops may occur continuously during the growth of a crop, but become only apparent when the rate of uptake is lower than the rate of loss [92].

In summary, denitrification losses could well have been overestimated, and there is an urgent need for techniques that can measure such losses directly under undisturbed field conditions.

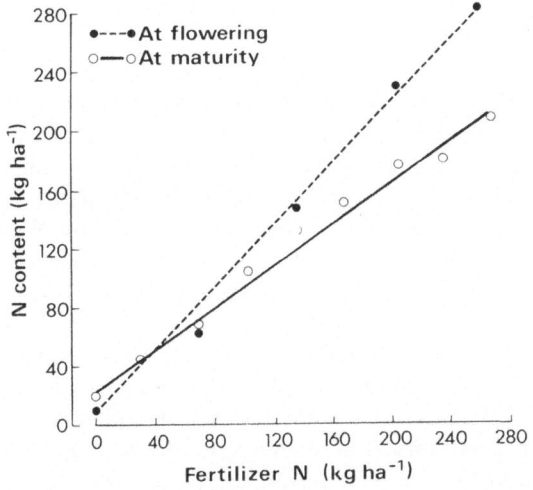

Fig. 3. N contents of rice at flowering (●) and at maturity (○), at different amounts of fertilizer N [3].

3.2.4. Wind erosion. The transfer of terrestrial material to the atmosphere through wind can cause substantial losses of N from an agrosystem. The removal of 1 mm of a topsoil containing 0.07% N would be equivalent to a loss of *c* 10 kg N ha^{-1}. However, reliable assessments of N losses due to wind erosion are lacking, probably because of the problems associated with obtaining reliable results.

It can be assumed that all of the wind component N present in the atmosphere is returned to the earth surface through dry and wet deposition. The assessment of the total N content of these depositions may therefore give a measure of the *maximum* N loss due to wind erosion. However, the deduction of the *actual* N loss due to wind erosion from this assessment remains a problem owing to (i) the presence in the depositions of N derived from processes other than wind erosion, and (ii) the difficulty of relating to its source any of this N in the depositions. The latter is due to the lateral transfer in the atmosphere of airborne particles of terrestrial origin caused by air movement. Possibly, the origin of the N from particular regions may be established by verification of ratios of certain elements or compounds that are specific to the region.

In conclusion, the assessment of the N contained in wet and dry deposition in a particular region is not likely to provide us with a proper assessment of N lost from that agrosystem through wind erosion. Possibly, flux measurements of airborne soil particles akin to those developed for ammonia by Denmead and various co-workers [14, 15, 16] could be considered.

4. The plant N pool: inputs and outputs

4.1. Inputs

4.1.1. Soil and fertilizer N. In most nitrogen balance studies, nitrogen in the soil and in plant roots is combined because of problems associated with separating them [92]. Consequently, when reference is made to nitrogen uptake by crops, this generally means the N content of the aboveground parts, or tops. For non-leguminous crops, uptake represents a transfer from the soil N pool to the plant N pool, and originates either from soil organic N that has been mineralized, or fertilizer N that has been placed on or in the soil or brought directly in contact with the plant.

The amount of soil N made available to plants in each growing season depends on soil type, land history and amount of rainfall, but is on average about 5.5% of the total organic N fraction in the soil for a semi-arid region such as northwest Australia [87]. In plant tops, the recovery of this N, together with that already available at the start of the growing season, can be as low as 20% and as high as 57% (Table 3). The high recovery by pearl millet and the low one by sorghum after a

Table 3. Plant nitrogen yields as percentage of total available soil nitrogen (nitrate in profile at start of season plus organic soil nitrogen mineralized during wet season) of non-leguminous crops following legumes [90]

After	Pearl millet	Grain sorghum	Sudan grass	Cotton	Weighted mean
Guar	42.7	43.1	56.8	43.1	46.4
Townsville stylo	37.9	47.4	42.6	42.6	42.6
Cowpea	54.0	47.3	56.0	48.7	51.3
Peanuts	35.4	41.9	50.2	38.6	41.4
Weighted mean	41.8	44.9	51.6	42.8	45.0
Bare fallow	46.2	19.9	25.3	25.6	

period of bare fallow is explained in terms of the differences in rooting depth between these species. During the bare fallow period, mineralized soil organic N accumulated as nitrate in the subsoil with a peak at about 1 to 1.5 m depth. Pearl millet can take up such nitrate from a much greater soil depth than sorghum [12, 95]. After five years of bare fallow in northwest Australia, one crop of millet recovered 203 kg N ha^{-1} and sorghum only 105 kg N ha^{-1} from the soil. However, when these two crops were sown in this same season (as for the previous measurements) on a 7-year old Townsville stylo (*Stylosanthes humilis*) pasture in which little or no nitrate had accumulated in the subsoil during the pasture phase, millet and sorghum took up 90 and 86 kg N ha^{-1} respectively [95]. In other words, the capacity of these two crops to take up nitrogen made available in the topsoil during the growing season was the same.

Clearly, the mode and time of application, the type of nitrogen fertilizer, and edaphic and climatic conditions play a major role in determining the recovery of fertilizer N. In northwest Australia, the recovery of fertilizer N by sorghum ranges from 14 to 74% under rainfed conditions [55]. However, a low recovery by plant tops does not necessarily mean that the unrecovered part has been lost from the system. For instance, during an above-average wet season, about one quarter of the 100 kg N ha^{-1} applied as ammonium sulphate to sorghum could be recovered as nitrate in the 60 to 120 cm soil layer at the end of the growing season [86]. This nitrate would still be available in the next wet season to a deep-rooting crop such as pearl millet. In this case, the low recovery of applied N by the sorghum crop was due to high amounts of rainfall leaching the nitrified applied ammonium. However, in the same region, a low recovery by a crop of sorghum due to lack of rain was also obtained when the application of 80 kg N ha^{-1} as ammonium sulphate was followed by small and infrequent showers. After three months 44% of the applied N could still be found unnitrified in the surface soil, and out of reach of plant roots, due to the predominantly dry conditions in that soil layer [86].

Under wetland conditions, the recovery of applied N by the tops has been notoriously low, but deep placement of ammoniacal nitrogen fertilizers has substantially increased this recovery [13]. Enhanced recovery could be due to a reduction in ammonia volatilization [98], and/or a reduction in nitrification and therefore in denitrification. The reduction in nitrification may be achieved by placing the fertilizer in the anaerobic zone that is present several cm below the soil surface. Such placement inevitably increases the local fertilizer concentration, which *per se* has an inhibiting effect on nitrification [96]. In addition, local high fertilizer N concentrations in the soil tend to stimulate root proliferation to such an extent that any fertilizer N moving away from the point of placement may be intercepted by the roots before it can be leached or lost in other ways [62].

4.1.2. N_2 fixation. Leguminous crops can derive their nitrogen from the soil and from the atmosphere. When the soil N supply is low, the atmospheric contribution may be as high as 80% of the total amount of N taken up [36]. In a semi-arid tropical region, the fixation of N_2 can depend variously on the presence of effective and specific *Rhizobium* strains, the amount of available N in the soil, and the soil water content. The results in Table 4 suggest that (i) where soybean was grown for the first time (in 1973), the effect of inoculation was dramatic, but no such effect was demonstrated with peanuts, (ii) fertilizer N had a depressing effect on N_2 fixation for both crops, and (iii) an inadequate supply of water reduces the grain yield of soybean, but not the proportion of N_2 fixed by the legume; for peanuts both were reduced (see also Fig. 4).

Table 4. The effect of three major environmental factors on the relative proportions of nitrogen coming from three sources (symbiotic fixation, soil and fertilizer) in aerial parts of two legumes and on their grain yield[a] [34]

Factors	Crop	Year	Treatment	Percentage of nitrogen in the aerial parts originating from			Grain yield (kg ha^{-1})
				N_2 fixation	Soil	Fertilizer	
Rhizobium	Soybean	1973	0	0	95	5	2020
	Soybean	1973	+	55	40	5	2320
	Peanuts	1974	0	44	54	2	1590
	Peanuts	1974	+	40	57	3	1620
	Peanuts	1975	0	66	32	2	1420
	Peanuts	1975	+	70	28	2	1500
N fertilization	Soybean	1979	s	58	38	4	840
	Soybean	1979	1	20	75	5	960
	Peanuts	1974	s	44	54	2	1590
	Peanuts	1974	1	20	68	12	1570
Precipitation	Soybean	1973	a	55	40	5	2320
	Soybean	1979	na	58	38	4	840
	Peanuts	1975	a	66	32	2	1420
	Peanuts	1974	i	44	54	2	1590
	Peanuts	1976	na	21	75	4	1130

[a] Work done within joint FAO/IAEA coordinated research program.
0, + = No inoculation or inoculation with effective rhizobium, respectively.
s, 1 = Small or large application of fertilizer, respectively.
na, i, a = Amount and distribution of precipitation is not adequate, intermediate or adequate, respectively.

There are many problems associated with the measurement of the *gross* amount of N_2 fixed by a legume over the whole period of its growth [48], and objections can be made to virtually all methods suggested [29]. The *net* contribution of legumes to the soil or to the soil—plant system can be measured slightly more accurately than the total amount fixed by determining (i) all N in the plant either remaining above the soil and/or being removed, plus (ii) all N in different soil layers before and after the introduction of the legume system. In this way, Henzell *et al.* [37] estimated that *Desmodium uncinatum* could make a net contribution of 112 kg N ha^{-1} yr^{-1}.

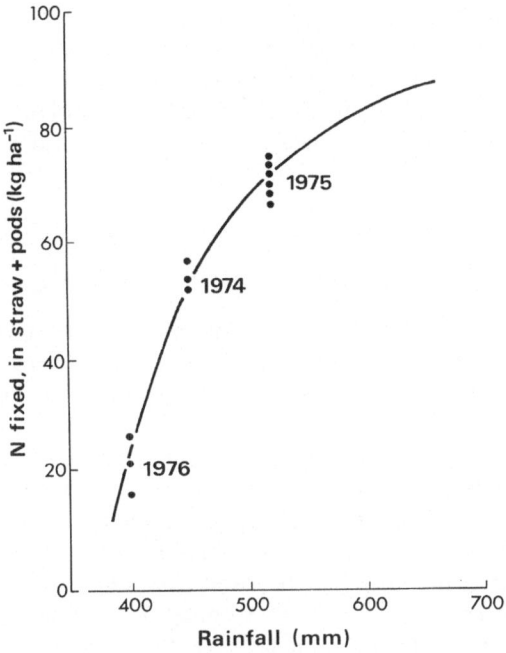

Fig. 4. Effect of rainfall on amount of N₂ fixed by peanuts, for the years 1974–76. Each point represents one plot at the Experimental Station in Bambey, Senegal (Ganry, unpublished results).

Although there is considerable evidence to suggest that fixation of N_2 takes place also in the phyllosphere of plants [67], such a contribution is small and probably not higher than about 1 kg N ha^{-1} crop^{-1} for cultivated annuals.

4.1.3. Ammonia absorption. Leaves can take up and metabolize ammonia from their surrounding atmosphere [40, 65]. Such uptake has been demonstrated when high external pressures of ammonia (30 to 20,000 nbar) were imposed. Under such conditions, the contribution of gaseous ammonia to the N content of plants is likely to be substantial. However, in most areas, the external partial pressure of ammonia is generally only 1 to 8 nbar [43]. Because in leaves massive release and refixation of ammoniacal N takes place in the photorespiratory carbon cycle [44, 100], Farquhar *et al.* [24] inferred that there is a finite partial pressure of ammonia in the intercellular spaces of leaves. When this pressure is equal to or above the ambient air no ammonia uptake takes place. As this internal pressure is generally close to that prevailing in the air of unpolluted areas, uptake of ammonia by plants is likely to be restricted to regions downwind of areas with high industrial activity, or where a dense plant community covers soils from which ammonia can volatilize. For the latter case, it has been shown that pasture species can 'absorb' virtually all ammonia released from decomposing plant material on the soil surface [14], and that rice plants take up some of the ammonia released from ammoniam sulphate applied to the flood-water [98]. In both cases, the actual rate of ammonia absorption was higher than could be expected from assimilation through the stomata alone, and an additional uptake *via* a water film on the plant surface was suspected.

4.2. Outputs

Outputs or losses from the plant N pool take place through a removal from the field or a return to the soil. Both will be discussed below, but reference is made to Table 5, which summarizes the processes involved in these two loss pathways.

4.2.1. Removal from the field. Nitrogen in plant material may be removed from the field through harvesting, grazing, or through losses into the atmosphere. The amount removed through harvesting will depend on the total amount of N taken up by the plants and its distribution between harvested and non-harvested plant parts. In general, the grain is the main part removed from crops such as grain sorghum, pearl millet, and wheat, with the straw, stubble, or litter being either ploughed back into the soil or burnt. The ratio between the N in the grain and the total amount of N in the tops is reasonably constant and in the order of 0.68 for winter wheat [2] and 0.65 for rice [89], although for wheat, values as high as 0.85 have been obtained [1]. Thus, when only the grain is removed from these two crops, a plant N loss of 65 to 68% is incurred.

The nitrogen content of different plant parts for some crops in some tropical regions is given in Table 6. For grain crops like pearl millet, peanuts, grain sor-

Table 5. Nitrogen losses from the plant pool

I. REMOVAL FROM THE FIELD				
Harvest	Grazing		Into atmosphere	
	Intentional	Unintentional	Solid	Gaseous
	*Cattle	*Birds	*Particulates	*Amines
	*Sheep	*Insects	*Pollen	*N_2
	*Goats	*Rats, etc		*N_2O, NO_x
		*Microorganisms		

II. RETURN TO THE SOIL			
Translocation	Leaching	Shedding	Return to soil
*to roots	*Rain	*Leaves	of non-harvested
*to soil	*Dew	*Flowers	part
	*Irrigation	*Fruits	
	*Spraying		

Table 6. Dry matter and nitrogen content of some tropical crops

Location	Crop	Treatment	Dry matter (kg ha⁻¹)	N (%)	N (kg ha⁻¹)	Reference
Senegal	Pearl millet grain straw	Rainfed, P & N	2560 6980	1.3 0.6	33 45	Ganry (unpublished)
Senegal	Peanuts kernels & straw shell & leaves	Rainfed P & starter N	4210 520	2.4 1.5	101 8	Ganry (unpublished)
Northwest Australia	Peanuts kernels shells trash	Rainfed & P	1120 560 1680	5.0 1.0 1.0	56 6 17	Wetselaar and Norman [94]
Northwest Australia	Grain sorghum grain stubble	Rainfed & P & N	2000 6700	1.5 0.5	30 33	Wetselaar and Norman [94]
Northwest Australia	Annual fodder crops cowpea guar Sudan grass sorghum bullrush millet	Rainfed & P & (N)	3920 4480 7840 10640 13440	2.0 2.0 0.7 0.75 1.0	78 90 55 80 134	Wetselaar and Norman [94]
Philippines	Rice grain stubble	Flooded, no N	4380 –	1.12 –	49 19	Khind and Ponnamperuma [45]
	grain stubble	Flooded, 150 N	6100 –	1.48 –	90 37	

ghum and rice, the amount of N lost on harvest will depend on whether the non-grain portion is used for other purposes such as fuel, cattle feed, etc., or whether it remains in the field. For fodder crops the total aboveground portion is lost from the plant N pool.

Losses through defoliation may be unintentional or intentional. Unintentional losses, such as occur when pests are in plague proportions, can induce substantial N removal, and NH_3 volatilization losses have been postulated [92] as occurring when plants are 'diseased' (through microbial oxidation by microorganisms of amino acids obtained in aerial parts of plants).

The magnitude of intentional losses through grazing will depend on the type of animal, its age and the intensity of grazing imposed. In northwest Australia, grazing intensities of beef cattle can vary from 2 to 12 ha beast^{-1} yr^{-1} on native pasture to 0.4 ha beast^{-1} yr^{-1} for an introduced Townsville stylo-birdwood grass (*Cenchrus setigerus*) pasture [59].

Nitrogen can be lost from plant material to the atmosphere through the shedding of pollen, loss of minerals that accumulate on leaf surfaces due to mechanical agitation [46], burning, fragmentation, loss of wax rodlets during rapid leaf expansion or through the production of airborne salt crystals that are generated during rapid transpiration [4]. The magnitude of such losses into the atmosphere is not known, but is likely to be low and probably less than 1 kg ha^{-1} crop^{-1}. Norman and Wetselaar [61] have shown that at least 93% of the N contained in air-dry pasture material will be lost during aerobic combustion.

Mechanisms involved in gaseous losses of N from plant tops have been reviewed in detail elsewhere [92]. When the partial pressure of ammonia in the atmosphere is below a critical level there is a net release of ammonia from plant leaves [25]. This critical level, called the ammonia compensation point, influences the molar flux of ammonia into the leaf, J, according to

$$J = g(n_a - \gamma)/P \tag{6}$$

g being the stomatal conductance to the diffusion of ammonia, n_a the ambient partial pressure of ammonia, γ the compensation point, and P the total atmospheric pressure.

Negative values for J, i.e. evolution of ammonia from the leaves, were observed during senescence of maize leaves [24]. This physiological period coincides with the flowering-to-maturity period during which the total N content of annual plant tops can decrease by as much as 75 kg ha^{-1} [92]. The partial pressure of ammonia in unpolluted air, i.e. in most atmospheres in the rural tropics, is about 1 to 8 nbar, and compensation points are about 2 to 6 nbar [25]. Therefore, an unrealistically high γ value of at least 100 nbar above ambient would have to be assumed to explain such high plant N loss on the basis of ammonia evolution alone [92].

It follows that the lower the partial pressure of ammonia in the ambient atmosphere the greater the loss of ammonia from the plant will be. This might partly explain why Stutte and co-workers [76, 83, 84] found ammonia losses of up to 1.6 kg N ha^{-1} d^{-1} at a leaf area index (LAI) of 5, since they subjected leaves to near zero partial pressure of ammonia.

When both the reduced and oxidized forms of N emitted from plants are determined, the ammonia form appears to be by far the most important one [83, 84]. This suggests that the total gaseous loss of reduced plus oxidized N from plants is likely to be low, because losses in the NH_3 form alone may already be assumed to be low and approximately of the order of 0.05 kg N ha^{-1} d^{-1} at LAI 5 [24]. The possible loss of gaseous N in the form of N_2 and HCN is discussed by Wetselaar and Farquhar [92], but this is likely to be small.

4.2.2. Return to the soil. The underground portion of most crop plants, excepting root crops and peanuts, remains in the soil after harvest and automatically becomes a part of the soil N pool. In principle, the N contained in plant tops can be returned to the soil through translocation into or *via* the roots, leaching out of plant parts by rain or irrigation water, shedding of plant material, and through decomposition of plant residues or litter not removed from the field after harvest.

The processes and amounts involved in a redistribution of plant N from the tops to the roots have been reviewed recently by Wetselaar and Farquhar [92], who indicate that for perennial plants such a translocation can definitely occur; there is no evidence for this process in annual plants. The authors conclude that the amount of N involved in root exudates is likely to be extremely small, but they caution that the lack of any soil N increase during the period when translocation of N from tops *via* the roots into the soil could be expected does not exclude the possibility of such a transfer followed by denitrification.

A return of N in tops to the soil *via* leaching is very likely to occur during rain, irrigation, spraying or dewfall in view of the evidence accumulated from throughfall measurements in forests [50]. For annual crops and pastures such evidence is sparse. When rice plants of high N content were sprayed each with 300 ml of distilled water per day for two days at ear initiation stage, the N content of the collected water was 46 ppm, equivalent to 0.56 kg N ha^{-1} per mm rainfall [77]. In addition, a nitrogen concentration of 102 ppm in dew was observed on the same rice plants [77]. In general, a transfer of N through leaching is therefore likely to be important in the tropics, where plants may experience intensive rainfall.

Other forms of plant N return to the soil are the spontaneous abortion of flowers and fruits, especially for a crop such as cotton, and the shedding of leaves. In pastures, all aboveground plant material that has not been consumed by herbivores is eventually returned to the soil surface. Legumes may be sown as green manure crops, converting N_2 into plant N; incorporation of this plant N into the soil provides extra nitrogen to the soil N pool. However, when some parts of the tops of legumes are removed, the net result is not always an increase of N in the soil, as can be seen from Table 7. Likewise, Pieri [63] reported a decrease in soil N after the cultivation of some legumes in West Africa. However, a temporarily favorable effect can be obtained when a non-legume is preceded by a legume [42, 57]. A beneficial effect from peanut cropping is mainly due to an accumulation of nitrate formed during the period of early growth in the wide interspace between the peanut rows [63, 90].

Table 7. Topsoil organic nitrogen content and amount of plant nitrogen removed and returned to the soil after 3 years of legumes [90]

	Soil org. N (kg ha^{-1})	Plant N removed (kg ha^{-1})	Plant N returned (kg ha^{-1})
After 3 years			
Guar	1720	50	202
Peanuts	1580	168	56
Townsville stylo	1640	230	34
Cowpea	1510	364	0
At zero time	1700	–	–

The fate of plant N returned to the soil will depend on the chemical composition of the plant material, climatic and edaphic factors, the presence and activity of microorganisms and fauna, and the position of the plant material in relation to the soil surface. In general, a high N content will increase the weight-loss rate of the plant material, while a high lignin content will retard this process [5]. Three phases in the nitrogen dynamics of litter decomposition can be distinguished [5]: (i) a leaching phase, representing a rapid decline in absolute N content (kg ha^{-1}) due to contact with water, (ii) an accumulation phase, when there is a net absolute increase in N content, and (iii) a release phase, consisting of a continuous net decrease in N content after the maximum N content has been reached. The rate and magnitude of these three phases will, no doubt, depend to a large extent on whether the plant material remains on the soil surface, or whether it is in close contact with the soil due to mixing of the litter material with the topsoil during cultivation. In the former case, losses due to ammonia volatilization into the atmosphere are possible [28] and are regarded by some authors [101] as a major source of loss of N in natural grasslands, when conditions for denitrification and leaching are not favourable. On the other hand, all litter N could be accounted for after two years of pasture growth, when ^{15}N-labelled pasture litter was incorporated into the top 0.5 cm of a clay loam [90, 97]. The actual release of this litter N for plant uptake depended on the C to N ratio of the litter, and varied from 6.5 to 17% during the first season.

Saprophytic grazers can consume microflora and excrete mineral forms of N, but their exact role in the turnover of litter N in different ecosystems is not known [101].

5. The animal N pool: inputs and outputs

In this section, we discuss vertebrate herbivores such as beef and dairy cattle and sheep that are introduced into the soil—plant system for the sole purpose of yielding animal products useful to mankind.

The animal N pool will be discussed in lesser detail than the other pools. This is certainly a reflection of the main interests of the authors. Nevertheless, an attempt is being made to indicate the major gain and loss pathways and their likely magnitude. Problems arising from the social behaviour patterns of many domesticated herbivores in relation to the measurement of changes of the size of the different N pools will not be discussed.

5.1. Inputs

5.1.1. Grazing. The amount of N consumed by grazing animals per unit area of land depends on the amount of consumable plant material on offer, its N content, and the stocking rate. The utilization of the total dry matter production can be as low as 10 to 20% for extensive grazing systems and as high as about 80% in rotationally grazed high production systems [27]. Because animals tend to graze selectively, their N intake is difficult to estimate, but it can be roughly calculated from known liveweight gains (Table 8). For native pasture in northwest Australia, the total intake is not higher than 6.7 kg N ha^{-1} yr^{-1}, which is close to the annual maximum N content of 8.5 kg ha^{-1} of the pasture. For introduced pastures in the same region, the intake is about 60 kg N ha^{-1} yr^{-1}, which is appreciably less than the total N content of these pastures of about 100 kg ha^{-1} [59]. N intake as high as 120 kg ha^{-1} yr^{-1} is possible on high quality pastures in tropical areas.

5.1.2. N_2 fixation. Increases of N content in the rumen over and above the amount of N brought in by feed has been suggested to be due to N_2 fixation, but such increases could possibly be due to endogenous additions [58]. Fixation of $^{15}N_2$ in rumen contents *in vitro* indicates that the contribution of N_2 fixation, if it occurs, has no major effect on the total amount of N available to the microorganisms in the rumen [39].

5.2. Outputs

5.2.1. Return to the soil. Of the nitrogen taken in by animals such as beef cattle, only about 10% is retained, the rest being excreted mainly as faeces plus urine, or shed as surface cells plus hair. The latter group is dependent on the weight of the animal according to

$$N_s = 0.02 \, W^{0.75} \tag{7}$$

where N_s is the amount of N lost through the shedding of surface cells and hair (g d^{-1}) and W is the liveweight (kg) [56]. With a liveweight of 100 kg ha^{-1}, this loss amounts to 1.3 kg ha^{-1} yr^{-1}.

When high quality diets are ingested, i.e. diets with a protein-N concentration of 1.5 to 2%, the excess N is excreted in the urine rather than in the faeces, with the ratio between the two being as high as 7 or as low as 3. However, when animals are on a low quality diet, having a crude-protein-N concentration of about 0.5%, as for

Table 8. Calculation of N intake and N excreted for approx. 2-year old beef cattle, based on known liveweight gains (LWG)

System	LWG[a] (kg beast^{-1})	Stocking rate[a] (beast ha^{-1})	LWG (kg ha^{-1})	N gain[c] (kg ha^{-1})	N intake[d] (kg ha^{-1})	N excreted Skin, etc.[e] (kg ha^{-1})	Faeces[f] (kg ha^{-1})	Urine[f] (kg ha^{-1})
Native pasture[a]								
Wet season (6 months)	104	0.25	26	0.522	5.22	0.05	1.4	3.25
Dry season (6 months)	−54	0.25	−13.5	−0.27	1–2(?)	0.05	0.18	0.04
Total	50	0.25	12.5	0.252	6.7	0.1	1.6	3.3
Introduced pasture[a]								
High qual. (6 months)	108	2.2	240	4.8	48	0.48	13	30.2
Low qual. (6 months)	30	2.2	66	1.32	13.2	0.13	3.5	8.3
Total	138	2.2	306	6.1	61.2	0.6	16.5	38.5
Humid tropics[b]								
High qual. (12 months)	–	–	600	12	120	1.2	32	75

[a] In northwest Australia [59].

[b] Simpson and Stobbs [70].

[c] Assumes that the N concentration of the LWG is between 1.6 (J.B. Coombe, personal communication) and 2.4% [70], i.e. 2% average.

[d] Assumes that the N gain represents 10% of the total N intake.

[e] Represents N losses through shedding of skin and hair, belching of ammonia, and ammonia loss through sweating, estimated at 1% of total intake (J.B. Coombe, personal communication).

[f] For high quality feed 80% of N is excreted via faeces and 70% via urine; for low quality feed these are 80 and 20% respectively (J.B. Coombe. personal communication).

beef cattle on native pastures during the late dry season in northwest Australia, cattle lose weight and excrete more N than they take in; most of this excreted N (80%) is in the faeces.

Virtually all (90%) N in urine is in the urea or amino-N form, the rest occurring as ammonia, allantoin, creatine, creatinine and hippuric acid [70]. Its fate will depend on processes described under 3.2.2., but it must be kept in mind that the return of urine N to the soil is highly localized. With 2.5 cows ha^{-1} only 15% of a pasture would be affected directly by urine per year [27] and one urine patch might therefore receive the equivalent of 300 to 500 kg N ha^{-1} [70]. Because of such high concentrations and the high pH developed, losses through ammonia volatilization can be substantial. Nevertheless, because urine N has the highest proportion of the amount of N returned by large herbivores, it plays a dominant role in the conversion of plant N, often biologically fixed by legumes, into available N in the soil.

The N compounds in faeces are not well known. About 10 to 20% occur in undigested residues and 50 to 60% in bacterial matter [70]. Because most of this N is insoluble, it will only become available after incorporation into the soil and mineralization by microorganisms. Therefore, the role of dung-feeding insects, such as certain dung beetles in burying faecal material is an important factor in the rate of faecal N cycling.

5.2.2. Removal from the field. Nitrogen stored in the animal N pool can be removed from the field through milk, beef, wool, and loss of ammonia through belching and sweating.

In many tropical areas the nature of the herbage tends to be more fibrous than in temperate areas. Since lactation yield for dairy cattle is usually determined by the intake of digestible energy rather than of digestible protein, milk production per cow tends to be lower in the tropics than in temperate climates [70]. On the basis of Stobbs' [75] data on milk production, and assuming an N concentration of 0.57%, the N removed in milk can vary from 6 kg ha^{-1} yr^{-1} for unfertilized tropical pastures to 114 kg ha^{-1} yr^{-1} for irrigated tropical pastures with a high fertilizer input. This N represents between 10 and 30% of the total amount of N ingested [70].

The amount of N removed from the field as beef can be estimated for different types of tropical pastures from Table 8. For low quality, lightly-stocked native pasture, this would be about 0.25 kg N ha^{-1} yr^{-1}, but can be as high as 12 kg N ha^{-1} yr^{-1} for high quality pastures. The takeoff of 300 kg ha^{-1} yr^{-1} of lamb would reduce the animal N pool by 7.5 kg N ha^{-1} yr^{-1}.

Greasy wool consists of about 70% protein, containing 16.4% N, and the removal of each kg of wool represents an output of 0.11 kg of N [70]. On average, one sheep adds 1 g N d^{-1} to its wool (J.B. Coombe, personal communication). With stocking rates varying from 0.2 to 30 head ha^{-1}, the annual takeoff of wool N would range from 0.07 to 11 kg ha^{-1}.

The amount of N removed *via* belching and sweating is not well known. The loss through belching is only likely to occur with high intakes of urea, e.g. from a urea lick, which would act as an ammonia source and also increase the pH in the rumen, which is normally between 6 and 7 (H. Dove, personal communication). The amount of N lost through sweating is generally regarded to be less than 1% of the total amount of N intake.

6. Examples of N balances

In the following, we shall give some examples of nitrogen balances made for different cropping systems in different parts of the tropics, to illustrate the type of measurements that have been made of inputs and outputs, and to indicate the changes that normally occur in the different pools. Although it is recognized that domesticated herbivores can play an important role in the cycling of nitrogen and in the maintenance of soil fertility, no N balances that include the animal N pool are presented, because of the lack of relevant data for the tropics.

In some cases, only changes in the size of the soil N pool have been determined, so that equation (1) applies. In other cases, the measurement of such changes are combined with measurements of *some* inputs and outputs. Based on equation (3) we can write

$$N_t = N_0 + \text{inputs} - \text{outputs} \tag{8}$$

or

$$N_t = N_0 + (I_k + I_u) - (O_k + O_u) \tag{9}$$

where I_k and I_u are known and unknown inputs, respectively and O_k and O_u are known and unknown outputs, respectively. Therefore,

$$I_u - O_u = N_t - N_0 - I_k + O_k \tag{10}$$

$I_u - O_u$ can be regarded as the *net* contribution of N to the soil−plant system, and represents in most cases a measure of the net amount of N added via N_2 fixation. Let $I_u - O_u = \text{Net } I_u$, then equation (10) becomes

$$\text{Net } I_u = N_t - N_0 - I_k + O_k \tag{11}$$

6.1. Casuarina (equation (1))

In the tropical arid climate of the northern coast of Cape Vert peninsula in West Africa, sand dunes have been planted to *Casuarina equisetifolia* for stabilization. After 13 years, ΔN for the soil and plant N pools were 229 and 531 kg ha^{-1}, respectively [20]. Thus, the total annual addition of N to the soil−plant system was 58.5 kg ha^{-1}, of which about 6 kg ha^{-1} originated from wet deposition. This suggests that on average 52.5 kg ha^{-1} yr^{-1} was contributed through biological N_2 fixation.

6.2. Peanuts and millet (equation (2))

For a 2-year cropping period of one season peanuts followed by one season millet in Senegal (Ganry, unpublished results), it is assumed that all inputs to and outputs from the soil N pool are known, in which case equation (2) will apply. The balance sheet (Table 9) suggests a total loss of 71 kg N ha^{-1} from the soil–plant system, on the assumption that inputs due to wet plus dry deposition and non-symbiotic N_2 fixation were negligible, which was not necessarily the case.

Table 9. N inputs and outputs, and soil N changes, for a 2-year period of peanuts–millet cropping in the semi-arid tropics of Senegal (Ganry, unpublished)

	N (kg ha^{-1} 2 yr^{-1})	
Inputs (I_k)		
Precipitation	Negligible	
Non-symbiotic N_2 fixation	Negligible	
Fertilizer	95	
Symbiotic N_2 fixation	82	
Σ_0^t inputs		177
Outputs (O_k)		
Removed through millet	33	
Removed through peanuts	109	
Leaching	20	
Denitrification and volatilization from soil N	50	
Denitrification and volatilization from plant residues	8	
Denitrification and volatilization from fertilizer N	28	
Σ_0^t outputs		248
Change of soil N pool (ΔN, equation (2))		−71

6.3. Different legumes (equations (11))

In the semi-arid tropics of northwest Australia, one bare fallow area and four different legumes were maintained in parallel for three consecutive years. The balance sheet is represented by Fig. 5. It assumes that, in relation to equation (11): (i) N = organic N in the topsoil + nitrate N in the whole soil profile, (ii) $N_0 = N$ value for bare fallow and $N_t = N$ value after three years of a legume, and (iii) I_k is ignored, representing only the N input *via* sowing of the legume, while the input *via* wet + dry deposition is zero, since N_0 and N_t received the same amount through this input, or $(N_0 + N \text{ in deposition}) - (N_t + N \text{ in deposition}) = N_0 - N_t$.

By making a N balance after the first year, actual values were found for Net I_u of 93, 90, 67, and 26 kg ha^{-1} yr^{-1} for the mean of the 2nd and 3rd year for Townsville stylo, guar (*Cyamopsis tetragonoloba*), cowpea, and peanuts, respectively [88]. These values are likely to represent the net contribution of the legume, *via* symbiotic N_2 fixation, to the soil–plant system.

26

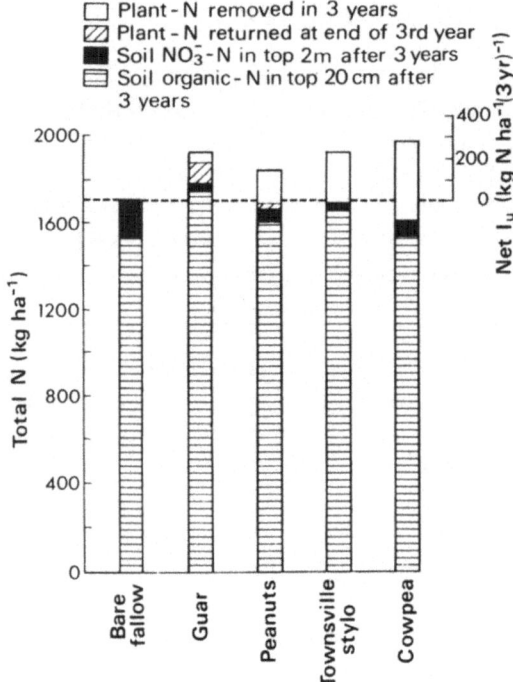

Fig. 5. Nitrogen balance after 3 years of legumes or bare fallow [88]; for Net I_u see equation (11), Section 6.

6.4. Rice (Equation (11))

In southern Senegal, rice was grown for nine consecutive years, without any application of nitrogen fertilizer, on a grey soil characterized by a temporary shallow groundwater table at the end of the wet season [69]. Over this cropping period there was no significant change in soil N content, i.e. $N_t - N_0 = 0$. The total known input was through wet deposition ($I_k = 10$), while the total amount of N removed in grain + straw was 460 kg ha^{-1}. Thus, according to equation (11), Net $I_u = 450$. This represents a net average gain of 50 kg N ha^{-1} yr^{-1}, perhaps through groundwater and N$_2$ fixation. A similar net gain (40 kg N ha^{-1} crop^{-1}) was found for unfertilized rice in central Thailand [26].

At the International Rice Research Institute at Los Baños, Philippines, 24 crops of three varieties were grown in 12 years without any nitrogen fertilizer [49]. The following values were found:

$$N_0 = 4130 \text{ kg N ha}^{-1}$$

$$N_t = 4440 \text{ kg N ha}^{-1} \text{ (mean of three varieties)}$$

$$I_k = 72 \text{ (rainwater)} + 120 \text{ (irrigation water) kg N ha}^{-1}$$

$$O_k = 1024 \text{ (grain)} + 368 \text{ (straw) kg N ha}^{-1}$$

Thus, according to equation (11), Net $I_u = 1510$ kg N ha^{-1}, which is equivalent to a net input of 126 kg N ha^{-1} yr^{-1} or 63 kg N ha^{-1} crop^{-1}.

All three balance sheets suggest that under wetland rice conditions, in the absence of fertilizer N, there is a net N gain for the soil—plant system. In the Senegal and Thailand experiments, the total nitrogen content in the soil did not change significantly, but in the Philippines there was a net increase of 13 kg N ha^{-1} crop^{-1}. All three examples indicate that the traditional method of wetland rice growing is a very stable one so far as nitrogen is concerned.

In addition, data given by Koyama and App [49] suggest that different varieties can have a significantly different effect on the changes in total soil N content. In the Thailand case [26], the rotation of rice with the legume mung bean indicates that this legume contributes between 58 and 107 kg N ha^{-1} crop^{-1} to the soil—plant system, but it has no positive effect on the soil N status, presumably because most of the N in the aboveground part of the legume is removed.

6.5. Rice (equation (5), ^{15}N)

In the central Plain of Thailand, ^{15}N-labelled ammonium sulphate was applied to an irrigated dry-season crop of rice, either on the surface or at depth, split or non-split, and at 50 or 100 kg N ha^{-1}. At 2, 4, 7 and 12 weeks after transplanting sub-plots were destructively sampled and the amount of ^{15}N was determined in different pools and sub-pools (Shaw, Oupathum, Thitipoca and Wetselaar, unpublished results). In addition, the NH$_3$ originating from the fertilizer, that had been volatilized was also determined [98]. The results are given in Table 10.

In this case, the unaccounted-for applied N could represent (i) denitrification losses, (ii) losses via plant tops, and (iii) losses through unknown pathways. Leaching losses to the subsoil were unlikely, because deep sampling did not indicate the presence of any ^{15}N compounds. Losses due to overflow or run-off were excluded, since for each subplot the fertilizer was applied within a frame, where the water level was strictly controlled.

The balance sheet suggests that at least some fertilizer N was lost through denitrification. In addition, there are strong indications that ^{15}N was lost from the plant tops between 7 and 12 weeks where 100 kg N ha^{-1} had been applied; this loss represents between 70 and 75% of the increase in unaccounted-for N during that period. At the same time, the total amount of ^{15}N in the soil (including roots) also decreased. It is therefore unlikely that the decrease in N content of the tops was due to a transfer of plant N into the roots or soil.

Table 10. Fate of fertilizer N (kg ha^{-1}) based on ^{15}N-labelled ammonium sulphate applied to transplanted rice, in the field, in the central plain of Thailand (Shaw, Oupathum, Thitipoca and Wetselaar, unpublished results). For experimental details see Wetselaar et al. [98]

Treatment[a]	Weeks after	Fertilizer N (kg ha^{-1})				
		Tops	Soil		NH$_3$ volatilized	Unaccounted for
			Mineral N	Org. N + root N		
S$_{50}$	2	2.7	10.5	11.7	4.4	20.7
	4	6.8	1.1	15.4	4.4	22.3
	7	7.3	0.2	16.7	4.4	21.4
	12	9.3	0.3	16.3	4.4	19.7
S$_{50+50}$	2	2.1[b]	9.8[b]	9.6[b]	3[b]	25.5[b]
	4	9.7[b]	1.2[b]	15.3[b]	3[b]	20.8[b]
	7	36.7	0.8	26.8	3	32.7
	12	20.1	0.5	21.2	3	55.2
S$_{100}$	2	3.5	56.4	13.1	4.3	22.7
	4	25.3	10.3	21.4	4.3	38.7
	7	24.6	0.4	23.8	4.3	46.9
	12	19.1	0.5	21.4	4.3	54.7
D$_{100}$	2	3.3	56.3	15.7	0.7	24.0
	4	25.6	17.3	18.9	0.7	37.5
	7	38.1	0.7	26.5	0.7	34.0
	12	30.2	1.1	23.4	0.7	44.6

^{a} S$_{50}$ = 50 kg N ha^{-1} on surface at transplanting time.
S$_{100}$ = 100 kg N ha^{-1} on surface at transplanting time.
S$_{50+50}$ = 50 kg N ha^{-1} on surface at transplanting time plus 50 kg N ha^{-1} on surface at panicle initiation.
D$_{100}$ = 100 kg N ha^{-1} in mudball, at 10 cm depth, one ball per plant.
^{b} For first 50 N application only.

6.6. Regional (equation (2))

For the Suchow district of the Jiangsu province of China, Zhu [102] drew up an N balance sheet for an average cropping system. In this district, the cropping systems are one winter crop (barley, rape, or wheat) plus one or two crops of irrigated wet-land rice per annum. Apart from chemical N fertilizers, organic ones such as farm-yard manure, compost, *Azolla* and milk vetch (*Astragalus* ssp.) are also used extensively.

Table 11. Nitrogen balance sheet, averaged over all agricultural fields of the Suchow district, for 1978 [102]

Input		Output $(kg N ha^{-1} yr^{-1})$	
Chemical fertilizers	246	Harvesting	204
Symbiotic fixation,		Leaching + run-off	2
milk vetch	12	Estimated loss from chemical fertilizers	123
azolla	2	Organic manures losses	8
Non-symbiotic fixation	?		
Aquatic plants	1		
Rice straw	11		
Night soil	10		
Pig manure	28		
Seeds	7		
Irrigation water	3		
Precipitation	23		
Total	343 + ?		337

Applying equation (2) to the results given in Table 11, $\Delta N = 6 + ? \, kg \, N \, ha^{-1} \, yr^{-1}$. This suggests that the agricultural systems in the district are on average reasonably stable as far as N is concerned. This could be due in part to the relatively high input of organic nitrogen.

7. Significance and conclusions

The management of an agrosystem should aim to, at least, maintain the N status of a soil by keeping all N losses to a minimum and bringing all N gains to a maximum, provided that the gains have no detrimental effects on the environment [99]. This aim cannot be achieved using only empirical procedures. For instance, although legumes may add N to the soil—plant system in the tropics (see 6.3.), a nitrogen balance study indicates that only under certain management conditions where legumes are employed is the N content of the soil maintained (Table 7). Monitoring of the soil N content will indicate the magnitude and direction of the net changes, but this cannot give the information required to correct the changes. Such information can be obtained through detailed N balance studies, involving measurement of as many inputs and outputs as possible. These studies will indicate (i) whether the soil–plant system as a whole gains or loses N over time, (ii) in which pool or sub-pool these positive or negative accumulations occur, and (iii) what the causes are for such accumulation.

In general, nitrogen balances are important means of determining which N loss and gain pathways are important and what their magnitudes are. They are therefore a useful tool in identifying research priorities, in testing potential improvements, and in assessing the net gains or losses due to certain management strategies such as the introduction of N_2 fixing species or associations. They are also likely to be a useful tool in testing of certain nitrogen simulation models.

The more the important inputs and outputs can be determined at the same time for the same agrosystem the higher the research efficiency becomes, since such results may verify the links between the different transfers and their interactions [91]. However, as has been shown in this chapter, such research could be improved by the development of appropriate methodology for certain gain or loss pathways; specifically these are denitrification, gross N_2 fixation, wind erosion, and gaseous losses from plant tops. In addition, realistic N assessments are required for wet and dry deposition and for leaching from plants.

References

1. Angus, J.F., Nix, H.A., Russell, J.S. and Kruizinga, J.E. 1980 Water use, growth and yield of wheat in a subtropical environment. Aust. J. Agric. Res. 31, 873–886.
2. Austin, R.B., Ford, M.A., Edrich, J.A. and Blackwell, R.D. 1977 The nitrogen economy of winter wheat. J. Agric. Sci., Camb. 88, 159–167.
3. Basinski, J.J. and Airey, D.R. 1970 Nitrogen response of IR8 rice at Coastal Plains Research Station in northern Australia. Aust. J. Exp. Agric. Anim. Husb. 10, 176–182.
4. Beauford, W., Barber, J. and Barringer, A.R. 1977 Release of particles containing metals from vegetation into the atmosphere. Science 195, 571–573.
5. Berg, B. and Staaf, H. 1981. Leaching, accumulation and release of nitrogen in decomposing forest litter. In: Terrestrial Nitrogen Cycles. Processes, Ecosystem Strategies and Management Impacts. Clark, F.E. and Rosswall, T. (eds.), Ecol. Bull. (Stockholm) 33, 163–178.
6. Birch, H.F. 1958. The effects of soil drying on humus decomposition and nitrogen availability. Plant Soil 10, 9–31.
7. Bremner, J.M. and Shaw, K. 1958 Denitrification in soil. II. Factors affecting denitrification. J. Agric. Sci. 51, 40–52.
8. Brotonegoro, S., Abdulkadir, S., Sukiman, H. and Partohardjono, J. 1981 Nitrogen cycling in lowland rice fields with special attention to the N_2 fixation. In: Nitrogen Cycling in South-East Asian Wet Monsoonal Ecosystems. Wetselaar, R., Simpson, J.R., and Rosswall, T. (eds.), Australian Academy of Science, Canberra, pp. 36–40.
9. Brown, A.D. 1981 Groundwater transport of nitrogen in rice fields in northern Thailand. In: Nitrogen Cycling in South-East Asia Wet Monsoonal ecosystems. Wetselaar, R., Simpson, J.R. and Rosswall, T., (eds.), Australian Academy of Science, Canberra, pp. 165–170.
10. Burford, J.R. and Bremner, J.M. 1975 Relationship between denitrification

capacities of soils and total, water-soluble, and readily decomposable soil organic matter. Soil Biol. Biochem. 7, 389–394.

11. Charreau, C. and Nicou, R. 1971 L'amélioration du profil cultural dans les sols sableux et sablo-argileux de la zone tropicale sèche Ouest-Africaine et ses incidences agronomiques. Bulletin agronomique, n° 23, Institut de Recherches Agronomiques Tropicales, Paris, 254 p.

12. Chopart, J.L. 1980 Etude au champ des systèmes racinaires des principales cultures pluviales au Sénégal (Arachide-Mil-Sorgho-Riz pluvial). Thèse Doctorat d'Université Institut National Polytechnique de Toulouse, No. 10.

13. Craswell, E.T. and Vlek, P.L.G. 1979. Fate of fertilizer nitrogen applied to wetland rice. In: Nitrogen and Rice, pp. 175–193. International Rice Research Institute Los Baños, Philippines.

14. Denmead, O.T., Simpson, J.R. and Freney, J.R. 1974. Ammonia flux into the atmosphere from a grazed pasture. Science 185, 609–610.

15. Denmead, O.T., Catchpoole, V.R., Simpson, J.R. and Freney, J.R. 1976 Experiments and models relating to the atmospheric diffusion of ammonia from plane and line sources. Proc. Symp. Air Pollution Diffusion Modelling, Canberra 1976, pp. 115–124. Australian Environment Council, Canberra.

16. Denmead, O.T., Simpson, J.R. and Freney, J.R. 1977 A direct field measurement of ammonia emission after injection of anhydrous ammonia. Soil Sci. Soc. Amer. J. 41, 1001–1004.

17. Denmead, O.T., Freney, J.R. and Simpson, J.R. 1979 Nitrous oxide emission during denitrification in a flooded field. Soil Sci. Soc. Amer. J. 43, 716–718.

18. Denmead, O.T., Freney, J.R. and Simpson, J.R. 1982. Dynamics of ammonia volatilization during furrow irrigation of maize. Soil Sci. Soc. Amer. J. (in press).

19. Döbereiner, J. and Boddey, R.M. 1981 Nitrogen fixation in association with Gramineae. In: Current Perspectives in Nitrogen Fixation, pp. 305–312. Gibson, A.H. and Newton, W.E. (eds.), Australian Academy of Science, Canberra.

20. Dommergues, Y. 1963 Evaluation du taux de fixation de l'azote dans un sol dunaire reboisé en filao (Casuarina equisetifolia). Agrochimica 7, 335–340.

21. Dommergues, Y. and Rinaudo, G. 1979 Factors affecting N_2 fixation in the rice rhizosphere. In: Nitrogen and Rice, pp. 241–260. International Rice Research Institute, Los Baños, Philippines.

22. Dommergues, Y., Balandreau, J., Rinaudo, G. and Weinhard, P. 1973 Non-symbiotic nitrogen fixation in the rhizospheres of rice, maize and different tropical grasses. Soil. Biol. Biochem. 5, 83–89.

23. Dowdell, R.J., Burford, J.R. and Crees, R. 1979 Losses of nitrous oxide dissolved in drainage water from agricultural land. Nature (London) 278, 342–343.

24. Farquhar, G.D., Wetselaar, R. and Firth, P.M. 1979 Ammonia volatilization from senescing leaves of maize. Science 203, 1257–1258.

25. Farquhar, G.D., Firth, P.M., Wetselaar, R. and Weir, B. 1980 On the gaseous exchange of ammonia between leaves and the environment: determination of the ammonia compensation point. Plant Physiol. 66, 710–714.

26. Firth, P., Thitipoca, H., Suthipradit, S., Wetselaar, R. and Beech, D.F. 1973 Nitrogen balance studies in the central plain of Thailand. Soil Biol. Biochem. 5, 41–46.

27. Floate, M.J.S. 1981 Effects of grazing by large herbivores on nitrogen cycling in agricultural ecosystems. In: Terrestrial Nitrogen Cycles. Processes, Ecosystem Strategies and Management Impacts. Clark, F.E. and Rosswall, T. (eds.), Ecol. Bull. (Stockholm) 33, 585–601.

32

28. Floate, M.J.S. and Torrance, C.J.W. 1970 Decomposition of the organic materials from hill soils and pastures. I. Incubation method for studying the mineralization of carbon, nitrogen and phosphorus. J. Sci. Food Agric. 21, 116—120.
29. Freney, J.R. and Denmead, O.T. 1981 Recent developments in methods for studying nitrogen cycle processes in the field. In: Nitrogen Cycling in South-East Asian Wet Monsoonal Ecosystems. Wetselaar, R., Simpson, J.R. and Rosswall, R. (eds.), Australian Academy of Science, Canberra, pp. 187—194.
30. Freney, J.R., Simpson, J.R. and Denmead, O.T. 1981 Ammonia volatilization. In: Terrestrial Nitrogen Cycles. Processes, Ecosystems Strategies and Management Impacts. Clark, F.E. and Rosswall, T. (eds.), Ecol. Bull. (Stockholm) 33, 291—302.
31. Freney, J.R., Denmead, O.T., Watanabe, I. and Craswell, E.T. 1981 Ammonia and nitrous oxide losses following application of ammonium sulphate to flooded rice. Aust. J. Agric. Res. 32, 37—45.
32. Ganry, F. and Guiraud, G. 1979 Mode d'application du fumier et bilan azoté dans un système mil-sol sableux du Sénégal. In: Isotopes and Radiation in Research on Soil-Plant Relationships. IAEA-SM-235/16, pp. 313—331. International Atomic Energy Agency, Vienna.
33. Ganry, F., Guiraud, G. and Dommergues, Y. 1978 Effect of straw incorporation on the yield and nitrogen balance in the sandy soil — Pearl millet cropping system in Senegal. Plant Soil 50, 647—662.
34. Ganry, F. and Wey, J. 1974, 1975, 1976, 1977 and 1980 Coordinated research programme on the use of isotopes in fertilizer efficiency studies on grain legumes. Research contract No. RC/1296-SEN of the joint FAO/IAEA Division. In: Reports presented at workshops held in Vienna.
35. Hanawalt, R.B. 1969 Environmental factors influencing the sorption of atmospheric ammonia by soils. Soil Sci. Soc. Amer. Proc. 33, 231—234.
36. Hauck, R.D. 1981 Nitrogen fertilizer effects on nitrogen cycle processes. In: terrestrial Nitrogen Cycles. Processes, Ecosystem Strategies and Management Impacts. Clark, F.E. and Rosswall, T. (eds.), Ecol. Bull. (Stockholm) 33, 551—562.
37. Henzell, E.F., Fergus, I.F. and Martin, A.E. 1966 Accumulation of soil nitrogen and carbon under a *Desmodium uncinatum* pasture. Aust. J. Exp. Agric. Anim. Husb. 6, 157—160.
38. Henzell, E.F., Fergus, I.F. and Martin, A.E. 1967 Accretion studies of soil organic matter. J. Aust. Inst. Agric. Sci. 33, 35—37.
39. Hobson, P.N. 1969 Annual report of studies in animal nutrition and allied sciences, vol. 25, p. 35. The Rosett Research Institute, Aberdeen. Cited by Nolan [58].
40. Hutchinson, G.L., Millington, R.J. and Peters, D.B. 1972 Atmospheric absorption by plant leaves. Science 175, 771—772.
41. International Rice Research Institute. 1979 Annual Report for 1978. The International Rice Research Institute, Los Baños, Philippines, 478 pp.
42. Jones, M.J. 1974 Effect of previous crop on yield and nitrogen response of maize at Samaru, Nigeria. Expl. Agric. 10, 273—274.
43. Junge, C.E. 1956 Recent investigations in air chemistry. Tellus 8, 27—139.
44. Keys, A.J., Bird, I.F., Cornelius, M.J., Lea, P.J., Wallsgrove, R.M. and Miflin, B.J. 1978 Photorespiratory nitrogen cycle. Nature (London) 275, 741—743.
45. Khind, C.S. and Ponnamperuma, F.N. 1981 Effects of water regime on growth, yield, and nitrogen uptake by rice. Plant Soil 59, 287—298.
46. Kingsley, A.F., Clogett, C.O., Klosterman, H.J. and Stoa, T.E. 1957 Loss of minerals from mature wheat and flax by simulated rain. Agron. J. 49, 37—39.

47. Knowles, R. 1981 Denitrification. In: Terrestrial Nitrogen Cycles. Processes, Ecosystem Strategies and Management Impacts. Clark, F.E. and Rosswall, T. (eds.), Ecol. Bull. (Stockholm) 33, 315–329.
48. Knowles, R. 1981 The measurement of nitrogen fixation. In: Current Perspectives in Nitrogen Fixation, pp. 327–333. Gibson, A.H. and Newton, W.E. (eds.), Australian Academy of Science, Canberra.
49. Koyama, T. and App, A. 1979 Nitrogen balance in flooded rice soils. In: Nitrogen and Rice, pp. 95–104. International Rice Research Institute, Los Baños, Philippines.
50. Likens, G.E., Bormann, F.H., Pierce, R.S., Easton, J.S. and Johnson, N.M. 1977 Biogeochemistry of a forested ecosystem. Springer-Verlag, New York, Berlin.
51. Lodge Jr., J.P., Machado, B.A., Pate, J.G., Sheesley, A.C. and Wartburg, A.F. 1974 Atmospheric trace chemistry in the American humid tropics. Tellus 26, 250–253.
52. Malo, B.A. and Purvis, E.R. 1964 Soil absorption of atmospheric ammonia. Soil Sci. 97, 242–247.
53. Murayama, S. 1977 Saccharides in some Japanese paddy soils. Soil Sci. Plant Nutr. 23, 479–489.
54. Myers, R.J.K. 1975 Temperature effects on ammonification and nitrification in a tropical soil. Soil Biol. Biochem. 7, 83–86.
55. Myers, R.J.K. 1978 Nitrogen and phosphorus nutrition of dryland grain sorghum at Katherine, Northern Territory 3. Effect of nitrogen carrier, time and placement. Aust. J. Exp. Agric. Anim. Husb. 18, 834–843.
56. National Research Council. 1976 Nutrient requirements of beef cattle. National Academy of Sciences, National Research Council, Washington, D.C., p.6.
57. Nicou, R. 1978 Etude de succession culturales au Sénégal – Résultats et méthodes. Agro. Trop. 33, 51–61.
58. Nolan, J.V. 1975 Quantitative models of nitrogen metabolism in sheep. In: Digestion and Metabolism in the Ruminant, pp. 416–431. McDonald, I.W. and Warner, A.C.I. (eds.), The University of New England Publishing Unit, Armidale, Australia.
59. Norman, M.J.T. 1966 Katherine Research Station 1956–64: A review of published work. CSIRO, Australia, Div. of Land Research, Tech. Pap. No. 28.
60. Norman, M.J.T. and Wetselaar, R. 1960 Performance of annual fodder crops at Katherine, N.T. CSIRO, Australia, Div. of Land Res. Regional Survey, Tech. Pap. No. 9.
61. Norman, M.J.T. and Wetselaar, R. 1960 Loss of nitrogen on burning native pasture at Katherine, N.T. J. Aust. Inst. Agric. Sci. 26, 272–273.
62. Passioura, J.B. and Wetselaar, R. 1972 Consequences of banding nitrogen fertilizers in soil II. Effect on the growth of wheat roots. Plant Soil 36, 461–473.
63. Pieri, C. 1982 La fertilisation potassique du mil Pennisetum et ses effets sur la fertilité d'un sol sableux. Revue de la Potasse, Berne (in press).
64. Ponnamperuma, F.N. 1978 Electrochemical changes in submerged soils and the growth of rice. In: Soils and Rice, pp. 421–441. International Rice Research Institute, Los Baños, Philippines.
65. Porter, L.K., Viets Jr., F.G. and Hutchinson, G.L. 1972 Air containing nitrogen-15 ammonia: foliar absorption by corn seedlings. Science 175, 759–761.
66. Pushparajah, E. 1981 Nitrogen cycle in rubber (*Hevea*) cultivation. In: Nitrogen Cycling in South-East Asian Wet Monsoonal Ecosystems. Wetselaar,

34

R., Simpson, J.R. and Rosswall, T. (eds.), Australian Academy of Science, Canberra, pp. 101–108.
67. Ruinen, J. 1974 Nitrogen fixation in the phyllosphere. In: The Biology of Nitrogen Fixation, pp. 121–167. Quispel, A. (ed.), Elsevier Scientific Publishing Company, Amsterdam, New York.
68. Samy, J. and Vamadevan, V.K. 1981 Sources of nitrogen and crop responses to fertilizer nitrogen in rice double-cropping systems in Malaysia. In: Nitrogen Cycling in South-East Asia Wet Monsoonal Ecosystems. Wetselaar, R., Simpson, J.R. and Rosswall, T. (eds.), Australian Academy of Science, Canberra, pp. 92–95.
69. Siband, P. 1976 Quelques réflexions sur les potentialités et les problèmes des sols gris de Casamance (Sénégal méridional). Agro. Trop. 31, 105–113.
70. Simpson, J.R. and Stobbs, T.H. 1981 Nitrogen supply and animal production from pastures. In: Grazing Animals, pp. 261–287. Morley, F.H.W. (ed.), Elsevier Scientific Publishing Company, Amsterdam, New York.
71. Singh, K.P. and Pandey, O.N. 1981 Cycling of nitrogen in a tropical deciduous forest. In: Nitrogen Cycling in South-East Asian Wet Monsoonal Ecosystems. Wetselaar, R., Simpson, J.R. and Rosswall, T. (eds.), Australian Academy of Science, Canberra, pp. 123–130.
72. Stanford, G., Dzienia, S. and Vanderpol, R.A. 1975 Effect of temperature on denitrification rate in soils. Soil Sci. Soc. Amer. Proc. 39, 968–970.
73. Stangel, P.J. 1979 Nitrogen requirement and adequacy of supply for rice production. In: Nitrogen and Rice, pp. 45–67. International Rice Research Institute, Los Baños, Philippines.
74. Stewart, W.D.P., Rowell, P., Ladha, J.K. and Sampaio, M.J.A.M. 1979 Blue-green algae (Cyanobacteria) – some aspects related to their role as sources of fixed nitrogen in paddy soils. In: Nitrogen and Rice, pp. 263–285. International Rice Research Institute, Los Baños, Philippines.
75. Stobbs, T.H. 1976 Milk production per cow and per hectare from tropical pastures. In: Proc., F.I.R.A. Seminario Internacional de Ganaderia Tropical. Acapulco, Mexico, pp. 129–146.
76. Stutte, C.A. and Weiland, R.T. 1978 Gaseous nitrogen loss and transpiration of several crop and weed species. Crop Sci. 18, 887–889.
77. Tanaka, A. and Navasero, S.A. 1964 Loss of nitrogen from the rice plant through rain and dew. Soil Sci. Plant Nutr. 10, 36–39.
78. Visser, S.A. 1964 Origin of nitrates in tropical rainwater. Nature (London) 201, 35–36.
79. Vlek, P.L.G. and Stumpe, J.M. 1978 Effects of solution chemistry and environmental conditions on ammonia volatilization losses from aqueous systems. Soil Sci. Soc. Amer. J. 42, 416–421.
80. Watanabe, I., Lee, K.K. and Alimagno, B.V. 1978 Seasonal change of N_2-fixing rate in rice field assayed by *in situ* acetylene reduction technique. I. Experiments in long term fertility plots. Soil Sci. Plant Nutr. 24, 1–13.
81. Watanabe, I., Craswell, E.T. and App, A. 1981 Nitrogen cycling in wetland rice fields in south-east and east Asia. In: Nitrogen Cycling in South-East Asian Wet Monsoonal Ecosystems. Wetselaar, R., Simpson, J.R. and Rosswall, T. (eds.), Australian Academy of Science, Canberra, pp. 4–17.
82. Weier, K.L. 1978 The fixation of nitrogen in association with roots of some tropical grasses. Soil Sci. Conf., Aust. Soil Sci. Soc. N.S.W. Branch, Armidale.
83. Weiland, R.T. and Stutte, C.A. 1979 Pyro-chemiluminescent differentiation of oxidized and reduced N forms evolved from plant foliage. Crop Sci. 19, 545–547.

84. Weiland, R.T. and Stutte, C.A. 1980 Concomitant determination of foliar nitrogen loss, net carbon dioxide uptake and transpiration. Plant Physiol. 65, 403–406.
85. Wetselaar, R. 1962 Nitrate distribution in tropical soils III. Downward movement and accumulation of nitrate in the subsoil. Plant Soil 16, 19–31.
86. Wetselaar, R. 1962 The fate of nitrogenous fertilizers in a monsoonal climate. In: Trans. Int. Soil Conf. Commisions IV and V, pp. 588–595. Neal, G.J. (ed.), Soil Bureau, Lower Hutt, New Zealand.
87. Wetselaar, R. 1967 Determination of the mineralization coefficient of soil organic nitrogen on two soils at Katherine, N.T. Aust. J. Exp. Agric. Anim. Husb. 7, 266–274.
88. Wetselaar, R. 1967 Estimation of nitrogen fixation by four legumes in a dry monsoonal area of north-western Australia. Aust. J. Exp. Agric. Anim. Husb. 7, 518–522.
89. Wetselaar, R. 1972 Physiological aspects of rice nutrition. Aust. Rice Res. Conf. 1972, Leeton, N.S.W., pp. 2(c) 1–2(c) 9.
90. Wetselaar, R. 1980 Nitrogen cycling in a semi-arid region of tropical Australia. In: Nitrogen Cycling in West African Ecosystems, pp. 157–168. Rosswall, T. (ed.), SCOPE/UNEP International Nitrogen Unit and the Swedish Academy of Science, Stockholm.
91. Wetselaar, R. 1981. Nitrogen inputs and outputs of an unfertilized paddy field. In: Terrestrial Nitrogen Cycles. Processes, Ecosystem Strategies and Management Impacts. Clark, F.E. and Rosswall, T. (eds.), Ecol. Bull (Stockholm) 33, 573–583.
92. Wetselaar, R. and Farquhar, G.D. 1980 Nitrogen losses from tops of plants. Adv. Agron. 33, 263–302.
93. Wetselaar, R. and Hutton, J.T. 1963 The ionic composition of rainwater at Katherine, N.T., and its part in the cycling of plant nutrients. Aust. J. Agric. Res. 14, 319–329.
94. Wetselaar, R. and Norman, M.J.T. 1960 Soil and crop nitrogen at Katherine, N.T. CSIRO, Australia, Div. of Land Res. and Regional Survey, Tech. Pap. No. 10.
95. Wetselaar, R. and Norman, M.J.T. 1960. Recovery of available soil nitrogen by annual fodder crops at Katherine, Northern Territory. Aust. J. Agric. Res. 11: 693–704.
96. Wetselaar, R., Passioura, J.B. and Singh, B.R. 1972 Consequences of banding nitrogen fertilizers in soil I. Effect on nitrification. Plant Soil 36, 159–175.
97. Wetselaar, R., Begg, J.E. and Torssell, B.W.R. 1974 The contribution of litter nitrogen to soil fertility in a tropical grass-legume pasture. Aust. Soil Sci. Conf., Melbourne, 1974, pp. 3(c) 13–3(c) 18.
98. Wetselaar, R., Shaw, T., Firth, P., Oupathum, J. and Thitipoca, H. 1977 Ammonia volatilization losses from variously placed ammonium sulphate under lowland rice field conditions in Central Thailand. Proc. Int. Sem. Soil Env. Fertility Manag. Intens. Agric., pp. 282–288. The Society of the Science of Soil and Manure, Tokyo.
99. Wetselaar, R., Denmead, O.T. and Galbally, I.E. 1981 Environmental problems associated with terrestrial nitrogen transformations in agrosystems in the wet monsoonal tropics. In: Nitrogen Cycling in South-East Asian Wet Monsoonal Ecosystems. Wetselaar, R., Simpson, J.R. and Rosswall, T. (eds.), Australian Academy of Science, Canberra, pp. 157–164.
100. Woo, K.C., Berry, J.A. and Turner, G.L. 1978 Release and refixation of ammonia during photorespiration. Carnegie Inst. Yearb. 77, 240–245.

101. Woodmansee, R.G., Vallis, I. and Mott, J.J. 1981 Grassland nitrogen. In: Terrestrial Nitrogen Cycles. Processes, Ecosystem Strategies and Management Impacts. Clark, F.E. and Rosswall, T. (eds.), Ecol. Bull. (Stockholm) 33, 443–462.
102. Zhu, Z.L. 1981 Nitrogen cycling and the fate of fertilizer nitrogen in rice fields of the Suchow district, Yiangsu province, China. In: Nitrogen Cycling in South-East Asian Wet Monsoonal Ecosystems. Wetselaar, R., Simpson, J.R. and Rosswall, T. (eds.), Australian Academy of Science, Canberra, pp. 73–76.

2. Nitrogen fixation by legumes in the tropics

A.H. GIBSON, B.L. DREYFUS and Y.R. DOMMERGUES

Introduction

Many tropical soils are limited in their ability to produce crops due to severe deficiency in soil nitrogen. Agronomists readily agree that the legumes, with their ability to reduce atmospheric nitrogen to a usable form of nitrogen, should be an important component of tropical agrosystems. Although some nodulated legumes have an outstanding potential for fixing N_2 (*Sesbania cannabina* and *Leucaena leucocephala* can fix up to 500 kg N ha^{-1} yr^{-1}, Table 1), many of them fail to achieve their potential in the field. Thus when biological or environmental conditions are not favorable, the N_2-fixing activity of *Leucaena leucocephala* may be as low as 13% of that found in a more favorable environment (Table 1). Other examples given in Table 1 confirm that the range of N_2-fixing rates of legumes is

Table 1. Examples of N_2 fixation rates (kg ha^{-1} crop^{-1}) by some tropical legumes grown for grains, forage or wood

Plant	Location	N_2 fixation rate	Method of estimation	References
Glycine max (soybean)	Senegal	165	Difference	[60]
	Senegal	200	Fertilizer equivalent	Mugnier (unpub.)
	India	102	A value	[56]
	n.i.	54–369	n.i.	[70]
Arachis hypogaea (groundnut)	Senegal	25–56	A value	Table 8
	n.i.	84–297	n.i.	[70]
	Israel	87–220	Acetylene technique	[137]
Vigna unguiculata (cowpea)	n.i.	24–240	n.i.	[70]
Centrosema pubescens (centro)	Africa	126–395		[115, 128]
Sesbania cannabina		542 yr^{-1}		[128]
Leucaena leucocephala (= glauca)		74–584 yr^{-1}		[4, 128]
Macroptilium atropurpureum		291		[128]

n.i. = not indicated.

Y.R. Dommergues and H.G. Diem (eds.), Microbiology of Tropical Soils and Plant Productivity. ISBN 978-94-009-7531-6.
© 1982 Martinus Nijhoff/Dr W. Junk Publishers, The Hague/Boston/London.

very large. Such variations are attributable to the impact of limiting factors that occur in the field in such a way that N_2 fixation is often seriously depressed.

The main objectives of this chapter are: (1) to identify the parameters that determine the N_2-fixing potential of the legumes; (2) to indicate the key limiting factors that operate in the tropics; (3) to suggest methods to reduce or eliminate the effect of these limiting factors.

1. Potential Symbiotic Nitrogen Fixation

The potential symbiotic N_2 fixation of a given legume is defined as the maximum activity of that legume when nodulated with the most effective Rhizobium strain and grown under the most favorable environmental conditions.

Total N_2 fixation is dependent upon two parameters, nodule weight and the specific N_2-fixing activity (N_2 fixed per gram nodule). While nodule weight shows a general increase during the growing season, there can be major fluctuations (Fig. 1). Futhermore, specific N_2-fixing activity, as estimated with the acetylene reduction technique, shows wide fluctuations (Fig. 2). These may be due to short-term changes in temperature, light intensity, moisture, or longer term effects involving these variables or stage of physiological development of the plants. It should be recognized that the acetylene reduction assay measures activity only during the course of the assay, and many observations are required to estimate overall rates of N_2 fixation [154]; in addition, it is necessary to know the appropriate $C_2H_2:N_2$ ratio to apply to convert the assay results to N_2 fixed.

One important feature of the nodulation pattern is the period taken for nodules to appear and commence N_2 fixation. This lag depends on a number of factors. One is the Rhizobium strain involved (e.g. Law and Strijdom [101]); those strains which nodulate first are active competitors against strains which nodulate later. The size of the population of rhizobia, the availability of soil moisture, the soil temperature, and the level of soil nitrogen are other important determinants affecting the speed of nodulation. The nodule pattern usually exhibits a maximum which occurs during the second half of the plant's growth. This maximum is followed by a decrease, often roughly sinusoidal, which indicates successive nodule turnovers. Such processes are most difficult to observe and consequently poorly documented.

The pattern of N_2 fixation by a legume nodule exhibits three phases — lag (during nodule initiation and early development), exponential (rapid nodule growth and development of nitrogenase activity), and senescence (breakdown of nodule tissue and decline in nitrogenase activity, which may be rapid or extend over a considerable period). With a growing plant, nodule initiation and development is occurring continually, but the overall pattern of N_2 fixation by the population of nodules on a plant resembles that of individual nodules, but over a longer time span. Droughting, and excess moisture, are two environmental factors known to promote nodule senescence. But of greater significance is the stage of physiological development. For example, numerous people have attributed the decline in nodule

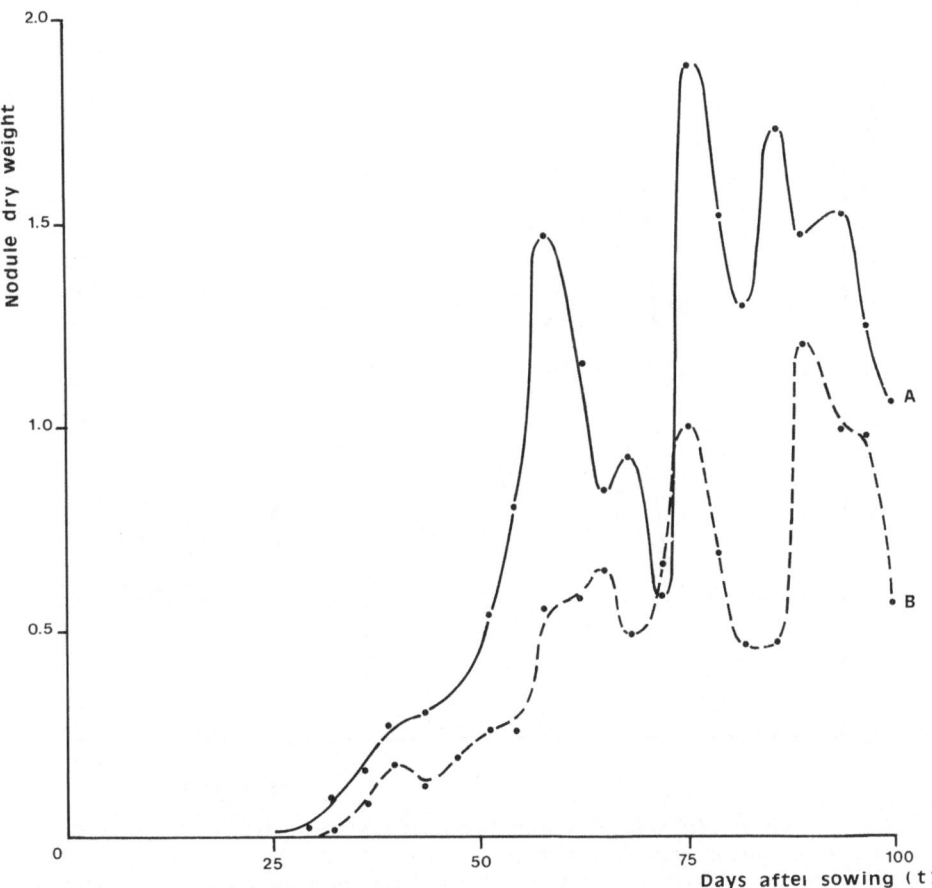

Fig. 1. Nodule weight of *Glycine max* cv. Jupiter grown at the ORSTOM station, Senegal during summer, the most favorable period for soybean growth. *Plot A*: soybean irrigated during the whole cycle; *Plot B*: soybean grown in an adjacent plot (same soil) but infested by nematodes; irrigation was interrupted between day 56 and 68. Each point is the mean of 10 replicates (unpublished data).

activity to the diversion of photosynthate supplies from the nodules to developing pods. In an attempt to define this more precisely, Lawn and Brun [102] suggested that the decline occurs when the growth rate of the pods equals that of the total plant shoot. Unfortunately, few other attempts have been made to examine this hypothesis, which could provide a basis for selecting the best species or varieties for an area. The strain of rhizobia has also been implicated as a factor determining the duration of N_2 fixation [30]. One of the major gaps in our knowledge of the symbiosis between legumes and rhizobia is that relating to host and bacterial factors responsible for normal nodule senescence, i.e. not induced by moisture excess or deficiency, high temperature etc.

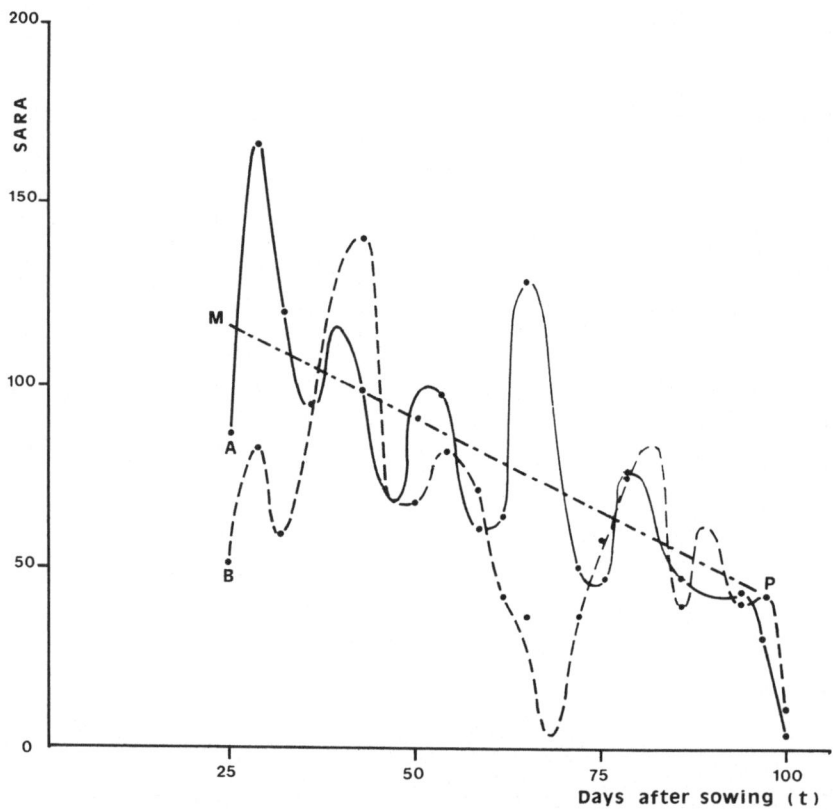

Fig. 2. Pattern of specific N_2-fixing activity (measured as μmoles C_2H_4 per g nodule (d.w.) per h, or SARA) of the soybean plants from plots A and B whose nodule pattern is represented in Fig. 1. The SARA of soybean from plot B exhibited a dramatic decrease when irrigation was interrupted (unpublished data).

The maximum specific N_2-fixing activity is an interesting feature of legumes. As direct methods of measurement are cumbersome, most determinations utilize the acetylene reduction assay. In general, meristematic-type nodules (e.g. Trifolium, Pisum and Phaseolus species) show significantly higher specific nitrogenase activities than the spherical or determinate-type nodules found on many tropical legumes (Table 2). An interesting exception is *Arachis hypogaea* [149] (Table 2) which has high specific activity. Vigna species have also shown high specific activity as young plants ($> 300\,\mu$moles/g nodule dry wt/hr) although the levels decline rapidly from two weeks after germination (Gibson, unpublished). An interesting observation in this program was that the rate of N increase in these plants growing in N-free media was constant, even though specific nitrogenase activity, determined by acetylene reduction assays, was declining. This indicates that caution should be exercised in extrapolating from acetylene reduction assays to the determination of N_2 fixation.

Under field conditions, many biological and environmental factors affect the

Table 2. Parameters of N$_2$ (C$_2$H$_2$) fixation in field-grown legumes

Plant	Location	Maximum nodule wt. (g plant)		Maximum specific activity (SARA) μmole C$_2$H$_4$ h^{-1} g^{-1}		References
		fresh	dry	fresh wt. basis	dry wt. basis	
Glycine max (soybean)						
cv. Jupiter	Senegal	–	1.2	–	200	[112]
cv. Clay	Minnesota	4.0	–	80	–	[102, 103]
cv. Chippewa	–id–	4.8	–	26	–	[102, 103]
Arachis hypogaea (groundnut)						
cv. 71-234	Israel	–	0.6–0.9	–	140–975	[137]
cv. 57-422	Senegal	–	0.27	–	400	[112]
cv. 28-206	–id–	–	0.15	–	400	[112]
cv. 55-437	–id–	–	0.09	–	400	[112]
cv. Kediri 71-1	India	–	–	–	600	[3]
Cowpea	Nigeria	–	0.3	60	–	[7]
Sesbania rostrata	Senegal					
stem		40.0	–	53	–	[44]
root		15.0	–	17	–	[44]
Acacia bivenosa	Senegal	–	–	–	40	[43]
Phaseolus vulgaris	Colombia	–	–	45	–	[73, 75]

N$_2$-fixing potential of legumes, through an effect on the nodulation pattern or on the specific nitrogenase activity of the nodules. The objective of the following sections is to attempt to define the major factors limiting N$_2$ fixation by legumes grown in the tropics.

2. Limiting Factors

2.1. Physical Factors

Relatively little is known of the effect of high soil temperatures on the survival of rhizobia under field conditions. Heating under moist conditions promotes death more rapidly than under dry conditions, and a 'wet' inoculant is more susceptible to higher temperature than a 'dry' inoculant [15]. The protection provided to *R. trifolii* and *R. leguminosarum* by montmorillonite clay against desiccation and heating, is not found within the slow-growing *Rhizobium* sp. [24], although this latter group appear capable of growth at higher temperatures than the fast-growers [15]. Adaptation to high soil temperatures has been advocated [178]. High temperature-tolerant strains of *Rhizobium* sp. have been isolated recently [47]; these isolates are also able to form effective nodules under high temperature conditions. The observation that moderate temperatures (35 °C) can induce plasmid loss, with consequent loss of nodulating ability [182] could have serious implications regarding the effectiveness and nodulating ability of strains of rhizobia under tropical conditions, although Eaglesham *et al.* [47] found no evidence for 'curing' in their study.

High temperatures reduce the longevity of commercial inoculants, and refrigeration is recommended during transportation and storage. The lack of adequate refrigeration in many tropical and sub-tropical countries can create severe problems in maintaining reasonable populations of rhizobia in inoculants.

Both nodulation and N_2 fixation are adversely affected by higher soil temperatures [64]. In general, temperatures above $30\,°C$ reduce the level of nodulation, although species such as *Acacia mellifera* [78], *Stylosanthes gracilis* and *Pueraria javanica* [157] make good growth up to $35\,°C$. As with the fast-growing *R. trifolii*, some *Rhizobium japonicum* strains, and those nodulating *Cicer arietinum*, are able to maintain high rates of N_2 fixation at temperatures under which many strains show low nitrogenase activity [30].

Moisture stresses, due to either deficiency or excess, are detrimental to the legume—Rhizobium symbiosis. Rhizobia show a rapid decline in viability under drying conditions, and this is exacerbated by cycles of wetting and drying [132, 135]. Nodulation does not occur when the soil is dry [10, 39] due to the failure of infection [180]. Desiccation severely depresses N_2 fixation by nodulated plants, due either to moisture loss from the nodules [158, 159] or the cessation of photosynthesis [85]. Prolonged desiccation leads to nodule loss (Fig. 1), with a consequent reduction in the level of N_2 fixation while new nodules form. Although the soil in the vicinity of the nodules may be dry, provided adequate moisture is supplied from the region below the nodule zone, reasonable rates of N_2 fixation will be maintained [86]. Soil moisture deficits are frequently associated with high soil temperatures, and the distinction between effects due to moisture stress and temperature is difficult in the field.

Waterlogging, especially when associated with poor soil structure, retards both nodulation and N_2 fixation even though some degree of adaptation is possible [141]. These effects are thought to be associated with low oxygen levels [28, 113, 134]. Excess water has pronounced effects on the growth of *Vigna unguiculata*, the effects being greater when such stress occurs early in plant growth [84, 114]. Legarda and Forsythe [104] found that production by *Phaseolus vulgaris* was retarded where the oxygen diffusion rate was less than $24–28\,g \times 10^{-8}$ cm^{-2} min^{-1}, a value higher than that considered by Loveday [107] and Gradwell [72] to retard the nodulation of Trifolium species. Some years ago, Masefield [110] made the interesting observation that the nodulation of a number of tropical legumes was vastly improved where the water table was high. Recent work in Australia has shown that soybeans grown with a water table maintained at 15 cm were highly productive, raising the possibility of 'paddy soybeans' [87]. From a study of survival of rhizobia during the paddy phase of rice cultivation, Rerkasem and Tongkumdee [138] concluded that introduced *R. japonicum* were more sensitive to flooding than naturally-occurring strains nodulating *Leucaena leucocephala*.

Both photoperiod and light intensity have been reported to affect nodulation and N_2-fixing activity [64]. Such an effect should not be overlooked with leguminous crops established under the canopy of trees, such as rubber, since in comparable situations nodulation and N_2 fixation are reduced under shaded conditions [161]. Similarly increasing plant density decreases N_2 fixation [74, 160]. Thus N_2 fixation of *Phaseolus vulgaris* cultivar P590 reached a maximum when the density was 8.5 plants per m^2 and rapidly decreased at higher densities (Fig. 3). One of the three cultivars studied was less sensitive to the harmful effect of increasing density. Expressing N_2 fixation on an area basis, N_2 fixation was shown to increase with density for all three cultivars.

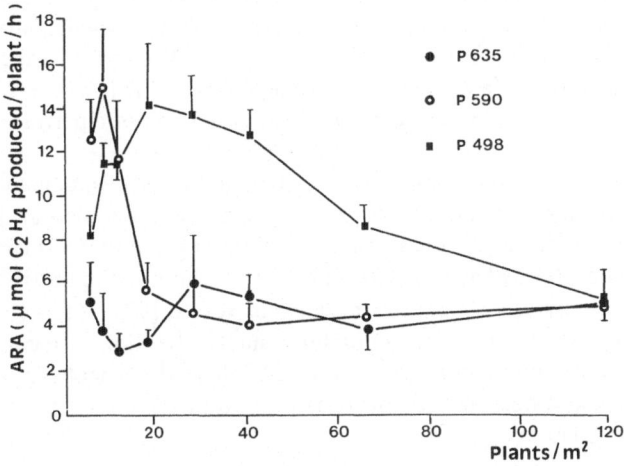

Fig. 3. Influence of planting density on N_2 fixation expressed as μmoles C_2H_4 produced per plant per h in three cultivars of *Phaseolus vulgaris*, 39 days after planting [74].

2.2. Nutritional factors

Nutritional factors may affect N_2 fixation by legumes by a direct effect on the infection and nodulation process, by influencing the effectiveness of the legume—Rhizobium symbiosis, or through an effect on the metabolism and growth of the plant independent of any effect on the symbiosis. Many plant nutrients have multiple functions, such that it is difficult to distinguish at which level they are acting. Among the nutritional factors which play a role in different stages of the symbiosis, soil acidity, phosphorus, molybdenum and combined N appear to be the most important in tropical conditions.

2.2.1. Soil acidity. Tropical legumes and their rhizobia are reputedly able to tolerate low pH better than the temperate legumes [125, 126]; however the usefulness of this generalization is now compromised by numerous exceptions [118, 119, 120, 121]. In West African soils, *Vigna unguiculata* and *Arachis hypogaea* are relatively tolerant of low pH (4—5) but as shown by Date [32], *Leucaena leucocephala* does not grow or nodulate in the low pH soils; it responds strongly to liming [88]. The complexity of the situation is indicated by a recent study with *Vigna radiata* [122] and *V. unguiculata* [93] in which strains of *Rhizobium* sp. demonstrated wide variation in acid tolerance, and their performance was influenced by the host cultivar. With soybeans growing in an acid soil in Nigeria, Bromfield and Ayanaba [19] observed that inoculation with strains of *Rhizobium japonicum* selected for symbiotic competence in acid soils resulted in yields up to $2\,t\,ha^{-1}$, and that liming had no effect on the response.

Acidity, calcium deficiency, and both manganese and aluminium toxicity tend to occur together, and determination of the precise cause of inadequate nodulation and N_2 fixation is frequently difficult. Toxicity by Mn and Al generally affects plant growth and the effect on nodulation is indirect [48] whereas low pH and calcium deficiency effects are more directed towards symbiotic performance.

2.2.2. Phosphorus. In many tropical soils, particularly acid soils, available phosphate is an important limiting factor for both growth and nodulation [118, 119]. Nodules often contain 2—3 times more phosphorus per unit dry matter than the roots [117]. Phosphorus has a beneficial effect on the nodulation of *Stylosanthes humilis*, nodule number and dry weight being increased [57]. It is well known that phosphate fertilization improves nodulation and N_2 fixation of legumes [2], but equally important is the maintenance of a well-balanced nutrient supply [58]. The effects of vesicular-arbuscular mycorrhizae on phosphate supply of the host plant are discussed in Chapter 8.

2.2.3. Combined nitrogen. The inhibition of nodulation and N_2 fixation by combined nitrogen has been long recognized [54, 173] but the nature of the inhibition is not clearly understood. Three phases are affected — root hair infection, nodule growth and nitrogenase activity. Current evidence suggests that an effect of nitrate on lectins, or specific recognition glycoproteins, on the root hair surface may be responsible for the lowered level of infection [37, 38]. Reduced nodule growth and nitrogenase activity appear to be a consequence of a lower supply of photosynthate to the nodules, due to the diversion of the photosynthate to assimilate the combined nitrogen [62, 67, 155]. Recent evidence for a localized effect of combined nitrogen was obtained with *Sesbania rostrata*, which shows profuse stem nodulation; these nodules are unaffected by supplying the roots with 3 m*M* nitrate, but root nodulation is severely retarded [42].

Due to fluctuations in nitrate and ammonium levels in soil, it is difficult to assess the significance of combined nitrogen under natural conditions. Tropical soils are frequently low in nitrogen [92], but high levels of nitrate may exist in non-fertilized fields for short periods following the dry season [76]. This could delay nodulation and hence the onset of N_2 fixation, as we have observed with *Arachis hypogaea* and cowpeas in Senegal, and be an important factor in limiting N_2 fixation. Against this, low levels of combined nitrogen can enhance nodulation and N_2 fixation [62, 63] after an initial delay. Obviously further work is required to determine the precise effects of combined nitrogen under field conditions and to seek ways of reducing any inhibition of symbiotic activity. The suggestion that strains of rhizobia vary in their response to the effects of combined nitrogen [31] should be explored in more detail.

2.2.4. Trace elements. Of the various trace elements shown to be essential for successful nodulation (boron, copper, molybdenum, cobalt [118]), molybdenum is most likely to be deficient in acidic tropical soils. In a survey of 41 Brazilian soils, 36 were found to have Mo deficiency [52]. In Australia, the incorporation of Mo in the superphosphate has meant the difference between success and failure in the establishment of legumes in many acidic soils. Under certain conditions, liming will alleviate Mo deficiency by releasing Mo sorbed to clay particles at low pH.

2.3. Biological factors

Biological factors directly limiting N_2 fixation by legumes may be broadly divided into two categories, namely those in the host plants and the bacteria affecting the specificity of nodule formation and symbiotic N_2 fixation, and those affecting the survival and activity of the Rhizobium in the soil.

2.3.1. Symbiotic compatibility between Rhizobium and host. Specificity between host legumes and rhizobia occurs at three levels, namely the ability to form nodules, the ability to fix N_2 and the degree to which symbiotic N_2 fixation meets the plant's nitrogen requirements. Despite this, there are groups of plants that tend to nodulate with groups of strains of rhizobia (the cross-inoculation group concept of Fred *et al.* [54]). Although most soils throughout the world possess rhizobia, many do not contain rhizobia for particular hosts. For example, the vast area of Cerrado soils in Brazil generally lacks strains of rhizobia capable of nodulating soybeans, while some soils in Australia contain strains that nodulate soybeans but only ineffectively. Similarly the very strain-specific legumes *Leucaena leucocephala, Lotononis bainesii, Trifolium semipilosum* and *T. ambiguum* required inoculation when introduced into Australia [32]. It is interesting to note, however, that in West Africa, Asian cultivars of soybeans are effectively nodulated by some native strains of Rhizobium belonging to the cowpea miscellany [20, 89].

Many tropical legumes belong to the cowpea cross-inoculation group, and nodulation failure is rare. Due to their ability to form nodules with a wide range of *Rhizobium* sp. strains, the cowpea group is often considered promiscuous. However this generalization can be dangerous as strong specificity often exists with regard to the symbiotic effectiveness of the associations formed [34]. Figure 4 provides examples of all three forms of specificity referred to above, such examples

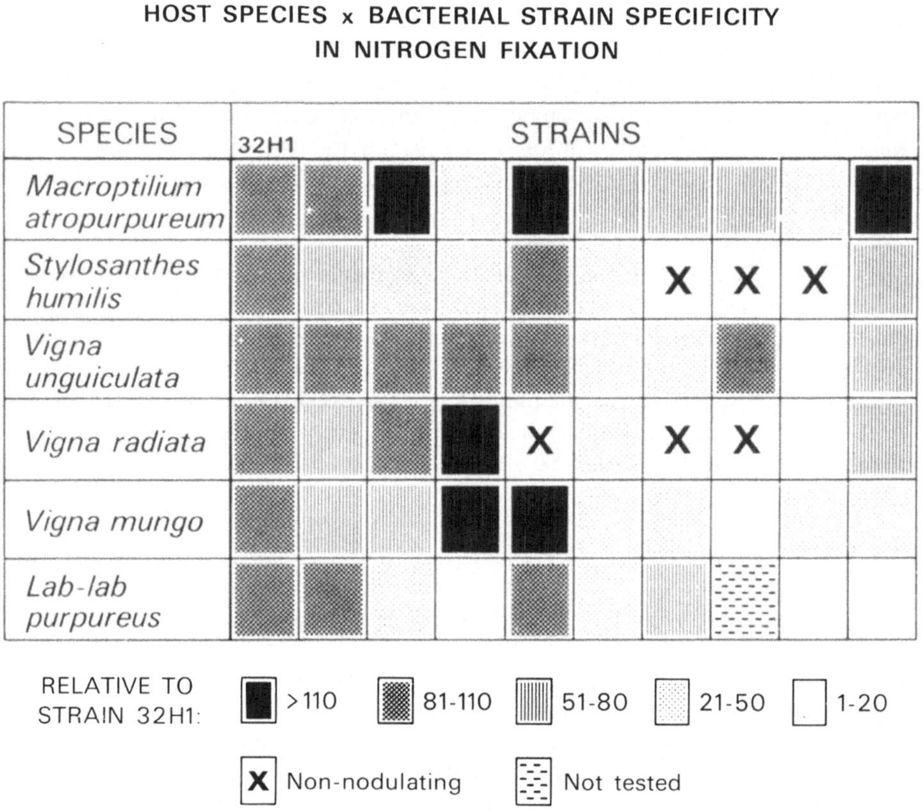

HOST SPECIES x BACTERIAL STRAIN SPECIFICITY
IN NITROGEN FIXATION

SPECIES	32H1	STRAINS								
Macroptilium atropurpureum										
Stylosanthes humilis						X	X	X		
Vigna unguiculata										
Vigna radiata				X		X	X			
Vigna mungo										
Lab-lab purpureus										

RELATIVE TO STRAIN 32H1: ▮ >110   ▨ 81-110   ▥ 51-80   ░ 21-50   ☐ 1-20

X Non-nodulating   ⋮ Not tested

Fig. 4. Host species X bacterial strain specificity in N_2 fixation [65].

occurring among 10 strains chosen for their symbiotic effectiveness with one or more hosts and six common legumes [65].

Hence it is important to exercise great care when introducing a species to a new area. A previous record of ready nodulation is no guarantee that the species will nodulate in the new environment. More importantly, appropriate nitrogen control treatments should be included with the introduced material in order to ascertain the extent to which the local population of rhizobia, and the inoculant bacteria, are able to develop symbiotic associations that will meet the plant's nitrogen requirements [66].

2.3.2. Ecology of Rhizobium in soil. Studies under the broad heading of the ecology of rhizobia are generally involved with two aspects, survival, or persistence, of both native and introduced populations, and competition between strains in forming nodules. Most of the biological limitations on the persistence of Rhizobium in the soil, including predation, parasitism and lysis, have been previously reviewed [133].

We shall here deal mainly with the survival of both native and newly introduced strains.

Surprisingly, relatively little information is available regarding changes in the populations of rhizobia in tropical soils, or in changes of individual strains in these populations. Recent work in Senegal has shown important fluctuations in the number of rhizobia nodulating *Acacia senegal*, a tree growing in sandy soils in the semi-arid region of the Sahel. During the dry season, when surface temperatures exceed 50 °C, very few rhizobia are found, but the numbers increase rapidly once the rainy season commences (Dreyfus, unpublished). While this may favor the introduction of improved strains, unless they are selected for their ability to withstand the extreme conditions, the benefits of inoculation are unlikely to extend beyond the year in which they are introduced. A major problem in studying individual strains is their identification, although the development of ELISA (enzyme-linked immunosorbent assay) and associated techniques [11, 97, 116, 179] and the use of antibiotic resistance markers in selected strains [23, 98] opens the way for the required detailed investigations without having to use host plants to 'sample' the soil population. Even so, such studies will be restricted to the measurement of changes in the population and further development of techniques will be required to study critical nutritional and biological factors influencing survival of strains, and of populations [133].

Possibly of greater significance is the competition between strains in forming nodules. The factors responsible for competition are not understood [146], although relative numbers of the component strains in a population are obviously important. While it appears that the more effective strains have a competitive advantage in forming nodules on some legumes [100, 108, 139], both the host plant and environmental factors can influence the relative competitiveness of strains on such species (e.g. Jones and Morley [91]). With tropical legumes, Franco and Vincent [51] concluded that competitiveness was a strain characteristic, and was not necessarily related to symbiotic effectiveness with the selected host.

Regarding inoculant strains, the extensive root systems of many tropical legumes pose the additional problem of achieving movement of the introduced strain along the lateral roots from the point of sowing. With soybeans, the one species studied in any detail, both host cultivar [25] and root temperature [175] have a marked effect on competition for nodule formation. Until the factors affecting competitiveness are more clearly understood, and methods devised to screen for appropriate attributes, it is likely that introduced strains will not persist for a significant period, at least as a component of the soil population that is forming the nodules, due to greater competitiveness of the naturally-occurring strains [32].

2.3.3. Attack by nematodes and nodule-feeding insects. Root-knot and cyst nematodes interfere with the nodulation of many legumes, including *Glycine max, Arachis hypogaea* and *Vigna unguiculata* [140, 164]. Soybeans are infected by both *Meloidogyne* species [8] and *Heterodera javanica* races [105]. Nematodes tend to prefer infecting nodules to roots, but in some cases, nodules have been found on the nematode-induced galls.

Root-lesion nematodes — such as *Pratylenchus* — have been shown to dramatically impede nodulation in many soils of Western Africa. By controlling the population of *P. sefaensis* by soil fumigation with 1,2-dibromo-3-chloropropane, the nodule weight on soybean increased from 1.7 to 2.8 g plant^{-1}, and the grain yield from 1.28 to 2.16 t ha^{-1} [60]. Similar effects have been found for groundnuts [59]. Besides their harmful effect upon nodulation, nematodes can also affect the specific N_2-fixing activity of the nodules (Table 3). Interestingly whereas the nematode infestation similarly affects nodulation of the three cultivars, its effect upon SARA significantly varies according to the cultivar.

Table 3. Specific nitrogenase activity (SARA) and nodule weight of three groundnut cultivars grown in nematode-infested soils in Senegal, expressed as percentage of the same characteristics measured on plants grown in a fumigated, nematode-free soil [112]

	cv. 55-437	cv. 28-206	cv. GH119-20
SARA	36%	90%	60%
Nodule weight	42%	32%	44%

The nodules on legumes are a favored place for *Sitona* sp. and *Rivellia* sp. to lay their eggs, with most of the damage arising from the feeding larvae. Diatloff [40] reported *Rivellia* sp. larvae in nodules on *Glycine wightii*, *Centrosema pubescens* and cowpeas, with up to 70% of the *G. wightii* nodules being infested. *Colaspis brunnea* is reported to attack soybean nodules [167] while at ICRISAT in India, chickpea nodules are reputedly attacked by insects [3].

3. Increasing N_2 Fixation by the Legumes

The foregoing observations and comments indicate that the N_2-fixing potential of legumes is far from being exploited. In this section, various approaches to increase symbiotic N_2 fixation by legumes will be discussed. These approaches range from the survey of hitherto unexploited legumes and the breeding of new cultivars better adapted to specific environments, through the selection of the most effective, and competitive, strains of rhizobia for those plants in their particular environment, to the use of management practices designed to minimize the impact of nutritional and environmental limitations on production by nodulated legumes.

3.1. Survey and utilization of unexploited N_2-fixing symbiotic systems

Up to now, the beneficial Rhizobium—legume symbiosis has only been partially explored. Of the 13,000 known species of legumes, only 100 are commercially grown. Most of the new genera and species which remain unstudied are tropical legumes. Already some species appear to have outstanding potential, such as *Leucaena leucocephala*, which not only shows a high N_2-fixing activity, but also offers one of the widest assortment of uses of all tropical legumes [4]. This legume tree can be used as forage for cattle, firewood, timber, reforestation, wind breaks and organic fertilizer. The winged bean, *Psophocarpus tetragonolobus*, a climbing legume with an active N_2-fixing capacity, has many uses [94]. Its leaves, flowers, pods, seeds and tubers have high nutritional values with high protein and vitamin contents. A recent publication by the National Academy of Sciences [5] lists numerous species with interesting nutritional characteristics and/or abilities to grow in harsh environments (Table 4). With few exceptions, little attempt has been made to domesticate these species. Little is known of their nodulation or N_2-fixing capabilities, although several belong to the subfamily Caesalpinioideae, a sub-family remarkable for the proportion of component species on which nodules have not been found. Growing such plants in the tropics should be of great value to meet the need of new sources of nitrogen and protein provided their nitrogen requirements can be obtained through effective nodulation.

Table 4. Interesting legumes with potential as food crops in tropical regions [5]

Root crops
Pachyrhizus erosus and *P. tuberosus* (yam bean)
Sphenostylis stenocarpa (African yam bean)
Vigna vexillata

Pulse crops
Voandzeia subterranea (Bambara groundnut)
Canavalia ensiformis (Jackbean), *C. gladiata* (Swordbean)
Lablab purpureus (Lablab bean)
Tylosema esculentum (or *Bauhinia esculenta*)[a] (Marama bean)
Vigna aconitifolia (moth bean), *V. umbellata* (Rice bean)
Lupinus mutabilis (Tarwi)
Phaseolus acutifolius (Tempary bean), *P. lunatus* (Lima bean)

Fruits
Ceratonia siliqua (Carob)
Tamarindus indicus[a] (Jamarina)

Forages
Acacia sp., *Prosopis* spp.,

---

[a] Sub-family Caesalpinioideae

Permanently or temporarily waterlogged soils, regarded in some areas as being favorable to legumes, are common in some tropical areas, and tropical aquatic legumes provide an opportunity to increase protein production in these soils. Genera such as Aeschynomene, Sesbania and Neptunia are known to grow in water-logged conditions, but only few attempts have been made to adapt them to agricultural usage. In Vietnam, planting *Sesbania aculeata, S. cannabina* and *S.*

grandiflora between flowering rice has produced $10 \, t \, ha^{-1}$ green matter, equivalent to $50 \, kg \, ha^{-1}$ fertilizer N (Dao The Tuan, personal communication). Similarly *Sesbania aculeata, Crotalaria juncea, C. usaramoensis* and *Aeschynomene americana* were found to be a suitable green manure incorporated to flooded rice soils within 4 to 6 weeks after sowing [168]. The incorporation of these last four species was equivalent to 20 to 40 kg N ha^{-1} applied as ammonium sulfate. Aquatic legumes, used as green manure in flooded paddy fields could thus favorably compare with the N_2-fixing fern, Azolla.

Within the aquatic legumes, stem-nodulated plants such as *Sesbania rostrata* [44, 45], and to a lesser degree *Aeschynomene indica* [181] and *Neptunia oleracea* [144], represent a further step in the adaptation of legumes to waterlogging. Due to its profuse stem nodulation, *Sesbania rostrata* has five to ten times more nodules than the best nodulated crops, which should confer on this species an outstanding potential for N_2 fixation in flooded soils (Fig. 5). Unlike most other legumes, *Sesbania rostrata* can grow on saline and alkaline waterlogged soils, and it is able to nodulate profusely in the presence of high levels of combined nitrogen in the soil. The occurrence of stem nodules has only been reported rarely, presumably due to the lack of interest by taxonomic botanists in N_2 fixation, and their possible belief that they were finding insect galls. Of interest, another stem-nodulated tree legume, *Aeschynomene elaphroxylon* (Dreyfus, unpublished) has been found growing in waterlogged soils in Senegal. Should the use of these species prove significant in raising the productivity of tropical agrosystems, the possibility of transferring the ability to produce dormant stem meristems (the site of infection by rhizobia, erroneously described as lenticels [44]) from such species to other legumes, or even non-legumes, should be explored.

3.2. Plant breeding

There is an increasing awareness of the importance of host genetics in the symbiosis between legumes and their rhizobia. Three levels of interest may be defined, although they are not mutually exclusive: (i) breeding to improve N_2 fixation; (ii) studies to describe genes responsible for the absence of nodules, or for ineffective nodulation; (iii) studies at the molecular biology level to understand and define genes and gene products responsible for various steps in nodule formation and development, and for the induction and maintenance of nitrogenase activity by the rhizobia.

These aspects have been reviewed recently [26, 65, 77, 129, 130, 169, 170], and will not be considered in detail in this section.

The most outstanding example of selection and breeding for improved N_2 fixation involves the heterogeneous *Trifolium pratense* [129]. Yield increases up to 40% over the parental population were achieved. In small part only, this was due to improved plant vigour independent of N_2 fixation. The selected material nodulated earlier, and with a greater volume of nodule tissue, but the efficiency expressed as C_2H_4 produced per g nodule dry wt in C_2H_2 reduction assays, was unaffected. Breeding for earlier, and improved, nodulation of *Trifolium ambiguum* led to the release of commercial material suitable for Australian conditions [82].

Fig. 5. (a) Three-month old stand of *Sesbania rostrata*; (b) with stem nodules; (c) stem-nodule section.

Selection and breeding for attributes other than N_2 fixation can also enhance symbiotic N_2 fixation. For example, *Phaseolus vulgaris* lines selected at CIAT, Colombia for later flowering and rapid pod development, fixed 2–3 times as much N_2 as earlier flowering lines, presumably due to the competitive effects of pod development on N_2 fixation (D.R. Laing, personal communication). This is an aspect of plant breeding that should be foremost in the minds of breeders and agronomists selecting unadapted material for use under tropical conditions.

Another aspect of selection and breeding is that many plant breeders depend on local populations of rhizobia, or even nitrogen fertilization, during their programmes. There is a distinct danger that the bred material, when tested in a different location, will produce poorly solely because the appropriate rhizobia are not present. This is demonstrated by the comparative failure of some *Glycine max* lines bred in certain US regions to yield as well when tested in a similar climatic region but with a different population of rhizobia in the soil. Hence it is vital that any breeding or selection programme involving legumes be undertaken in conjunction with trained legume bacteriology scientists.

Attributes that may be amenable for selection, with a view to improving symbiotic performance, include ability to nodulate and to fix N_2 in the presence of moderate levels of combined nitrogen, early nodulation, continued N_2 fixation during pod development, and the ability to maintain reasonable rates of N_2 fixation during moderate drought and/or temperature stress [83]. Vesicular-arbuscular mycorrhizae infection is another characteristic worthy of study, with the possibility that genetic variation within host material could be exploited to improve symbiotic performance.

Much publicity has been given to the possibility of transferring the *nif* (nitrogen fixation) genes to the plant genome. Consideration of the problems involved, particularly in locating the genes such that nitrogenase can function [68, 131], suggests that this approach to increasing N_2 fixation will be very difficult. More feasible could be the transfer of those genes in legumes, or the nodulating non-legumes, responsible for the establishment of symbiotic associations with Rhizobium or Frankia [36, 68, 169]. Such an approach will require a closer understanding of the physiological processes involved in the development and function of nodules than we presently possess. However, it is possible that many species already possess some of the necessary genetic information and the transfer of only three or four genes may be adequate to confer nodule-forming ability on such species.

3.3. Microbial approach

3.3.1. Strain selection. As indicated in 2.3.2, a successful inoculant strain of Rhizobium must be effective in N_2 fixation with the selected hosts, competitive in nodule formation on these hosts, and in many situations, show indefinite persistence as a significant component of the Rhizobium population. Effectiveness studies, made under a range of environmental and nutritional conditions, and involving different species and cultivars, are most efficiently done under laboratory or greenhouse conditions [66]. Most of the strains now included in commercial inoculants around the world have been selected from laboratory or greenhouse trials, albeit under optimal conditions in most instances. Date [32] and Ayanaba [6] each record successful strain selections for inoculant use on forage legumes and soybeans, respectively, growing under tropical conditions. In other work, Silvestre [152] observed an increase in soybean yield from 240 to 1440 kg ha^{-1} following inoculation, while in Senegal, inoculation of *Vigna unguiculata* with CB756 significantly hastened and increased nodulation when the soil population of *Rhizobium* sp. was low.

The need to check inoculant strains against all species for which it is recommended is paramount. Although the cowpea cross-inoculation group is diverse, there are a number of species with specific requirements (Table 5). For example, 13 Acacia species were shown to fall into three groups according to their effectiveness response patterns with a large group of fast- and slow-growing strains [43]. Further-

Table 5. Promiscuous and specific cross-inoculation groups of tropical legume species

Nodulation group	Rhizobia growth rate Characteristics	Legume species
Promiscuous	Slow-growing strains	*Acacia albida* *Arachis hypogaea* *Cajanus cajan* *Clitoria* sp. *Crotalaria* sp. *Glycine wightii* *Indigofera* sp. *Lablab purpureus* *Macroptilium atropurpureum* *Phaseolus lunatus* *Psophocarpus tetragonolobus* *Pueraria* sp. *Stylosanthes guyanensis* (most cultivars) *Stylosanthes humilis* *Teramnus* sp. *Vigna radiata* *Vigna unguiculata* *Voandzeia subterranea* *Zornia* sp.
Specific	Slow-growing strains	*Adesmia* spp. *Centrosema* spp. *Coronilla* spp. *Desmodium intortum* *Lotononis* spp. *Lotus* sp. *Stylosanthes* spp.
	Fast-growing strains	*Acacia raddiana* *Acacia senegal* *Albizia julibrissin* *Andura* sp. *Leucaena leucocephala* *Psoralea* sp. *Samanea* sp. *Sesbania sesban* *Sesbania rostrata*

more, the strain selected for an inoculant should be the most effective strain available for that species (Fig. 4).

There are many situations in which there is no response to inoculation, particularly where legumes in the cowpea cross-inoculation groups are involved [6]. While this may be due to the presence of highly effective and competitive strains in the soil, it may be due to (i) the failure of the inoculant strains to form nodules, (ii) nutritional or environmental factors limiting symbiotic development, and (iii) nutritional or environmental factors limiting plant production *per se*. For all field studies, it is important that control treatments with adequate fertilizer nitrogen be included, in order to help define any major limitations and determine potential productivity of that crop. However, in order to make the assessment realistic, N fertilizer should not be applied until the inoculated treatments have nodulated [66]. It is also important to determine whether the inoculant strains have formed nodules, either by visual inspection or, more usually, by serological or other procedures used to identify particular strains [46, 148, 172].

The significance of competition between strains, and the lack of knowledge of the fundamental basis of strain competition, was discussed above (2.3.2). An example of host effects on strain competition, and its significance with regard to N_2 fixation and plant growth, is shown in Table 6. Whereas strain 61B9 forms a more effective symbiosis on both *Macroptilium atropurpureum* and *Vigna unguiculata* than does strain QA922, in a mixed inoculant, QA922 dominates the response on *V. unguiculata*, whereas 61B9 has the stronger effect with *M. atropurpureum*.

Table 6. Shoot weight of *Macroptilium atropurpureum* and *Vigna unguiculata* when inoculated with three strains of *Rhizobium* sp., either singly or in pairs, with equal numbers of each strain – 10 replicates (B.L. Dreyfus and A.H. Gibson, unpublished)

	Macroptilium atropurpureum (mg plant^{-1})[a]			*Vigna unguiculata* (g plant^{-1})[b]		
	CB756	61B9	QA922	CB756	61B9	QA922
CB756	98	110	87	3.0	3.3	2.9
61B9	–	107	108	–	2.7	0.6
QA922	–	–	82	–	–	0.4

[a] Uninoculated, 15 mg. LSD = 15.7, $P = 0.05$.
[b] Uninoculated, 0.4 g. LSD = 0.5, $P = 0.05$

The selection of strains of rhizobia able to persist in soils remains a major problem, especially for the harsh conditions frequently encountered in tropical soils. Date and Halliday [33] have shown that special procedures are required to isolate some acid-tolerant strains for *Stylosanthes* spp. while Eaglesham *et al.* [47] have shown that selection of high temperature-tolerant strains of *Rhizobium* sp. is possible. Numerous studies from Alexander's laboratory in Cornell University (e.g. Ramirez and Alexander [136]) have indicated that, in general, predation is not a serious problem. Nor have studies on bacteriophage action consistently shown that this significantly affects field populations of rhizobia, although lysogenic phages may have a significant effect in mixed strain inoculants [9, 109, 147]. Of potential interest is the possibility of conferring on inoculant strains characteristics enabling them to survive in hostile environments. However, before the application of such bacterial breeding techniques, it will be necessary to identify the genetic basis of the attributes.

3.3.2. Inoculation techniques. The preparation of inoculants prior to inoculation involves different steps (strain selection, culture, preparation of carriers, the mixing of culture broth and carrier, maturation, and storage) which have been reviewed in detail recently [22, 35, 165]. In this section we shall focus our attention (1) on the form of carriers that can be used, and (2) on techniques of inoculation.

Legume inoculants have been prepared commercially since late in the 19th century. Their quality partly depends upon the suitability of the carrier, with the best results usually obtained with peat or soil high in organic matter [163]. Many substitutes for peat have been proposed, such as bentonite, lignite, cellulose powder, various powdered crops residues, etc., but peat has proved generally superior to these carriers. However the composition of peat is highly variable, and rigorous testing of new deposits is essential before the material is used commercially. Due to variation in peat quality, even from a single source [162], there is appeal in having a synthetic carrier of constant quality. Initial tests with entrapping rhizobia in a polyacrylamide gel have been promising [41] and current efforts are directed to adapting this approach for use at the farm level. Where possible, peats should be sterilized by γ-radiation or autoclaving [142] or partially sterilized by heating [22], and care should be taken to adjust both the moisture level and pH if high counts are to be maintained during prolonged storage. Frozen inoculants are prepared for soybeans in the USA, but difficulties in providing adequate storage conditions in many tropical countries precludes consideration of this type of commercial inoculant.

The classical procedure for inoculating seed is to prepare a slurry containing an adhesive, such as gum arabic [17]; coating the inoculated seed with finely divided $CaCO_3$, rock phosphate, or bentonite, confers a degree of protection from immediate desiccation. Other techniques involve the use of peat granules [13, 53], or other inoculant material (e.g. marble chips), sown with the seed. Alternatively 'drenching', or spraying suspended peat inoculant into the furrow below the seed, has achieved a degree of popularity for large-scale sowings, especially as there are considerable labour savings compared with preparing inoculated seed [18, 145]. This and the granule method have other advantages over conventional inoculation methods in that the bacteria are protected from potentially-damaging fungicide or insecticide seed dressings, the bacteria are not lifted above the soil during (epigeal) germination, and it is possible to achieve high rates of nodulation.

3.3.3. Vesicular-arbuscular mycorrhizae. Since legumes have a restricted root system, they are 'relatively poor foragers for phosphate' [123] and since many tropical soils are phosphorus deficient, a positive response to root infection by vesicular-arbuscular mycorrhizae (VAM) can be expected. Thus, it is not surprising that VAM and phosphorus application have similar favorable effects not only on plant growth but also on nodulation and specific N_2-fixing activity in many tropical soils. Besides beneficial nutritional effects, VAM may be of importance in semi-arid conditions, such as providing increased resistance to drought, or in alleviating some forms of soil toxicity (see Chapter 8).

3.4. Soil management

Assuming that the crop selected is capable of a high level of production in a particular environment, and that the inoculant strain is capable of forming an effective symbiosis with that host, it is then necessary to ensure that all other major limitations are minimized by the adoption of appropriate management practices (Table 7).

Table 7. Major factors limiting symbiotic N_2 fixation in West Africa; proposed means of controlling their effect

Limiting factor	Control
1. Moisture stress	— irrigation, mulching — search for drought resisting cultivars and *Rhizobium* strains — stimulating VAM infection
2. Nematodes[a]	— fumigation by nematicides — biological and integrated control
3. Soil acidity and toxicity	— liming — addition of organic matter (FYM; green manure; compost) — selections of acid-resistant strains
4. Mineral deficiencies, especially phosphorus	— addition of phosphorus — stimulating VA mycorrhizal infection — determination of other deficiencies and their rectification
5. Inadequacy of native *Rhizobium* populations and competition between native and introduced strains	— inoculation with strains selected for effectiveness, competitiveness and persistence

[a] Some pests and diseases may become serious in some circumstances

3.4.1. *Mulching*. Surface mulches of straw, sugarcane 'bagasse', or even grass-clippings, are recommended to conserve soil moisture, reduce soil temperature, decrease the nitrate concentration in the soil and to control competing weeds. Masefield [110, 111] reported a 3-fold increase in the nodulation of cowpeas following mulching. Significant increases in N_2 fixation by soybeans were found when straw was incorporated into soil, with the effects being comparable to those achieved with CO_2 fertilization of the canopy [151]. However, results obtained using 'bagasse' in Brazil were variable, with the yield and nodulation of *Glycine wightii* and *Teramnus uncinatus* being increased and that of *Medicago sativa* decreased; *Phaseolus vulgaris* showed improved nodulation but severely depressed grain yield [127]. It is obvious that mulching can be beneficial but careful experimentation is required to determine the most effective forms and rates of incorporation.

3.4.2. Irrigation and drainage. As previously discussed (2.1), both drought and waterlogging reduce nodulation and N_2 fixation. Irrigation can provide marked yield improvement where the soil is not waterlogged for long periods after application [150]. Care should be taken to ensure that the irrigation water is not alkaline. The poor nodulation of *Arachis hypogaea* at Bambey, Senegal was attributed to water with pH 8.0, due to a high Mg-salts level. The need to drain crop land should be obvious from poor nodulation and plant growth, but less obvious will be waterlogging effects in pastures where the growth of associated grasses will tend to obscure, and accentuate, the poor growth of the legume component due to waterlogging.

3.4.3. Liming. Tropical legumes vary in their response to liming [118, 119, 120, 121] and caution must be exercised in the use of this practice. While it raises pH, and alleviates Mn and Al toxicities, ill-managed liming can induce deficiencies of Mg, Cu, Zn and B [1]. Due to the high cost of liming, a more appropriate strategy could be the selection of legumes, and their rhizobia, that are able to tolerate acid conditions [143].

3.4.4. Nitrogen fertilizer. The effects of nitrogen fertilizer on the nodulation and N_2 fixation by legumes are complex, and vary with the form and concentration of the fertilizer, the species (and possibly the strains of rhizobia), the environmental conditions, and the level of available N in the soil [31, 171]. The data in Table 8 show depression of N_2 fixation at high rates of application, and either no effect or stimulation at an intermediate rate. Low initial levels of fertilizer N can promote extensive nodulation and N_2 fixation by overcoming any N deficiency induced before nodulation occurs [63]; they may also enable less effective strains to form a more effective symbiosis than they would in the absence of combined N.

As a generalization, well-nodulated legumes grow and yield at a similar level to plants supplied with adequate mineral nitrogen, provided they do not suffer N deficiency in the early stages of growth. Above a certain level of supplied N, further increases reduce the level of N_2 fixed; below that level, the full N_2-fixing potential may not be achieved. The aim, therefore, should be to supplement soil N with the

58

Table 8. Effect of nitrogen application on N_2 fixation (estimated by A value method) by groundnut

Rates of application of nitrogen fertilizer at seeding (kg ha^{-1})	kg N_2 fixed per ha		
	Senegal[a]		Ghana [99]
	(1974)	(1975)	(1975)
15	52.0	67.5	60.7
30	56.0	75.1	69.3
60	25.0	–	–

[a] Ganry (1976), unpublished data

required dose of 'starter' N. As an example, trials in southern Brazil at a number of centers have indicated the benefits of 4–8 kg N ha^{-1} as a starter dose for soybeans (Jardim Freire and Kolling, personal communication). Similar results have been obtained in Minnesota, USA (Ham, personal communication), but in Illinois, USA, only one response to fertilizer N was reported from 132 trials [176]. By using different forms of N fertilizer, the response was enhanced, or depressed [81]. Although it is generally recognized that combined N retards nodulation and N_2 fixation, it is obvious that by correct application of fertilizer N, N_2 fixation may be promoted. Only through a series of trials on different soils, with different crops, and in different agricultural cropping systems, will the most appropriate level of application be determined [64]. Other interesting approaches include deepbanding [177] or the use of slow-release fertilizers [50]; caution should be adopted in the use of nitrification inhibitors [156].

3.4.5. Other fertilizers. After nitrogen, phosphorus is most likely to be the limiting nutrient in tropical soils. A striking example of the effect of P fertilization was reported by Graham and Rosas [75], who observed a linear increase in nodule mass (9-fold), and specific nitrogenase activity of *Phaseolus vulgaris* over the range 0–315 kg triple superphosphate ha^{-1}. None of the 30 cultivars showed reasonable nodulation or N_2 fixation at low P. However, some *Stylosanthes* [21] and *Lupinus* [69] species make better growth with low P levels and it would be of interest to know the physiological basis of this characteristic. Sulphur, and less commonly potassium, may also be limiting in acidic, well-leached tropical soils. As indicated by Munns [119] and Gates and Muller [58], a balanced nutrition is essential to maximum production, and suitable trials should be undertaken to ensure that nutrients are not limiting production.

3.4.6. Micronutrients. In alkaline soils, Mn, Fe, B and Zn are less available than in acid soils, while in acid soils, Mo availability is reduced [79]. In many tropical regions, trace elements have received little attention, although it is possible that the deficiency of any one, but particularly Mo, B, Fe or Co, could severely limit productivity.

3.4.7. Organic matter. It is unfortunate that many tropical soils have very low organic matter levels in view of the generally beneficial effects of organic matter on the physical structure, and the chemical nature, of soil. Another attribute of organic matter is that it has been reported to enhance legume nodulation and N_2 fixation [14, 30, 90, 124]. Various reasons may be advanced for these effects. They could be providing nutrients for the rhizobia or protecting them against harmful environmental factors. They could be promoting plant growth [49], or they could be exerting their effect through their influence on soil structure, by helping retain moisture in the soil, and by reducing phosphate fixation to clays [27]. Regardless of the actual effect(s), and this is most worthy of investigation, agronomic practices should be directed towards increasing the organic matter content of those tropical soils low in organic matter.

An interesting effect of different types of organic matter on nitrogen fixation by *Arachis hypogaea* has been observed (Table 9). The amount of N_2 fixed was determined by the 'isotope dilution' method and from the differences in total N of nodulated and non-nodulated legumes (see Appendix). Compost and farmyard manure (FYM) increased N_2 fixation 30–50%, as also found by Masefield [111], whereas millet straw had no effect. Under some conditions, green manuring may depress N_2 fixation (Ganry, unpublished).

Table 9. Aftereffects of different types of organic matter on N_2 fixation in peanuts, expressed in mg N_2 fixed per plant (Ganry, Guiraud and Dommergues, unpublished)

Treatments	Estimation of N_2 fixation by 2 methods[a]	
	^{15}N	Difference
Control	101	97
Incorporation of pearl millet straw	111	101
Incorporation of compost	142	158
Incorporation of FYM	136	146

[a] See Appendix

3.4.8. Multiple cropping systems involving legumes. These systems involve two or more crops grown on the same field in one year. The cropping may be sequential, or concurrent (i.e. intercropping).

Sequential cropping involving legumes and rice is common in paddy rice fields in Asia. In China and Vietnam, milk vetch (*Astragalus sinensis*) and *Sesbania* species respectively are commonly used, and in Thailand, soybeans are being investigated [138]. The legume may be used as a green manure crop, or as a grain crop, although in this latter case, the benefits are likely to be reduced as the grain is removed.

Intercropping is very popular in Africa and India. The most important legume in Africa, *Vigna unguiculata*, is grown with other crops in 98% of its cultivated area, while in Nigeria 95% of the groundnuts are intercropped [61, 70]. Although the yield of individual crops is usually decreased, the overall yield is often higher.

Furthermore, N_2 fixation by the legume may be enhanced compared with the single-cropped legume, either in absolute amounts, or when expressed as a proportion of total plant N (Table 10). These experiments were done in lysimeters containing 78 kg soil uniformly labelled with ^{15}N. Presumably the cereal has utilized most of the available N with the consequence that N_2 fixation by the legume is favored. Another example of intercropping involves maize (*Zea mays*) and the climbing bean (*Phaseolus vulgaris*) which uses the maize for support [73].

Table 10. Percent of total N_2 fixed symbiotically in two leguminous plants grown as pure crops or in association with pearl millet (Ganry, Guiraud and Dommergues, unpublished data)

Plant species	Pure crop	Intercropping
Soybean	88	97
Groundnut	75	90

4. Concluding Remarks

The legume is not so much a plant as it is a symbiotic association, and it will be as a symbiotic association that the legumes will realize their potential in helping to overcome the severe nitrogen deficiency limiting production in many tropical soils. Hence it is important that as much attention be given to the rhizobia, both as free-living components of the soil microflora and as partners in the N_2-fixing symbiosis, as is given to the agronomy, selection, breeding, pathology and physiology of the more obvious partner in the symbiosis. In addition to selecting effective, competitive and persistent strains for particular hosts, and to devising appropriate procedures for producing inoculants, and applying them, consideration must be given to management practices that minimize the effect of nutrient deficiencies and toxicities, environmental stresses, and biological agents that affect the association. It is obvious from the limited data in Table 1, and the wide range of species listed in Tables 4 and 5, that the legumes possess an exciting potential to help overcome nitrogen deficiency in many types of agrosystems. However it will only be through the combined efforts of scientists in many disciplines that this potential will be achieved.

References

1. Andrew, C.S. 1978 Legumes and acid soils. In: Limitations and Potentials for Biological Nitrogen Fixation in the Tropics, pp. 135–160. Dobereiner, J. Burris, R.H. and Hollaender, A.E. (eds.), Plenum, New York.
2. Andrew, C.S. and Robins, M.F. 1969 The effect of phosphorus on the growth and chemical composition of some tropical pasture legumes. I. Growth and critical percentages of phosphorus. Australian Journal of Agricultural Research 20, 665–674.

3. Anon. 1977 Annual Report, International Crops Research Institute for the Semi-arid Tropics, 1976–1977, Hyderabad, India.
4. Anon. 1977 Leucaena. Promising forage and tree crop for the tropics, National Academy of Sciences, Washington.
5. Anon. 1979 Tropical legumes: Resources for the Future. National Academy of Sciences, Washington, 331 pp.
6. Ayanaba, A. 1977 Towards better use of inoculants in the humid tropics. In: Biological Nitrogen Fixation in Farming Systems of the Tropics, pp. 181–187. Ayanaba, A. and Dart, P.J. (eds.), Wiley-Interscience, Chichester.
7. Ayanaba, A. and Lawson, T.L. 1977 Diurnal changes in acetylene reduction in field-grown cowpeas and soybeans. Soil Biology Biochemistry 9, 125–129.
8. Balasubramanian, M. 1971 Root-knot nematodes and bacterial nodulation in soybeans. Current Science 40, 69–70.
9. Barnet, Y.M. 1980 The effect of rhizobiophages on populations of *Rhizobium trifolii* in the root zone of clover plants. Canadian Journal of Microbiology 26, 572–576.
10. Beadle, N.C.W. 1964 Nitrogen economy in arid and semi-arid plant communities. III. The symbiotic nitrogen-fixing organisms. Proceedings of the Linnean Society of New South Wales 89, 273–286.
11. Berger, J.A., May, S.N., Berger, L.R. and Bohlool, B.B. 1979 Colorimetric enzyme-linked immunosorbent assay for the identification of strains of *Rhizobium* in culture and in the nodules of lentils. Applied and Environmental Microbiology 37, 642–646.
12. Bergersen, F.J. 1980 Measurement of nitrogen fixation by direct means. In: Method for Evaluating Biological Nitrogen Fixation, pp. 65–110. Bergersen, F.J. (ed.), Wiley, Chichester.
13. Bezdicek, D.F., Evans, D.W., Abede, B. and Witters, R.E. 1978 Evaluation of peat and granular inoculum for soybean yield and N fixation under irrigation. Agronomy Journal 70, 865–868.
14. Bhardwaj, K.K.R. and Gaur, A.C. 1968 Humic acid in relation to inoculation of crops. *Agrochimica 12*, 100–108.
15. Bowen, G.D. and Kennedy, M. 1959 Effect of high soil temperatures on *Rhizobium* spp. Queensland Journal of Agricultural Science 16, 177–197.
16. Bremner, J.M. 1977 Use of nitrogen-tracer techniques for research on nitrogen fixation. In: Biological Nitrogen Fixation in Farming Systems of the Tropics, pp. 335–352. Ayanaba, A. and Dart, P.J. (eds.), Wiley-Interscience, Chichester.
17. Brockwell, J. 1977 Application of legume seed inoculants. In: A Treatise on Dinitrogen Fixation. IV – Agronomy and Ecology, pp. 277–309. Hardy, R.W.F. and Gibson, A.H. (eds.), Wiley-Interscience, New York.
18. Brockwell, J., Gault, R.R., Chase, D.L., Hely, F.W., Zorin, M. and Corbin, E.J. 1980 An appraisal of practical alternatives to legume seed inoculation: Field experiments on seed bed inoculation with solid and liquid inoculants. Australian Journal of Agricultural Research 31, 47–60.
19. Bromfield, E.S.P. and Ayanaba, A. 1980 The efficacy of soybean inoculation on acid soil in tropical Africa. *Plant and Soil 54*, 95–106.
20. Bromfield, E.S.P. and Roughley, R.J. 1980 Characterization of rhizobia isolated from nodules on locally adapted *Glycine max* grown in Nigeria. Annals Applied Biology 95, 185–190.
21. Burt, R.L., Williams, W.T., and Grof, B. 1980 Stylosanthes – Structure, adaptation and utilization. In: Advances in Legume Science, pp. 553–558. Summerfield, R.J. and Bunting, A.H. (eds.), Royal Botanic Gardens, Kew.

22. Burton, J.C. 1979 New developments in inoculating legumes. In: Recent Advances in Biological Nitrogen Fixation, pp. 380–405. Subba Rao, N.S. (ed.), Oxford and I.B.H. Publishers, New Delhi.
23. Bushby, H.V.A. 1981 Quantitative estimation of rhizobia in soils and rhizospheres. In: Current Perspectives in Nitrogen Fixation, p. 432. Gibson, A.H. and Newton, W.E. (eds.), Australian Academy of Science, Canberra.
24. Bushby, H.V.A. and Marshall, K.C. 1977 Some factors affecting the survival of root-nodule bacteria on desiccation. Soil Biology Biochemistry 9, 143–147.
25. Caldwell, B.E. 1969 Initial competition of root nodule bacteria on soybeans in a field environment. Agronomy Journal 61, 813–815.
26. Caldwell, B.E. and Vest, H.G. 1977 Genetic aspects of nodulation and dinitrogen fixation by legumes: The macrosymbiont. In: A Treatise on Dinitrogen Fixation. III – Biology, pp. 557–576. Hardy, R.W.F. and Silver, W.S. (eds.). Wiley-Interscience, New York.
27. Charreau, C. 1974 Soils of tropical dry and dry-wet climatic areas of West Africa and their use and management. Agronomy Mimeo 74–26, Agronomy Dept. Cornell University, Ithaca.
28. Criswell, J.G., Havelka, U.D., Quebedeaux, B. and Hardy, R.W.F. 1977 Effect of rhizosphere pO_2 on nitrogen fixation by excised and intact nodulated soybean roots. Crop Science 17, 39–43.
29. Dart, P.J. 1981 Genetic variability in the nodulation of groundnut. In: Current Perspectives in Nitrogen Fixation, p. 216. Gibson, A.H. and Newton, W.E. (eds.), Australian Academy of Science, Canberra.
30. Dart, P., Day, J., Islam, R. and Dobereiner, J. 1976 Symbiosis in tropical grain legumes: some effects of temperature and the composition of the rooting medium. In: Symbiotic Nitrogen Fixation in Plants, pp. 361–384. Nutman, P.S. (ed.), Cambridge University Press, London.
31. Dart, P.J. and Wildon, D.C. 1970 Nodulation and nitrogen fixation by *Vigna sinensis* and *Vicia atropurpurea*: the influence of concentration, form, and site of application of combined nitrogen. Australian Journal of Agricultural Research 21, 45–56.
32. Date, R.A. 1977 Inoculation of tropical pasture legumes. In: Exploiting the Legume-Rhizobium Symbiosis in Tropical Agriculture, pp. 293–311. Vincent, J.M., Whitney, A.S. and Bose, J. (eds.), Univ. Hawaii College of Agriculture Miscellaneous Publication No. 145.
33. Date, R.A. and Halliday, J. 1979 Selecting Rhizobium for acid, infertile soils of the tropics. Nature 277, 62–63.
34. Date, R.A. and Halliday, J. 1980 Relationships between Rhizobium and tropical forage legumes. In: Advances in Legume Science, pp. 597–601. Summerfield, R.J. and Bunting, A.H. (eds.), Royal Botanic Gardens, Kew.
35. Date, R.A. and Roughley, R.J. 1977 Preparation of legume seed inoculants. In: A Treatise on Dinitrogen Fixation. IV – Agronomy and Ecology, pp. 243–275. Hardy, R.W.F. and Gibson, A.H. (eds.), Wiley – Interscience, New York.
36. Davey, M.R. and Cocking, E.C. 1979 Tissue and cell cultures and bacterial nitrogen fixation. In: Recent Advances in Biological Nitrogen Fixation, pp. 281–324. Subba Rao, N.S. (ed.), Oxford and IBH Publishers, New Delhi.
37. Dazzo, F.B. and Brill, W.J. 1978 Regulation by fixed nitrogen of host-symbiont recognition in Rhizobium-clover symbiosis. Plant Physiology 62, 18–21.
38. Dazzo, F.B., Hrabak, E.M., Urbano, M.R., Sherwood, J.E., and Truchet, G.

1981 Regulation of recognition in the Rhizobium-clover symbiosis. In: Current Perspectives in Nitrogen Fixation, pp. 292—295. Gibson, A.H. and Newton, W.E. (eds.), Australian Academy of Science, Canberra.

39. Diatloff, A. 1968 Effect of soil moisture fluctuation on legume nodulation and nitrogen fixation in a black earth soil. Queensland Journal Agriculture and Animal Science 24, 315—321.

40. Diatloff, L. 1965 Larvae of *Rivellia* sp. attacking the root nodules of *Glycine javanica* L. Journal of the Entomological Society of Queensland 4, 86.

41. Dommergues, Y.R., Diem, H.G. and Divies, C. 1979 Polyacrylamide-entrapped Rhizobium as an inoculant for legumes. Applied and Environmental Microbiology 37, 779—781.

42. Dreyfus, B. and Dommergues, Y.R. 1980 Non-inhibition de la fixation d'azote atmosphérique par l'azote combiné chez une légumineuse à nodules caulinaires, *Sesbania rostrata*. C.R. Académie Sciences, Paris D, 291, 767—770.

43. Dreyfus, B. and Dommergues, Y.R. 1981 Nodulation of *Acacia* species by fast- and slow-growing tropical strains. Applied Environmental Microbiology 41, 97—99.

44. Dreyfus, B. and Dommergues, Y.R. 1981 Nitrogen-fixing nodules induced by *Rhizobium* on the stem of the tropical legume *Sesbania rostrata*. FEMS Microbiology Letters 10, 313—317.

45. Dreyfus, B.L. and Dommergues, Y. 1981 Stem nodules on the tropical legume, *Sesbania rostrata*. In: Current Perspectives in Nitrogen Fixation, p. 471. Gibson, A.H. and Newton, W.E. (eds.), Australian Academy of Science, Canberra.

46. Dudman, W.F. 1977 Serological methods and their application to dinitrogen fixing organisms. In: A Treatise on Dinitrogen Fixation IV — Agronomy and Ecology, pp. 487—508. Hardy, R.W.F. and Gibson, A.H. (eds.), Wiley-Interscience, New York.

47. Eaglesham, A., Seaman, B., Ahmad, H., Hassouna, S., Ayanaba, A., and Mulongoy, K. 1981 High-temperature tolerant 'cowpea' rhizobia. In: Current Perspectives in Nitrogen Fixation, p. 436. Gibson, A.H. and Newton, W.E. (eds.), Australian Academy of Science, Canberra.

48. Edwards, D.G. 1977 Nutritional factors limiting nitrogen fixed by rhizobia. In: Biological Nitrogen Fixation in Farming Systems of the Tropics, pp. 189—204. Ayanaba, A. and Dart, P.J. (eds.), Wiley-Interscience, Chichester.

49. Flaig, W. 1972 An introductory review on humic substances: aspects of research on their genesis, their physical and chemical properties, and their effect on organisms. Proceedings International Meeting on Humic Substances, Nieuwersluis, Pudoc, Wageningen. pp. 19—42.

50. Focht, D.D. and Verstraete, W. 1977 Biochemical ecology of nitrification and denitrification. In: Advances in Microbial Ecology. I, pp. 135—154. Alexander, M. (ed.), Plenum, New York.

51. Franco, A.A. and Vincent, J.M. 1976 Competition amongst rhizobial strains for the colonization and nodulation of two tropical legumes. Plant and Soil 45, 27—48.

52. Franco, A.A., Peres, J.R.R. and Nery, M. 1978 The use of *Azotobacter paspali* N_2-ase (C_2H_2 reduction activity) to measure molybdenum deficiency in soils. Plant and Soil 50, 1—11.

53. Fraser, M.E. 1975 A method of culturing *Rhizobium meliloti* on porous granules to form a pre-inoculant for lucerne seed. Journal of Applied Bacteriology 39, 345—351.

64

54. Fred, E.B., Baldwin, I.L. and McCoy, E. 1932 Root Nodule Bacteria and Leguminous Plants. University of Wisconsin Studies in Science No. 5. Univ. Wisconsin Press, Madison.
55. Fried, M. and Broeshart, H. 1976 An independent measurement of the amount of nitrogen fixed by a legume crop. Plant and Soil 43, 707–711.
56. Fried, M. and Mistry, K.B. 1979 Application of isotope techniques in research into the chemistry and availability of plant nutrients. In: Soil Research in Agroforestry, pp. 471–522. Mongi, H.O. and Huxley, P.A. (eds.), ICRAF, Nairobi.
57. Gates, C.T. 1974 Nodule and plant development in *Stylosanthes humilis* H.B.K.: Symbiotic response to phosphorus and sulphur. Australian Journal of Botany 22, 45–55.
58. Gates, C.T. and Muller, W.J. 1979 Nodule and plant development in the soybean, *Glycine max* (L.) Merr.: Growth response to nitrogen, phosphorus and sulfur. Australian Journal of Botany 27, 203–215.
59. Germani, G., Diem, H.G., and Dommergues, Y. 1980 Influence of 1, 2-dibromo-3-chloropropane fumigation on nematode population, mycorrhizal infection, N_2 fixation and yield of field-grown groundnut. *Revue Nématol. 3*, 75–78.
60. Germani, G., unpublished data.
61. Gibbons, R.W. 1980 Adaptation and utilization of groundnuts in different environments and farming systems. In: Advances in Legume Science, pp. 483–493. Summerfield, R.J. and Bunting, A.H. (eds.), Royal Botanic Gardens, Kew.
62. Gibson, A.H. 1974 Consideration of the growing legume as a symbiotic association. Proceedings of the Indian National Science Academy, 40B, 741–767.
63. Gibson, A.H. 1976 Recovery and compensation by nodulated legumes to environmental stress. In: Symbiotic Nitrogen Fixation in Plants, pp. 385–403. Nutman, P.S. (ed.), Cambridge University Press, London.
64. Gibson, A.H. 1977 The influence of the environment and managerial practices on the legume-Rhizobium symbiosis. In: A Treatise on Dinitrogen Fixation. IV – Agronomy and Ecology, pp. 393–450. Hardy, R.W.F. and Gibson, A.H. (eds.), Wiley-Interscience, New York.
65. Gibson, A.H. 1980 Host determinants in nodulation and nitrogen fixation. In: Advances in Legume Science, pp. 69–76. Summerfield, R.J. and Bunting, A.H. (eds.), Royal Botanic Gardens, Kew.
66. Gibson, A.H. 1980 Methods for legumes in glasshouses and controlled environment cabinets. In: Methods for Evaluating Biological Nitrogen Fixation, pp. 139–184. Bergersen, F.J. (ed.), Wiley-Interscience, Chichester.
67. Gibson, A.H. and Pagan, J.D. 1977 Nitrate effects on the nodulation of legumes inoculated with nitrate-reductase-deficient mutants of *Rhizobium*. Planta 134, 17–22.
68. Gibson, A.H., Scowcroft, W.R. and Pagan, J.D. 1977 Nitrogen fixation in plants: An expanding horizon? In: Recent Developments in Nitrogen Fixation, pp. 387–417. Newton, W.E., Postgate, J.R. and Rodriguez-Barrueco, C. (eds.), Academic Press, London.
69. Gladstones, J.S. 1970 Lupins as crop plants. Field Crop Abstracts 23, 124–147.
70. Gomez, A.A. and Zandstra, H.G. 1977 An analysis of the role of legumes in multiple-cropping systems. In: Exploiting the Legume-Rhizobium Symbiosis in Tropical Agriculture, pp. 81–95. Vincent, J.M., Whitney, A.S. and Bose, J.

(eds.), Univ. Hawaii College of Agriculture Miscellaneous Publication No. 145.

71. Gorbet, D.W. and Burton, J.C. 1979 A non-nodulating peanut. Crop Science 19, 727–728.
72. Gradwell, M.W. 1969 Soil oxygen and the growth of white clover. New Zealand Journal of Agricultural Research 12, 615–629.
73. Graham, P.H. and Rosas, J.C. 1978 Plant and nodule development and nitrogen fixation in climbing cultivars of *Phaseolus vulgaris* L. grown in monoculture, or associated with *Zea mays* L. Journal of Agricultural Science, Cambridge 90, 311–317.
74. Graham, P.H. and Rosas, J.C. 1978 Nodule development and nitrogen fixation in cultivars of *Phaseolus vulgaris* L. as influenced by planting density. Journal of Agricultural Science, Cambridge 90, 19–29.
75. Graham, P.H. and Rosas, J.C. 1979 Phosphorus fertilization and symbiotic nitrogen fixation in common bean. Agronomy Journal 71, 925–926.
76. Greenland, D.J. 1959 Nitrate fluctuations in tropical soils. Journal of Agricultural Science, Cambridge 50, 82–92.
77. Gresshoff, P.M., Carroll, B., Mohapatra, S.S., Reporter, M., Shine, J. and Rolfe, B.G. 1981 Host factor control of nitrogenase function. In: Current Perspectives in Nitrogen Fixation, pp. 209–212. Gibson, A.H. and Newton, W.E. (eds.), Australian Academy of Science, Canberra.
78. Habish, H.A. 1970 Effect of certain soil conditions on nodulation of *Acacia* spp. Plant and Soil 33, 1–6.
79. Hallsworth, E.G. 1972 Factors affecting the response of grain legumes to the application of fertilizers. In: Uses of Isotopes for Study of Fertilizer Utilization by Legume Crops, pp. 1–16. International Atomic Energy Agency, Vienna.
80. Ham. G.E. 1977 The acetylene-ethylene assay and other measures of nitrogen fixation in field experiments. In: Biological Nitrogen Fixation in Farming Systems of the Tropics, pp. 325–334. Ayanaba, A. and Dart, P.J. (eds.), Wiley-Interscience, Chichester.
81. Hardy, R.W.F., Burns, R.C., and Holsten, R.D. 1973 Applications of the acetylene-ethylene assay for measurement of nitrogen fixation. Soil Biology Biochemistry 5, 47–81.
82. Hely, F.W. 1972 Genetic studies with wild diploid *Trifolium ambiguum* M. Bieb. with respect to time of nodulation. Australian Journal of Agricultural Research 23, 437–446.
83. Holl, F.B. and Larue, T.A. 1976 Genetics of the legume plant host. In: Proceedings of First International Symposium on Nitrogen Fixation, pp. 391–399. Newton, W.E. and Nyman, C.J. (eds.), Washington State Univ. Press, Pullman.
84. Hong, T.D., Minchin, F.R. and Summerfield, R.J. 1977 Recovery of nodulated cowpea plants (*Vigna unguiculata* (L.) Walp.) from waterlogging during vegetative growth. Plant and Soil 48, 661–672.
85. Huang, C.-y, Boyer, J.S. and Vanderhoef, L.N. 1975 Limitation of acetylene reduction (nitrogen fixation) by photosynthesis in soybean having low water potentials. Plant Physiology 56, 228–232.
86. Hume, D.J., Criswell, J.G. and Stevenson, K.R. 1976 Effects of soil moisture around nodules on nitrogen fixation by well-watered soybeans. Canadian Journal of Plant Science 56, 811–815.
87. Hunter, M.N., de Jabrun, P.L.M. and Byth, D.E. 1980 Response of nine soybean lines to soil moisture conditions close to saturation. Australian Journal of Experimental Agriculture and Animal Husbandry 20, 339–345.

66

88. Hutton, E.M. and Andrew, C.S. 1978 Comparative effects of calcium carbonate on growth, nodulation, and chemical composition of four *Leucaena leucocephala* lines, *Macroptilium lathyroides* and *Lotononis bainesii.* Australian Journal of Experimental Agriculture and Animal Husband. 18, 81—88.

89. Jara, P., Dreyfus, B. and Dommergues, Y.R. 1981 Is it necessary to inoculate soybean in West African soils? In: Proc. 8th North American *Rhizobium* Conference, Winnipeg, 2—7 August, 1981. (in press).

90. Johnson, H.S. and Hume, D.J. 1972 Effects of nitrogen sources and organic matter on nitrogen fixation and yield of soybeans. Canadian Journal of Plant Science 52, 991—996.

91. Jones, D.G. and Morley, S.J. 1981 Marker techniques, competitive ability and host-strain preference in the clover-Rhizobium symbiosis. In: Current Perspectives in Nitrogen Fixation, p. 429. Gibson, A.H. and Newton, W.E. (eds.), Australian Academy of Science, Canberra.

92. Kang, B.T., Nangju, D. and Ayanaba, A. 1977 Effects of fertilizer use on cowpea and soybean nodulation and nitrogen fixation in lowland tropics. In: Biological Nitrogen Fixation in Farming Systems in the Tropics, pp. 217—232. Ayanaba, A. and Dart, P.J. (eds.), Wiley-Interscience, Chichester.

93. Keyser, H.H., Munns, D.N. and Hohenberg, J.S. 1979 Acid tolerance of rhizobia in culture and in symbiosis with cowpea. Soil Science Society of America Journal 43, 719—722.

94. Khan, T.N. and Eagleton, G.E. 1980 The winged bean (*Psophocarpus tetragonolobus*). In: Advances in Legume Science, pp. 383—392. Summerfield, R.J. and Bunting A.H. (eds.), Royal Botanic Gardens, Kew.

95. Knowles, R. 1980 Nitrogen fixation in natural plant communities and soils. In: Methods for Evaluating Biological Nitrogen Fixation, pp. 557—582. Bergersen, F.J. (ed.), Wiley, Chichester.

96. Knowles, R. 1981 The measurement of nitrogen fixation. In: Current Perspectives in Nitrogen Fixation, pp. 327—333. Gibson, A.H. and Newton, W.E. (eds.), Australian Academy of Science, Canberra.

97. Kremer, R.J. and Wagner, G.H. 1978 Detection of soluble *Rhizobium japonicum* antigens in soil by immunodiffusion. Soil Biology Biochemistry 10, 247—255.

98. Kuykendall, L.D. and Weber, D.F. 1978 Genetically-marked *Rhizobium* identifiable as inoculum strains in nodules of soybean plants grown in fields populated with *Rhizobium japonicum.* Applied and Environmental Microbiology 36, 915—919.

99. Kwakye, P.K. and Ofori, C.S. 1977 The use of isotopes in fertilizer efficiency studies on groundnuts (*Arachis hypogaea*) and cowpeas (*Vigna unguiculata*). Research Contract No. 1230/RB, Soil Research Institute, Kwadaso-Kumasi, Ghana 25pp.

100. Labandera, C.A. and Vincent, J.M. 1975 Competition between an introduced strain and native Uruguayan strain of *Rhizobium trifolii.* Plant and Soil 42 327—347.

101. Law, I.J. and Strijdom, B.W. 1974 Nitrogen-fixing and competitive abilities of a Rhizobium strain used in inoculants for *Arachis hypogaea.* Phytophylactica 6, 221—228.

102. Lawn, R.J. and Brun, W.A. 1974 Symbiotic nitrogen fixation in soybeans. I. Effect of photosynthetic source-sink manipulations. Crop Science 14, 11—16.

103. Lawn, R.J. and Brun, W.A. 1974 Symbiotic nitrogen fixation in soybeans.

III. Effect of supplemental nitrogen and intervarietal grafting. Crop Science 14, 22–25.

104. Legarda, L. and Forsythe, W. 1978 Soil water and aeration and redbean production. II. Effect of soil aeration. *Turrialba 28*, 175–178.

105. Lehman, P.S., Huisingh, D. and Barker, K.R. 1971 The influences of races of *Heterodera glycines* on nodulation and nitrogen-fixing capacity of soybean. Phytopathology 61, 1239–1244.

106. Liu, M.-c. and Hadlee, H.H. 1976 Effects of a non-nodulating gene (rj_1) on seed protein and oil percentages in soybeans with different genetic backgrounds. Crop Science 16, 321–325.

107. Loveday, J. 1963 Influence of oxygen diffusion rate on nodulation of subterranean clover. Australian Journal of Science 26, 90–91.

108. Marques Pinto, C., Yao, P.Y. and Vincent, J.M. 1974 Nodulating competitiveness amongst strains of *Rhizobium meliloti* and *Rhizobium trifolii*. Australian Journal of Agricultural Research 25, 317–329.

109. Marshall, K.C. 1956 A lysogenic strain of *Rhizobium trifolii*. *Nature 177*, 92.

110. Masefield, G.B. 1957 The nodulation of annual leguminous crops in Malaya. Empire Journal of Experimental Agriculture 25, 139–150.

111. Masefield, G.B. 1965 The effect of organic matter in soil on legume nodulation. *Experimental Agriculture 1*, 113–119.

112. Meyer, J., Germani, G., Dreyfus, B., Saint-Macary, H., Ganry, F., and Dommergues, Y.R. 1982 Estimation de l'effet de deux facteurs limitant sur la fixation de l'azote (C_2H_2) par l'arachide et le soja. Oléagineux (in press).

113. Minchin, F.R. and Pate, J.S. 1975 Effects of water, aeration, and salt regime on nitrogen fixation in a nodulated legume: Definition of an optimum environment. Journal of Experimental Botany 26, 60–69.

114. Minchin, F.R., Summerfield, R.J., Eaglesham, A.R.J., and Stewart, K.A. 1978 Effects of short-term waterlogging on growth and yield of cowpea (*Vigna unguiculata*). Journal of Agricultural Science, Cambridge 90, 355–366.

115. Moore, A.W. 1960 Symbiotic nitrogen fixation in a grazed tropical grass-legume pasture. *Nature 185*, 638.

116. Morley, S.J. and Jones, D.G. 1980 A note on a highly sensitive modified ELISA technique for Rhizobium strain identification. Journal of Applied Bacteriology 49, 103–109.

117. Mosse, B., Powell, C.L., and Hayman, D.S. 1976 Plant growth responses to vesicular-arbuscular mycorrhiza. 9. Interactions between vesicular-arbuscular mycorrhiza, rock phosphate and symbiotic nitrogen fixation. New Phytologist 76, 331–342.

118. Munns, D.N. 1977 Mineral nutrition and the legume symbiosis. In: A Treatise on Dinitrogen Fixation. IV – Agronomy and Ecology, pp. 353–391. Hardy, R.W.F. and Gibson, A.H. (eds.), Wiley-Interscience, New York.

119. Munns, D.N. 1977 Soil acidity and related factors. In: Exploiting the Legume-Rhizobium symbiosis in Tropical Agriculture, pp. 211–236. Vincent, J.M., Whitney, A.S. and Bose, J. (eds.), Univ. Hawaii College of Agriculture Miscellaneous Publication No 145.

120. Munns, D.N. and Fox, R.L. 1976 Depression of legume growth by liming. Plant and Soil 45, 701–705.

121. Munns, D.N. and Fox, R.L. 1977 Comparative liming requirements of temperate and tropical legumes. Plant and Soil 46, 533–548.

122. Munns, D.N., Keyser, H.H., Fogle, V.W., Hohenberg, J.S., Righetti, T.L., Lauter, D.L., Zaroug, M.G., Clarkin, K.L., and Whitacre, K.W. 1979

Tolerance of soil acidity in symbioses of mung bean with rhizobia. Soil Science Society of America Journal 71, 256–260.

123. Munns, D.N. and Mosse, B. 1980 Mineral nutrition of legume crops. In: Advances in Legume Science, pp. 115–125. Summerfield, R.J. and Bunting, A.H. (eds.), Royal Botanic Gardens, Kew.

124. Myskow, W. 1970 Humus substances of different origin and their influence on the symbiosis of red clover with Rhizobium. Zaklad Mikrobiologia 39, 39–54.

125. Norris, D.O. 1958 Lime in relation to the nodulation of tropical legumes. In: Nutrition of the Legumes, pp. 164–180. Hallsworth, E.G. (ed.), Butterworths, London.

126. Norris, D.O. 1959 Legume bacteriology in the tropics. Journal of the Australian Institute of Agricultural Science 25, 202–207.

127. Norris, D.O., Lopes, E.S. and Weber, D.F. 1970 The use of organic matter and lime pelleting in field testing of Rhizobium strains under tropical conditions. Pesquisas Agropecuarias Brasileira 5, 129–146.

128. Nutman, P.S. 1976 IBP field experiments on nitrogen fixation by nodulated legumes. In: Symbiotic Nitrogen Fixation in Plants, pp. 211–237. Nutman, P.S. (ed.), Cambridge Univ. Press, London.

129. Nutman, P.S. 1980 Adaptation. In: Nitrogen Fixation, pp. 335–354. Stewart, W.D.P. and Gallon, J.R. (eds.), Academic Press, London.

130. Nutman, P.S. 1981 Hereditary host factors affecting nodulation and nitrogen fixation. In: Current Perspectives in Nitrogen Fixation, pp. 194–204. Gibson, A.H. and Newton, W.E. (eds.), Australian Academy of Science, Canberra.

131. Orme-Johnson, W.H. 1977 Biochemistry of nitrogenase. In: Genetic Engineering for Nitrogen Fixation, pp. 317–331. Hollaender, A. et al. (eds.), Plenum Press, New York.

132. Osa-Afiana, L.O. and Alexander, M. 1979 Effect of moisture on the survival of Rhizobium in soil. Soil Science Society of America Journal 43, 925–930.

133. Parker, C.A., Trinick, M.J. and Chatel, D.L. 1977 Rhizobia as soil and rhizosphere inhabitants. In: A Treatise on Dinitrogen Fixation. IV – Agronomy and Ecology, pp. 311–352. Hardy, R.W.F. and Gibson, A.H. (eds.), Wiley – Interscience, New York.

134. Pate, J.S. 1976 Physiology of the reaction of the nodulated legumes to environment. In: Symbiotic Nitrogen Fixation in Plants, pp. 335–360. Nutman, P.S. (ed.), Cambridge University Press, London.

135. Pena-Cabriales, J.J. and Alexander, M. 1979 Survival of Rhizobium in soils undergoing drying. Soil Science Society of American Journal 43, 962–966.

136. Ramirez, C. and Alexander, M. 1980 Evidence suggesting protozoan predation on Rhizobium associated with germinating seeds and in the rhizosphere of bean (Phaseolus vulgaris L.). Applied and Environmental Microbiology 40, 492–499.

137. Ratner, E.I., Lobel, R., Feldhay, H. and Hartzook, A. 1979 Some characteristics of symbiotic nitrogen fixation, yield, protein and oil accumulation in irrigated peanuts (Arachis hypogaea). Plant and Soil 51, 373–386.

138. Rerkasem, B. and Tongkumdee, D. 1981 Rhizobium symbiotic development in rice based multiple cropping systems. In: Current Perspectives in Nitrogen Fixation, p. 435. Gibson, A.H. and Newton, W.E. (eds.), Australian Academy of Science, Canberra.

139. Robinson, A.C. 1969 Competition between effective and ineffective strains of Rhizobium trifolii in the nodulation of Trifolium subterraneum. Australian Journal of Agricultural Research 20, 827–841.

140. Robinson, P.E. 1961 Root-knot nematodes and legume nodules. Nature 189, 506–507.

141. Roughley, R.J. 1980 Environmental and cultural aspects of the management of legumes and Rhizobium. In: Advances in Legume Science, pp. 97–103. Summerfield, R.J. and Bunting, A.H. (eds.), Royal Botanic Gardens, Kew.

142. Roughley, R.J. and Vincent, J.M. 1967 Growth and survival of *Rhizobium* spp. in peat culture. Journal of Applied Bacteriology 30, 362–376.

143. Sanchez, P.A. and Isbell, R.F. 1979 A comparison of the soils of tropical Latin America and tropical Australia. In: Pasture Production in Acid Soils of the Tropics, pp. 25–53. Sanchez, P.A. and Tergas, L.E. (eds.), CIAT, Cali, Colombia.

144. Schaede, R. 1941 Die knöllchen der adventiven wasserwurzeln von *Neptunia oleracea* und ihre bakterien symbiose. *Planta 31*, 1–21.

145. Schiffman, J. and Alper, Y. 1968 Inoculation of peanuts by application of Rhizobium suspension into the planting furrows. Experimental Agriculture 4, 219–226.

146. Schmidt, E.L. 1978 Ecology of the legume root nodule bacteria. In: Interactions between Non-pathogenic Soil Microorganisms and Plants, pp. 269–303. Dommergues, Y.R. and Krupa, S.V. (eds.), Elsevier, Amsterdam.

147. Schwinghamer, E.A. and Brockwell, J. 1978 Competitive advantage of bacteriocin- and phage-producing strains of *Rhizobium trifolii* grown in mixed culture in nutrient broth and in peat. Soil Biology Biochemistry 10, 383–387.

148. Schwinghamer, E.A. and Dudman, W.F. 1980 Methods for identifying strains of diazotrophs. In: Methods for Evaluating Biological Nitrogen Fixation, pp. 337–365. Bergersen, F.J. (ed.), Wiley-Interscience, Chichester.

149. Sen, D. and Weaver, R.W. 1980 Nitrogen fixing activity of rhizobial strain 32H1 in peanut and cowpea nodules. Plant Science Letters 18, 315–318.

150. Shimsi, D., Schiffmann, J., Kost, Y., Bielorai, H. and Alper, Y. 1967 Effect of soil moisture regime on nodulation of inoculated peanuts. Agronomy Journal 59, 397–400.

151. Shivashankar, K., Vlassak, K. and Livens, J. 1976 Effect of straw and ammonium nitrate on growth, nitrogen fixation and yield of soybeans. Z. Pflanzenern. Bodenk. 3, 357–360.

152. Silvestre, P. 1970 Travaux de l'I.R.A.T. sur le soya. Ford Foundation Grain Legume Seminar, Ibadan, Nigeria (June 22–26, 1970).

153. Silvester, W.B. 1981 Measurement of nitrogen fixation. In: Current Perspectives in Nitrogen Fixation, pp. 334–337. Gibson, A.H. and Newton, W.E. (eds.), Australian Academy of Science, Canberra.

154. Sloger, C., Bezdicek, D., Milberg, R., and Boonkerd, N. 1975 Seasonal and diurnal variation in $N_2(C_2H_2)$ – fixing activity in field soybeans. In: Nitrogen Fixation by Free-living Micro-organisms, pp. 271–284. Stewart, W.D.P. (ed.), Cambridge University Press, London.

155. Small, J.G.C. and Leonard, O.A. 1969 Translocation of C^{14}-labelled photosynthate in nodulated legumes as influenced by nitrate nitrogen. American Journal of Botany 56, 187–194.

156. Smith, S.E. and Kelley, B.C. 1981 Effects of N-Serve on nitrogen fixation in *Trifolium* and a photosynthetic bacterium, *Rhodopseudomonas capsulata*. In: Current Perspectives in Nitrogen Fixation, p. 474. Gibson A.H. and Newton, W.E. (eds.), Australian Academy of Science, Canberra.

157. Souto, S.M. and Dobereiner, J. 1970 Soil temperature effects on nitrogen fixation and growth of *Stylosanthes gracilis* and *Pueraria javanica*. Pesquisas Agropecuarias Brasileira 5, 365–371.

70

158. Sprent, J.I. 1972 The effects of water stress on nitrogen-fixing root nodules. IV. Effects on whole plants of *Vicia faba* and *Glycine max*. New Phytologist 71, 603—611.

159. Sprent, J.I. 1976 Water deficits and nitrogen-fixing root nodules. In: Water Deficits and Plant Growth, Vol. IV, pp. 291—315. Kozlowski, T.T. (ed.), Academic Press, New York.

160. Sprent, J.I. and Bradford, A.M. 1977 Nitrogen fixation in field beans (*Vicia faba*) as affected by population density, shading, and its relationship with soil moisture. Journal of Agricultural Science, Cambridge 88, 303—310.

161. Sprent, J.I. and Silvester, W.B. 1973 Nitrogen fixation by *Lupinus arboreus* grown in the open and under different aged stands of *Pinus radiata*. New Phytologist 72, 991—1003.

162. Steinborn, J. and Roughley, R.J. 1974 Sodium chloride as a cause of low numbers of Rhizobium in legume inoculants. Journal of Applied Bacteriology 37, 93—99.

163. Strijdom, B.W. and Deschodt, C.F. 1976 Carriers of rhizobia and the effects of prior treatment on the survival of rhizobia. In: Symbiotic Nitrogen Fixation in Plants, pp. 151—168. Nutman, P.S. (ed.), Cambridge Univ. Press, London.

164. Taha, A.H.Y. and Raski, D.J. 1969 Interrelationships between root-nodule bacteria, plant-parasitic nematodes, and their leguminous host. Journal of Nematology 1, 201—211.

165. Thompson, J.A. 1980 Production and quality control of legume inoculants. In: Methods of Evaluating Biological Nitrogen Fixation, pp. 489—533. Bergersen, F.J. (ed.), Wiley — Interscience, Chichester.

166. Turner, G.L. and Gibson, A.H. 1980 Measurement of nitrogen fixation by indirect means. In: Methods for Evaluating Biological Nitrogen Fixation, pp. 111—138. Bergersen, F.J. (ed.), Wiley, Chichester.

167. Turnispeed, S.G. 1973 Insects. In: Soybeans, pp. 542—572. Caldwell, B.E. (ed.), American Society of Agronomy, Madison.

168. Vachhani, M.V. and Murty, K.S. 1964 Green manuring for rice. Indian Council of Agricultural Research Report, Ser. No 17, 50 pp.

169. Verma, D.P.S. 1980 Expression of host genes during symbiotic nitrogen fixation. In: Genome Organization and Expression in Plants, pp. 439—452. Leaver, C.J. (ed.), Plenum, New York.

170. Verma, D.P.S., Legocki, R.P. and Auger, S. 1981 Expression of nodule-specific host genes in soybean. In: Current Perspectives in Nitrogen Fixation, pp. 205—208. Gibson, A.H. and Newton, W.E. (eds.), Australian Academy of Science, Canberra.

171. Vincent, J.M. 1965 Environmental factors in the fixation of nitrogen by the legume. In: Soil Nitrogen, pp. 384—435. Bartholomew, W.V. and Clark, F.C. (eds.), American Society of Agronomy, Madison.

172. Vincent, J.M. 1970 A Manual for the Practical Study of Root-Nodule Bacteria. IBP Handbook No 15. Blackwell Scientific Publications, Oxford, 164 pp.

173. Vines, S.H. 1888 On the relation between the formation of tubercles on the roots of Leguminoseae and the presence of nitrogen in the soil. Annals Botany (London) 2, 386—389.

174. Weber, C.R. 1966 Nodulating and non-nodulating soybean isolines. II. Response to applied nitrogen and modified soil conditions. Agronomy Journal 38, 46—49.

175. Weber, D.F. and Miller V. 1972 Effect of soil temperature on *Rhizobium*

japonicum serotype distribution in soybean nodules. Agronomy Journal 64, 796–798.

176. Welch, L.F., Boone, L.V., Chambliss, C.G., Christiansen, A.T., Mulvaney, D.L., Oldham, M.G. and Pendleton, J.W. 1973 Soybean yields with direct and residual fertilization. Agronomy Journal 65, 547–550.
177. Wetselaar, R., Passioura, J.B. and Singh, B.R. 1972 Consequences of banding nitrogen fertilizers in soil. I. Effects on nitrification. Plant and Soil 36, 159–175.
178. Wilkins, J. 1967 The effects of high temperatures on certain root-nodule bacteria. Australian Journal of Agricultural Research 18, 299–304.
179. Wollum, A.G. and Miller, R.H. 1980 Density centrifugation method for recovering *Rhizobium* spp. from soil for fluorescent-antibody studies. Applied and Environmental Microbiology 39, 466–469.
180. Worrall, V.S. and Roughley, R.J. 1976 The effect of moisture stress on infection of *Trifolium subterraneum* L. by *Rhizobium trifolii* Dang. Journal of Experimental Botany 27, 1233–1241.
181. Yatazawa, M. and Yoshida, S. 1979 Stem nodules in *Aeschynomene indica* and their capacity of nitrogen fixation. Physiologia Plantarum 45, 293–295.
182. Zurkowski, W. and Lorkiewicz, Z. 1979 Plasmid-mediated control of nodulation in *Rhizobium trifolii*. Archives of Microbiology 123, 195–201.

Appendix. Principles of the Main Methods for Estimating Nitrogen Fixation Under Field Conditions

In the field, legumes receive their nitrogen from two sources, soil (+ fertilizer) N, and through N_2 fixation. Two of the methods to differentiate between these sources depend on total N determination in the nodulated plant, and the comparison of this value with that for non-nodulated plants, or with plants in different N-fertilizer rate treatments. The third method depends on ^{15}N, either added to the soil or measured as a change in natural abundance, while the fourth utilizes the ability of nitrogenase to reduce the surrogate substrate, acetylene, to ethylene, which is then determined by gas chromatographic procedures.

1. The difference method

During, or at the end of, the growing season, the total plant N in a non-nodulated legume, or a non-legume, is subtracted from that in the nodulated legume [174], the assumption being that both sets of plants remove the same amount of soil nitrogen. This method is most effective when the control is of the same species as the test plant. Non-nodulating isolines are available for a number of soybean cultivars [106] while non-nodulating *Arachis hypogaea* lines have been developed recently [29, 71]. The alternatives are to use soils free of the appropriate rhizobia (these are often very difficult to find, and maintain) or to use a non-legume, which may be criticized because the rooting pattern and growth habit is unlikely to be the same as that of the legume.

2. Fertilizer equivalent method

The growth of the nodulated plant is compared with a series of N-fertilizer 'rate of application' treatments applied to non-nodulated plants of the same species, or such plants very heavily inoculated with a known ineffective strain. This approach was used successfully during the International Biological Programme to determine N_2 fixation by a number of legumes in a wide range of environments [128].

3. ^{15}N-tracer methods

[15]N-tracer methods are regarded as the most reliable as the measure of N_2 fixation is direct [16]. The principal method used involves adding ^{15}N-labelled fertilizer and comparing the dilution of the label in nodulated and non-nodulated plants (isotope dilution method; [56]) although attention is being given also to the 'natural abundance method'.

Non-nodulated plants assimilate soil (S_1) and fertilizer (F_1) nitrogen, whereas nodulated plants assimilate these forms (S_2 and F_2) plus atmospheric N_2 (Y). The ^{15}N-labelled fertilizer is added to soil at a rate not exceeding 30 kg N ha^{-1} (and preferably less) in order to minimize effects on N_2 fixation. If s_1 and s_2 are the fractions of total plant N derived from the soil N, and f_1 and f_2 the percentage derived from fertilizer N, and y the percentage originating from N_2 fixation, then:

$$S_1 + F_1 = 1 \tag{1}$$

and

$$S_2 + F_2 + Y = 1 \tag{2}$$

The assumption is that both nodulated and non-nodulated plants will take up nutrient from each source in proportion to the amount available. Thus

$$F_1/S_1 = F_2/S_2 \tag{3}$$

Introducing equation (1) into equation (3) gives

$$S_2 = \frac{F_2}{F_1}(1 - F_1) \tag{4}$$

$$Y = 1 - \frac{F_2}{F_1} \tag{5}$$

As the percentages f_1 and f_2 have the same denominator (atom % [15]N excess in the labelled fertilizer, $a \% e$):

$$y = \frac{a \% e \text{ in nodulated plant}}{a \% e \text{ in non-}N_2\text{-fixing plant}} \tag{6}$$

The amount of N_2 fixed is then calculated by multiplying total N in the plants by Y ($\Sigma N \cdot Y$); it can also be calculated by using the 'A' value transformation proposed by Fried and Broeshart [55].

A variation on this method involves incorporating ^{15}N into the soil organic matter (^{15}N fertilizer plus glucose, sucrose or cellulose incubated in the soil), with all the ^{15}N then being available following mineralization. As Ham [80] points out, the assumption made is that inorganic nitrogen available to the plant by mineralization of the labelled organic matter represents the mineralizable fraction of soil nitrogen.

The 'natural abundance method' depends on the difference in the level of ^{15}N in atmospheric N_2 and that in the pool of soil nitrogen. Modern mass spectrometers are highly sensitive and are able to detect the small differences in the ^{15}N constitution of plants fixing N_2 and those assimilating their nitrogen in a combined (NO_3, NH_3) form, which has a slightly higher level of ^{15}N than atmospheric N_2. This and the isotope dilution method are described in greater detail by Knowles [95], and the methods for determining ^{15}N are described by Bergersen [12].

4. The acetylene reduction technique

This indirect assay of measuring N_2 fixation is simple to use, relatively cheap to establish and operate, and has been a major factor facilitating the large increase in N_2 fixation research in recent years. The assay only measures activity during the assay period. While this is generally satisfactory for comparative purposes (but see below), it requires that frequent assays must be made to construct daily and seasonal profiles for estimating nitrogen fixation over a long period [7, 154]. It is also essential to determine a conversion ($C_2H_2 : N_2$) ratio by doing other measurements (e.g., ^{15}N, total N). Many regard the theoretical ratio as 3:1, as the reduction of acetylene to ethylene requires one pair of electrons whereas N_2 to $2NH_3$ requires 3 pairs of electrons. However this ignores H_2 evolution, the electrons for which are also used to reduce acetylene, and should raise the theoretical ratio to 4:1, or higher. While published values average around 3.8–4.0:1 [81], the range is 0.6–6.2:1, and it is not possible to specify a suitable conversion factor for any situation without doing the essential calibrations for those conditions. Recent work has shown that strains of rhizobia may differ in the $C_2H_2 : N_2$ ratio, while the level of desiccation of the nodules also affects the ratio (Gibson and Sprent, unpublished), requiring that care must be exercised in comparisons of treatments in which these factors are variables.

Other considerations in the use of the technique were discussed recently by Ham [80], Knowles [95, 96], by Turner and Gibson [166], and by Silvester [153]. In brief, the technique is valuable, but great caution must be exercised, particularly in the interpretation of the data.

3. Heterotrophic N_2 fixation in paddy soils

T. YOSHIDA and G. RINAUDO

1. Introduction

1.1. Maintenance of N fertility of paddy soils

The yield of rice crops in many South Asian countries affects the food supply of almost half the world's population. Nitrogen is the element that often becomes a limiting factor for crop production. In many parts of Asian rice-growing countries, chemical fertilizers are not yet commonly used and rice yields greatly depend on the natural fertility of paddy soils. Ancient farmers practiced the burnt-land type of agriculture to produce their food, mainly upland rice. They had to move from land to land every 2 to 3 years because the soil fertility did not last long. However, the farmers growing rice in flooded soil have been growing it for more than a thousand years in the same rice paddies. The continuous supply of elements, particularly nitrogen, despite its removal over the years by rice crops, is considered caused by the fixation of atmospheric N_2 by microorganisms in paddy soils [41]. But it is not yet certain which microorganisms contribute to the fixation of atmospheric nitrogen.

Long-term NPK experiments at the various prefectures in Japan have demonstrated the continuous natural supply of nitrogen in paddy soils. Konishi and Seino [57] showed that at least about 20 kg N ha^{-1} is annually fixed from the atmosphere in paddy fields that received no fertilizer (Treatment None, Table 1). The addition of phosphorus and potassium increases nitrogen enrichment in the paddy field (Treatment PK, Table 1).

Table 1. Nitrogen balance in rice growing paddy field for 22 years (1929–1950) Ishikawa Agricultural Experiment Station, Japan [57]

Treatment	Gains or losses of nitrogen[a] (kg ha^{-1})	
	Without Ca	With Ca
None	+ 420	+ 651
PK (−N)	+ 741	+ 838
NK (−P)	− 507	− 372
NP (−K)	− 57	+ 82
NPK	+ 39	+ 95

[a] Positive values were considered as gains through atmospheric N_2 fixation. The calculation was done as follows: (N taken up by plants) − (N added as fertilizers + lost soil N)

Y.R. Dommergues and H.G. Diem (eds.), Microbiology of Tropical Soils and Plant Productivity. ISBN 978-94-009-7531-6.
© 1982 Martinus Nijhoff/Dr W. Junk Publishers, The Hague/Boston/London.

Matsuo and Takahashi [75], in a long-term experiment in Japan, calculated the nitrogen balance in paddy fields without nitrogen application and estimated the average natural supply of nitrogen in such fields at 49 to 91 kg ha^{-1} year^{-1} and 70 kg ha^{-1} year^{-1} for all the districts in Japan.

In Hokkaido, Japan, a 41-year rice experiment without nitrogen fertilizer showed average 2.8 t ha^{-1} grain yield with no decline in soil fertility [100]. The amount of nitrogen removed from the paddy was 53 kg ha^{-1} year^{-1}. Koyama and App [58] reported an input of 35–50 kg N crop^{-1} into flooded rice soils in temperate and tropical regions.

In a Maahas clay area at the experimental field of the International Rice Research Institute (IRRI), Philippines, net amount of soil nitrogen released was estimated at 57 kg ha^{-1} in fields submerged for 110 days during the dry season and 77 kg ha^{-1} in fields submerged for 94–98 days during the wet season [82, 101]. The input and output of natural nitrogen in the rice-growing soil seem well balanced. Input of natural nitrogen such as nitrogen in rain or irrigation water did not account for the balance. However, if the soil mineralized 57 kg ha^{-1} every cropping season, its total nitrogen (2,800 kg ha^{-1}, on the assumption of 0.14% soil N) would be compensated for in about 50 croppings unless the soils gain nitrogen by some mechanisms. Nevertheless without nitrogen fertilizer the rice yield in different paddies located in the Philippines has remained at several tons ha^{-1} over the past many years [48] (Table 2).

1.2. Measurement of N_2 fixation in the field

There are three major methods of studying the biological N_2 fixation in rice soils: the Kjeldahl technique, the N-15 isotope technique, and the acetylene reduction (C_2H_2-C_2H_4) technique. A direct estimate of N_2 fixation is obtained by determining the increase in total fixed N in samples. Total nitrogen is determined by the Kjeldahl method, which is adequate for a system that is capable of vigourous N_2 fixation, but is considered unreliable for a system that shows a small percentage of increase in fixed nitrogen. However, the method is satisfactory if length of exposure to N_2 results in a reasonably high percentage change in total nitrogen in the system used.

An experiment by App et al. [4] using this technique explains the natural nitrogen fertility of flooded rice fields and indicates that the rice plants significantly improve the nitrogen economy of such soils (Table 3).

More specifically, a direct and probably more accurate method for measuring N_2 fixation would be the N-15 isotope technique. But this tracer technique requires the use of expensive $^{15}N_2$, particularly if it is used on a large scale, as in a growth chamber or a paddy field. It also needs analysis with a mass spectrometer, a rather expensive piece of equipment and one requiring some laborious procedures in sample preparation. Therefore, most of the research on N_2 fixation in rice soils using the N-15 isotope technique has been restricted to experiments in the

Table 2. Yields of a traditional variety and two modern varieties grown without nitrogen fertilizer [48]

Yield (t ha^{-1})

Wet season	IRRI			Maligaya station			Bicol station			Visayas station			All stations		
	Peta	IR8	IR20	Peta	IR8	IR20	Peta	IR8	IR20	Peta	IR8	IR20	Peta	IR8	IR20
1968	3.9	4.1	3.7	4.1	3.7	4.0	3.8	5.1	4.5	2.4	3.0	3.1	3.6	4.0	3.8
1969	2.9	5.6	5.2	3.8	5.2	4.4	3.4	4.6	4.0	4.1	4.0	5.2	3.6	4.8	4.7
1970	2.8	4.9	4.6	2.8	4.0	4.3	0.0	2.8	2.4	3.5	2.2	3.5	2.3	3.5	3.7
1971	1.3	3.6	3.8	2.9	2.5	3.2	3.1	3.2	2.9	4.6	3.9	3.8	3.0	3.3	3.4
1972	2.9	3.3	4.0	3.2	3.6	3.6	3.8	4.2	3.8	3.4	3.2	3.5	3.3	3.6	3.7
Average	2.8	4.3	4.3	3.4	3.8	3.9	2.8	4.0	3.5	3.6	3.3	3.8	3.2	3.8	3.9

Table 3. Nitrogen balance sheet for flooded soil planted to rice determined by the Kjeldahl method [4]

Planted to rice	Exposed to light	Supplemental treatment	Mg N			N balance (X + Y − Z)	N gain as % of crop N
			Crop ± S.E. (X)	Change in soil ± S.E. (Y)	Misc. Inputs (Z)		
			Experiment A (4 crops)				
+	−	None	997±15 (a)	−795±39 (a)	21	181** (a)	18
+	−	Stubble removed	1042±22 (a)	−875±66 (a)	21	148* (a)	14
−	−	None	0 (b)	−243±40 (b)	0	−243** (b)	–
			Experiment B (6 crops)				
+	+	None	1175±11 (a)	−604±70 (a)	27	544** (a)	46
+	−	None	1148±16 (a)	−961±50 (b)	27	160* (b)	14
−	+	None	0 (b)	193±82 (c)	24	169 n.s (b)	–
			Experiment C (6 crops)				
+	+	None	1203±12 (b)	−789±61 (a)	27	387** (a)	32
+	+	P, Fe	1211±43 (b)	−895±77 (b)	92	723** (b)	60
+	+	P, Fe, algae	1273±33 (b)	−261±80 (b)	98	914** (b,c)	72
+	+	P, Fe azolla	1681±21 (a)	−421±67 (b)	106	1153** (c)	69

Note: Significantly different from zero at the 5% and 1% level represented by * and ** . Means followed by the same letter not significantly different at the 5% level by Duncan Multiple Range Test. S.E. means standard error of mean laboratory. The isotope technique is direct and, perhaps, much less subject to sampling errors than the Kjeldahl technique, however. Recently, the N_2-fixing activities of the rice root zone were measured by ^{15}N-labeled dinitrogen in water culture conditions [52]. A similar experiment using the tracer technique in flooded soil conditions [37, 133] indicated that the N-15 isotope technique is sensitive enough to detect within 1 or 2 weeks the N_2-fixing activities in the rice soil-plant system under *in situ* conditions.

The acetylene reduction method which is simple, sensitive and inexpensive, greatly enhanced N_2 fixation studies [22, 45]. Because it indirectly measures potential biological N_2 fixation, it has been considered unreliable for critical experiments aimed at estimating N_2 fixation. But N_2 fixation has been studied by the C_2H_2-C_2H_4 technique under *in situ* conditions in paddy fields [3, 60, 62, 63, 119].

Alimagno and Yoshida [3] used a device for *in situ* acetylene ethylene assay of submerged rice soils. Similarly, the N_2-fixing activities in the rice rhizosphere were measured with a metal cylinder attached to a plastic bag for C_2H_2-C_2H_4 assay in *in situ* field conditions [62]. The N_2-fixing activity was markedly higher in planted areas of the field than in non-planted areas between plant rows. Lee and Watanabe [60] found that stirring the soil-water system within the assay device maximized the recovery of the evolved ethylene.

1.3. Contributions of heterotrophic and photosynthetic organisms

It appears from the foregoing review that the continuous supply of nitrogen despite its removal by rice crops, is caused by microorganisms in paddy fields. It is interesting now to know which microorganisms actually take important part in the N_2 fixation in paddy fields. The blue-green algae have been receiving more attention than the bacteria in relation to the N_2 fixation in rice soils. Much work has been done on N_2 fixation by blue-green algae in rice soils particularly, in Japan and India (refer to Chapter 5 of this book). Little work regarding the study on the photosynthetic bacteria in rice soils has been reported [56, 77, 123].

The photosynthetic microorganisms may have a great potential in the N_2 fixation in paddy fields if the various ecological factors are suitable for the growth and maintenance of the N_2-fixing inhabitants. A greenhouse experiment aimed at determining the role of the photosynthetic microorganisms in rice soils showed that N_2-fixing activity is much higher in flooded conditions under light than under dark [130]. But there were no significant differences, as measured by the weights of straw and filled grain and by nitrogen uptake in rice between the light and dark treatments during two continuous croppings.

In tests using the *in situ* device for assaying photosynthetic N_2 fixation in paddy fields, the estimated amounts of N_2 fixation ranged from 2.3 to 5.7 kg N ha^{-1} in one location and 18.5 to 33.3 kg N ha^{-1} in another, in the Philippines [3]. Inoculating paddy fields with N_2-fixing blue-green algae such as *Anabaena* sp., *Nostoc* sp., *Aulosira fertilissima*, or *Tolypothrix tenuis* benefited rice growth and yield [113] in India, although it did not significantly affect growth and yield of rice plants in both field experiments at Los Baños, Philippines [2].

Rouquerol [98] reported that *Azotobacter* and *Clostridium* are responsible for the N_2 fixation in a rice field in southern France. Okuda and Yamaguchi [77] studied the distribution of photosynthetic N_2-fixing bacteria in paddy soils and reported that the occurrence of the bacteria appears important in the N_2 fixation in paddy soils in Japan. Becking [15] reported that *Beijerinckia* is distributed widely in tropical regions, mainly in lateritic soil, and more abundantly than *Azotobacter* in soil with low pH. However, no published paper gives a quantitative estimation of N_2 fixation by these heterotropic N_2-fixing bacteria. Jensen [53] or Alexander [1] consider the frequently estimated annual gains of 20–50 kg N ha^{-1} by heterotrophic N_2-fixing bacteria too high because these bacteria must consume at least 1–2.5 tons of organic materials of the same nutritive value as glucose.

Many papers report that the addition of organic materials to flooded soil stimulates the N_2-fixing activities of the soils [14, 67, 68, 72, 85, 89, 90]. But none of these studies reported quantitative estimates of N_2 fixation in the paddy field.

Atmospheric N_2 was recently found to be fixed by heterotrophic N_2-fixing bacteria in the rhizosphere of the rice plant [17, 34, 40, 47, 62, 63, 66, 111, 128, 129]. The rice rhizosphere has some unique characteristics, which include the air-transporting root tissues and the aerobic-anaerobic interface between the oxidative root tissue and the anaerobic soil [126]. The N_2-fixing activity increases as the plant ages, reaching a maximum at mid-reproductive stage. Rice plants rapidly translocate into plant parts the N_2 fixed in the rhizosphere [52, 133]. The amount of N_2 fixed in the rice rhizosphere by heterotrophic bacteria has been considered significant for rice production without fertilizers [4, 128, 133] Wada *et al.* [114] reported that the most important site of N_2 fixation was the reduced horizon of the paddy field, and the role of the rhizosphere in the estimation of N_2 fixation in paddy fields could not be neglected (Fig. 1.). Matsuguchi [72] reported that heterotrophic N_2 fixation shared more than 60 percent of the total N_2 fixed in the field. Direct measurement of N_2 fixation by the Kjeldahl technique gave an additional evidence of simultaneous N_2 fixation by heterotrophic and photo-trophic organisms as shown in Fig. 2 [4, 49].

1.4. Conclusion

Scientists have long recognized the maintenance of nitrogen fertility in rice paddies. The mechanism involved in this phenomenon is believed to be the biological N_2 fixation. The evidence supporting the theory has increased in the last several decades. Major biological agents contributing to nitrogen enrichment are, apparently, photosynthetic microorganisms, including the N_2-fixing blue-green algae and heterotrophic bacteria in rice paddies. The magnitude of the contribution of each N_2 fixer to the maintenance of nitrogen fertility of rice paddies may depend on environmental conditions and has not yet been clearly determined. How-

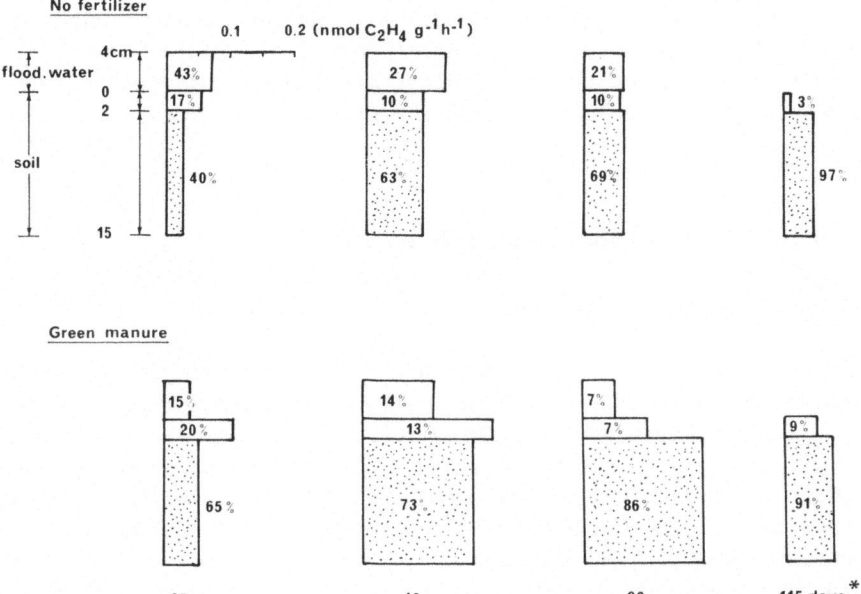

Fig. 1. Distribution of N_2-fixing activity in soil at different sites in Konosu [114]. * days after transplanting.

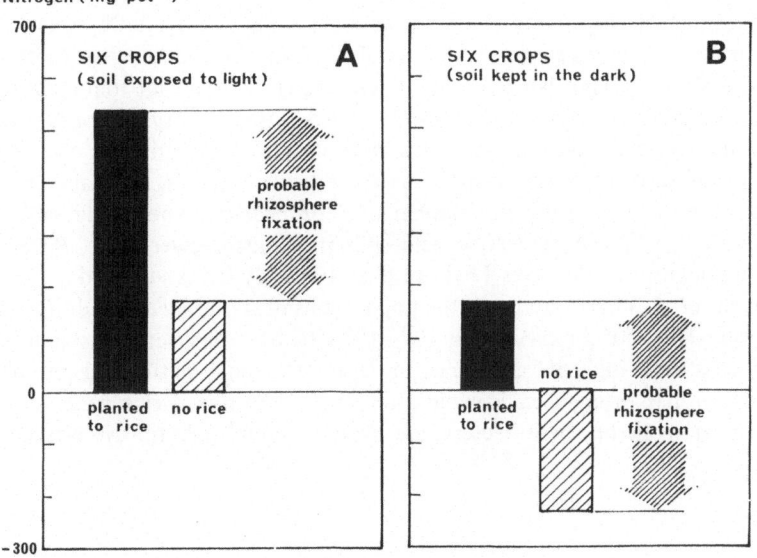

Fig. 2. Direct measurement of nitrogen balance suggests significant N_2 fixation in the root zone of paddy rice as a complement to that by blue-green algae (soil exposed to light: A) or in the absence of the blue-green algae (soil kept in the dark: B) [49].

ever, the role in the nitrogen enrichment of both phototrophic and heterotrophic N_2-fixing microorganisms appears important; probably that of the former is more important in the earlier stage of the rice-growing season and that of the latter is more important in the reproductive stage of the rice plant.

2. Microbiology of heterotrophic N_2 fixation

2.1. Microorganisms involved

2.1.1. Qualitative studies. A wide range of N_2-fixing bacteria occurs in rice soils: *Pseudomonas, Azotomonas, Azotobacter, Beijerinckia, Flavobacterium, Arthrobacter, Bacillus, Clostridium* [10]. *Desulfovibrio, Desulfotomaculum* and methane-oxidizing bacteria [19] have also been frequently found in paddy soils. *Derxia* was found particularly prevalent in waterlogged soils [27]. Recently, different strains of *Azospirillum* [39, 59, 76, 102], *Enterobacter* sp. [44, 118], *Pseudomonas, Arthrobacter, Spirillum, Vibrio* (Rinaudo, unpublished), and possibly *Achromobacter* [116, 121] have been isolated from washed or surface-sterilized rice roots. Such a large range of N_2-fixing heterotrophs suggests that heterotrophic N_2 fixation adapts to widely diverse soil environments.

2.1.2. Quantitative estimations. According to Ishizawa *et al.* [50] and Balandreau *et al.* [9], aerobic N_2 fixers are more abundant than anaerobic N_2 fixers in rice soils and in the rice rhizosphere. Counts of N_2 fixers on washed rice roots [111] revealed a similar trend. Ishizawa and Toyota [51] pointed out that *Azotobacter* was distributed more widely in wetland rice soils than in dryland ones, and also more widely in non-volcanic soils than in volcanic ones, although the bacterial population in most soils was as sparse as 10^3 g (soil)$^{-1}$ or less. In alkaline rice soils such as those in Egypt [69] the number of *Azotobacter* can be considerable $10^3 - 10^7$ g (soil)$^{-1}$, while in neutral to acid soils it is much lower [50]. *Beijerinckia* occurs abundantly in rice soils [16]. In over 40 paddy fields surveyed in Thailand, Matsuguchi *et al.* [74] found that the populations of *Azotobacter*, and *Clostridium* which were $10^0 - 10^4$ g (soil)$^{-1}$ and $10^3 - 10^6$ g (soil)$^{-1}$, respectively, tended to be high in soils with high pH and large amounts of organic matter and phosphorus. *Azospirillum* has often been isolated from surface sterilized rice roots. Unfortunately, most work on these bacteria are qualitative and quantitative estimates are lacking.

2.2. Distribution of N_2-fixing microorganisms

2.2.1. Location of N_2-fixing bacteria in non-rhizosphere soil. Little attention has been paid to heterotrophic N_2 fixers in non-rhizosphere soil. However, a considerable amount of organic matter (stuble and root litter from crop residues) is left in the field after harvest. Recent studies in Japan suggest that: (1) organic debris are active microsites for non-symbiotic N_2 fixation [115], (2) rice straw decomposition and subsequent heterotrophic N_2 fixation proceed not only in the surface layer, where aerobic or microaerophilic conditions facilitate both processes but also in the lower plow layer [72].

2.2.2. Location of N_2-fixing bacteria in the rhizosphere. After the rediscovery of *Spirillum lipoferum* [28], one of the most interesting problems was the location of N_2 fixers in or on root tissues and its establishment in association with the plants.

This point was recently discussed by Diem and Dommergues [23]. Intracellular colonization may occur but is generally discrete: Lakshmi *et al.* [59] observed only some *Azospirillum* cells within rice root hairs. In contrast, intercellular colonization of root cells, often reported, seems mainly restricted to older parts of roots (nodal zone) where the cells are generally damaged [6, 25, 78]. Umali-Garcia *et al.* [112] reported that *Azospirillum* cells were observable within the middle lamella of the walls of living root cells. That observation may be correlated with the ability of most *Azospirillum* strains to use the pectin as sole carbon source [24, 108]. Asanuma *et al.* [6] found that rhizoplane microorganisms of rice seedlings were distributed in every region except the root tip zone, and mainly at ruptured sites of epidermal cells and cell junctions. They were also dispersed and they adhered to mucigel, the bare epidermal surface, and root hairs. The ratio of bacterial coverage on the root surface was surprisingly small: 1–9% of the total root surface.

2.2.3. Distribution of N_2-fixing microorganisms in different sectors of the rhizosphere. The rhizosphere can be divided into three sectors: (1) the rhizosphere soil comprising the region of soil immediately surrounding the plant roots and the organisms living therein, (2) the rhizoplane formed by the root surface and the microorganisms living on it, and (3) the endorhizosphere formed by the root cortical tissue involved and colonized by heterotrophic microorganisms [32]. Brown [21] considered many N_2-fixing bacteria (e.g.) *Azotobacter* as not true rhizosphere bacteria. Colonization by *Azotobacter*, when it occurred, was limited to the rhizosphere soil, and very few cells were found on the root surface. In contrast, Döbereiner and Ruschel [30] reported that *Beijerinckia* grew better on the root surface of waterlogged rice than in the surrounding soil. The results obtained by Döbereiner's group [29] suggested that plants with the C_4 photosynthetic pathway are preferentially infected by *Azospirillum lipoferum* while C_3 plants are infected by *A. brasilense*. In flooded rice, all *Azospirillum* isolates from surface-sterilized roots (15 min in 1% chloramine T) belonged to the species *A. brasilense*, 96% of them being nir⁻ (nitrite reductase⁻), while in non-planted soil only 50% of the isolates belonged to *A. brasilense* nir⁻.

Watanabe *et al.* [121] studied the distribution of aerobic heterotrophic bacteria among the three sectors of the rhizosphere of rice cultivar IR26. Bacterial isolates were taken from the plates used for these counts, and checked for nitrogenase activity. Similar studies including actinomycetes counts have been done in Senegal with rice cultivar Moroberekan [97]. The results are summarized in Table 4. The percentage of bacterial isolates that were nitrogenase positive increased from rhizosphere soil to endorhizosphere in both the 'rice cultivar IR26-Philippine soil' and 'rice cultivar Moroberekan-Senegalese *sol-Dior*' systems. The percentages of endorhizosphere isolates showing nitrogenase activity (81% and 50%, respectively) were very high. Surprisingly, when Moroberekan was grown on Senegalese *sol gris*, the percentage of nitrogenase-positive isolates was lower in the endorhizosphere than in the rhizosphere soil (16% and 53%, respectively), whereas the proportion of actinomycetes decreased from endorhizosphere to rhizosphere soil (7.2% and 1.7% respectively). On the basis of these data Roussos *et al.* [97] suggested that actinomycetes might affect the rhizosphere colonization by N_2-fixing microorganisms.

A comparison of the results obtained in a Philippine soil and a Senegalese *sol gris* revealed other important differences: (1) all isolates from the endorhizosphere of the Philippine soil were presumably *Achromobacter*, whereas the most important N_2-fixing group in Senegalese *sol gris* was constituted by a pink pigmented strain belonging probably to the genus *Arthrobacter*; (2) Watanabe and Barraquio [116] found that glucose-utilizing N_2 fixers were more numerous than malate-using N_2 fixers, whereas Roussos *et al.* [97] observed that malate was more widely used by isolates from all three samples; (3) nutritional requirements (vitamins or amino acids) of N_2-fixing isolates from the Philippine soil were more marked than those of the Senegalese *sol gris*.

2.3. Conclusion

Miscellaneous bacteria other than *Azotobacter, Clostridium,* or *Azospirillum* thrive in the rice rhizosphere. Many are microaerophilic. Obligate or facultative anaerobes are present but usually less abundant. According to the different studies reported here, the rhizosphere population of N_2 fixers varies widely both qualitatively and quantitatively.

The high proportion of N_2-fixing bacteria among other rhizosphere microorganisms suggests that the rice rhizosphere could be a potential site of active N_2 fixation. However we shall see hereafter that the expression of this potential is limited by environmental factors or agents, such as actinomycetes, that impede this activity. Besides, one should be aware that the occurrence of N_2-fixing populations as high as 10^7 to 10^8 microorganisms g (dry root)$^{-1}$ does not imply that the root surface is covered by a sheath of bacteria since the bacteria coverage on the root surface never exceeds 10%.

Table 4. Colonization of rhizosphere soil, rhizoplane and endorhizosphere of three rice-soil systems

Rice-soil system	Total microflora	% of total microflora	
		N_2-fixing bacteria	Actinomycetes
Rice cv IR26-Philippine Soil [116]			
Rhizosphere soil	2.3×10^7	2.4	n.d.
Rhizoplane	1.5×10^7	76	n.d.
Endorhizosphere	1.1×10^8	81	n.d.
Rice cv Moroberekan-Senegalese sol gris [97]			
Rhizosphere soil	2.0×10^8	53	1.7
Rhizoplane	1.8×10^8	25	4.1
Endorhizosphere	1.2×10^8	16	7.2
Rice cv Moroberekan-Senegalese sol Dior[a]			
Rhizosphere soil	2.4×10^9	17	0.5
Rhizoplane	1.6×10^9	28	0.1
Endorhizosphere	2.4×10^7	50	0.1

Number of microorganisms expressed per g dry soil rhizosphere soil or dry root (rhizoplane or endorhizosphere); n.d. = not determined
[a] From Rinaudo (unpublished data)

3. Factors affecting heterotrophic N_2 fixation in paddy soils

The ecological aspects of N_2 fixation by rhizosphere microorganisms have been recently reviewed [8, 12, 23, 35, 55]. The review of Dommergues and Rinaudo [35] particularly concerns the rice rhizosphere. N_2 fixation in the rice rhizosphere depends primarily upon the plant and the rhizospheric N_2-fixing bacteria. Climatic and soil factors also directly or indirectly affect the rhizospheric microorganisms through the plant.

3.1. Effect of the plant

3.1.1. Plant genotype. Several reports confirm that the plant's genotype influences its association with N_2-fixing bacteria in its rhizosphere [42, 47, 61, 63, 91]. According to Lee *et al.* [63] the levels of N_2 fixation measured by the acetylene reduction activity (ARA) are highly correlated with the rice dry root weight at heading stage, the amount of organic material supplied by roots to the rhizosphere probably being one of the main factors responsible for such correlation.

In a first experiment, Dommergues and Rinaudo [35] found that various mutants of two rice varieties (Cesariot and Cigalon) obtained by gamma irradiation exhibited rhizospheric ARAs ranging from 500 to 5,320 (Cesariot) and 860 to 7,370 (Cigalon) nmol C_2H_4 g (dry root)$^{-1}$ h^{-1} when grown in a non-sterile Camargue soil (Table 5). A second experiment was set up with part of the same plant material, the original Cigalon variety, and two rice mutants L, which exhibited the lowest ARA, and H, which exhibited the highest ARA in the first experiment, but growing in a non-sterile Senegalese *sol gris*. Unlike in the first experiment, the ARA of the three rice genotypes did not differ significantly (Table 5). The discrepancies in results between the two experiments were attributed to differences in the soil chemical and biological properties. These data lead to the conclusion that the results of screening of rice genotypes for their ARA should be interpreted with utmost caution and their generalization to undefined environmental conditions avoided.

Table 5. Rhizospheric ARA of rice cv Cigalon and two mutants grown in alluvial soil from Camargue and a Senegalese *sol gris* (3-week old rice seedlings)

Rice 'Cigalon'	ARA (nmol C_2H_4 g (dry root)$^{-1}$ h^{-1} ± S.E.)	
	Camargue soil (first experiment)	Senegalese *sol gris* (second experiment)
Original cv	4079±1383	2589± 466
Mutant L	861± 456	3682±1254
Mutant H	7368±1971	2330± 595

S.E. = standard error of mean

3.1.2. Variations with time. Since climatic conditions vary with time, variations in the rate of some plant physiological processes and, consequently, in N_2- fixing activity associated with the plants, can be expected. Actually diurnal variations were reported by Balandreau *et al.* [11], Trolldenier [111], Rinaudo *et al.* [94] and Boddey and Ahmad [18]. Early reports on the variation in ARA throughout the rice-growing season indicated maximal activity at heading stage [63, 128]. *In situ* assays [18, 94, 120] confirmed this finding.

3.2. Climatic factors

3.2.1. Light intensity. When the plants are still young, the energy-yielding substrates available to N_2 fixers are mainly made up by root exudates. Since root exudation depends upon light intensity, shading 10-day old rice seedlings dramatically decreased rhizospheric ARA [34]. This decrease was attributed to a reduction of the carbon supply to the rhizosphere as ARA was similarly drastically reduced when 14- or 21-day old plants were decapitated. The diurnal fluctuations in rhizopheric ARA of rice seedlings reported by Rinaudo *et al.* [94] were thought to result from the close relationship of ARA to the photosynthetic activity of the plant. However, such relationship was not observed in more mature plants. Lee *et al.* [62] reported that decapitation of the shoot had little effect on *in situ* ARA within 24 hours. When a plant is aging, root lysates and root litter possibly provide N_2 fixers with an extra supply of energy-yielding compounds that greatly enlarge the carbon pool.

Light intensity may directly or indirectly affect the heterotrophic N_2 fixation in the non-rhizosphere soil. The direct effect is mainly related to the growth and activities of photosynthetic N_2-fixing bacteria. The indirect effect is the stimulation or inhibition of N_2-fixing heterotrophs in soil by the growth of photosynthetic N_2 fixers. The photosynthetic bacteria have been considered as important N_2 fixers in paddy soils [123]. The photosynthetic bacteria would have greater advantage than the blue-green algae in fixing N_2 under low light intensity and anaerobic environment.

The indirect effect of photosynthetic microorganisms on N_2 fixation can be associated with other heterotrophic N_2 fixers in paddy soils, but little information on their behaviour in the field is available [123].

3.2.2. Temperature. Relatively low temperatures limit: (1) photosynthesis, translocation, and exudation, thereby reducing the supply of energy substrate required by N_2 fixers [10] and (2) the activity of N_2 fixers. So, it is logical to assume that increased temperatures up to $30-35\,^{\circ}C$ would enhance rhizosphere ARA, although few data prove this assumption.

The effect of temperature on the non-rhizosphere soil is more related to the geographical factor than to the photosynthetic activity of plant and subsequent energy supply from roots to which it is greatly related on rhizospheric soil. The biological N_2-fixing activities are generally considered much higher in the tropical regions than in the temperate regions because of the temperature differences. A heterotrophic N_2-fixing bacterium, *Beijerinckia*, is reportedly distributed mainly in tropical regions [15]. However, its occurrence may not be due to the temperature differences but to some other soil factors such as pH or competition between microbes as reported by Becking [16].

3.3. Soil factors

3.3.1 Variation with soil type. Rinaudo *et al.* [93, 94, 95, 96] and Garcia *et al.* [38] showed that the same rice variety grown in different non-sterile soils exhibited different N_2-fixing abilities. For example, the ARA of 17-day old seedlings (Sefa 319 G) was 32 times higher when grown in Camargue soil (France), than when grown in Boundoum soil (Senegal). These differences in ARA could obviously be attributed to soil-related factors, indicating that soil is an essential part of the environment, which has often been overlooked. Of course further studies would be necessary to elucidate the roles of chemical and biological soil characteristics responsible for the variations in ARA such as those mentioned above.

3.3.2. Gaseous nitrogen. Flooded rice soil seems to have disadvantages for the biological N_2 fixation because of the limited supply of N_2 gas into the soil.

Takai *et al.* [106] reported that the amount of N_2 gas dissolved in flooded soils was generally constant for 2 weeks after flooding and the amount of methane gas increased considerably thereafter. If the flooded soils are planted to rice, N_2 gas would be supplied into soils through the plant's air-transporting system [131, 132]. In the absence of the rice plant, however, the partial pressure of N_2 gas in flooded soil may be inadequate to promote the N_2 fixation by heterotrophic N_2 fixers. The atmospheric N_2 gas diffuses into paddy water and further into the surface soil of the rice field. According to Magdoff and Bouldin [68], the concentration of N_2 diffused from the atmosphere approaches theoretically zero at a depth of about 6.6 cm below the surface of flooded soil.

Alternate flooding and drying provides favorable conditions for a high rate of N_2 fixation mainly because of the increase in the partial pressure of N_2 gas [68].

3.3.3. Inorganic nitrogen. Repression of nitrogenase synthesis by various forms of combined nitrogen in all types of N_2-fixing organisms (except derepressed mutants) is well established. Yoshida *et al.* [130] noted that 160 ppm of added combined nitrogen completely inhibited ARA in a rice soil system. Applied nitrogen, either as ammonium or nitrate, significantly inhibits the N_2-fixing activites of rhizosphere soil at 50 ppm N [66]. Trolldenier [111] found that the ARA of excised rice roots was significantly affected by an even lower application of combined nitrogen (10 ppm urea-N). However, the effect of combined nitrogen is more complex than expected. According to Balandreau *et al.* [9], the application of up to 40 ppm

NH_4-N ($120 \, kg \, N \, ha^{-1}$) caused no inhibition of ARA in the rice rhizosphere; on the contrary, ARA slightly increased probably because of the increase of exudate input into the soil. Similarly, Trolldenier [111] found that the application of up to 140 kg N ha^{-1} eventually caused an increase in rhizospheric ARA. Since rice plants absorb combined nitrogen all the more rapidly as they develop, nitrogenase repression occurs only at the early stages of growth. Starter doses of nitrogen fertilizers can therefore benefit plant growth without hindering biological N_2 fixation.

3.3.4. pH and inorganic elements other than nitrogen. Laboratory experiments indicate that the growth of pure culture of N_2 fixers is pH dependent. Most field surveys confirm these data: *Azotobacter* is generally restricted to soils of neutral and alkaline pH whereas *Beijerinckia* is more tolerant to acidic conditions [15]. *Azospirillum* species occur mainly in soils of pH 5.6–7.2 [31]. In contrast, the effects of pH on the overall N_2-fixing activity in the rhizosphere are unknown. A field survey [38] suggested that pH effect is less marked than expected.

The status of other elements besides mineral nitrogen may also affect N_2 fixation through alteration of plant exudation, about which little is known, however. Phosphorus content must be an important limiting factor, especially in poor soils, as reported for legumes. In long-term fertility assays in Thailand, Watanabe and Cholitkul [117] observed that when rice plants responded to phosphorus addition, N_2-fixing activity associated with rice was enhanced.

Trolldenier [110] found that potassium deficiency caused a decline in oxygen content and in redox potential, which might provide better conditions for N_2 fixation, but have detrimental effects on plant growth.

3.3.5. Soil oxygen and water regime. The soil water content does not directly affect N_2 fixation but controls it by affecting the rates of gas exchanges. Therefore, the effect of oxygen and that of the water regime on N_2 fixation cannot be dissociated.

The N_2-fixing enzymes are principally anaerobic and any system including them has an oxygen-scavenge system. Oxygen in air transported by the air-transporting system of flooded rice would be mostly consumed by respiration of root tissue which would provide favorable conditions for N_2-fixing bacteria in the rhizosphere.

Rice et al. [90] showed that the occurrence of an aerobic-anaerobic interface was of great importance for N_2 fixation in soil-straw mixtures: the products of the degradation of hemicellulose and cellulose in the 2 mm thick aerobic zone diffused across the interface to support N_2 fixation by *Clostridia* in the anaerobic zone. There are some similarities between such a model system and the rhizosphere of flooded rice, which is the site of an aerobic-anaerobic interface resulting from air diffusion from the leaves to the roots. Rice plants growing in anaerobic water-saturated soil can supply oxygen to the root, thus creating a gradient in oxygen concentration around its roots [5, 65, 83]. It is likely that in this gradient, a micro-zone with the oxygen concentration optimum for the development and activity of each of the various N_2-fixing strains that are diversely protected against oxygen, will exist (Rinaudo, unpublished data). Since this protection is generally not well developed in most free-living N_2-fixing bacteria, except *Azotobacter*, it is not surprising that rhizosphere ARA was constantly reported to be more pronounced when the plants were grown in waterlogged than in upland conditions [95, 128]. The addition of organic materials into paddy soils stimulates the respiration of aerobic microorganisms and enhances it to create an anaerobic environment in the paddy soil by expelling molecular oxygen and lowering the oxidation-reduction process. The quantity of oxygen supplied by diffusion through the floodwater into the soil is small and enough only in the few millimeters of surface soil. Magdoff and Bouldin [68] showed evidence supporting the hypothesis that N_2 fixation is enhanced when the products of anaerobic decomposition of cellulose are subjected to aerobic condition by such processes as diffusion, mixing, and drying.

3.3.6. Energy supply. Fixation of significant amounts of N_2 is dependent upon a suitable supply of carbon and energy. When plants are still young, root exudates and root lysates, mainly epidermal and cortical cells [70], are the major sources of energy substrates for rhizospheric microorganisms. With plant aging, root litter possibly provides N_2 fixers with an extra supply of energy-yielding compounds [35]. There is good evidence that the amounts of organic material released into the soil by the roots of actively growing plants are increased by root damage [7] and the presence of soil microorganisms [13].

The amounts of available substrates in the rhizosphere of rice grown under non-sterile conditions is poorly documented. At IRRI, the amount of exudates from IR8 and IR22 was evaluated by the isotope technique: the proportions of the total assimilated ^{14}C released into the soil were 1.9% and 2.2% at flower initiation, and 3.2% and 6.7% at harvest. But this amount of carbon indicated only carbon excreted from roots and remained in the rhizosphere soil, and did not include carbon mineralized by rhizosphere bacteria (IRRI Ann. Rep. 1973).

Data obtained by Martin [70] show a transfer of photosynthetate to the root system of field-grown wheat between early tillering and flowering that is equivalent to about 1,800 kg C ha^{-1}. Allowing for 30% of the root carbon occurring in liquid and other compounds, which is not readily available for microbial decomposition, there would be some 1,000 kg C ha^{-1} available to the microflora and microfauna.

'There is some doubt about the amount of carbon required for the heterotrophic fixation of N_2, but a common estimate is 30–40 kg C kg^{-1} N_2 fixed. If the amount of carbon and N_2 located in the root system of flooded rice is of the same order of magnitude as for wheat, one can obtain an upper limit of heterotrophic N_2 fixation equivalent to 25–30 kg N ha^{-1}. The actual amount of N_2 fixed per 1,000 kg available C ha^{-1} would be much less because no allowance has been made for root respiration, and that will induce competition for carbon among the different microbial populations, many of which may not be capable of N_2 fixation.' (K. Martin's comments on the communication presented by Dommergues and Rinaudo [35] at the Symposium 'Nitrogen and Rice' 1979, p. 259.)

In rice, N_2 fixers make up a rather large percentage of the total microflora so that one can assume that a significant part of the energy flowing out of the rhizosphere can be used for N_2 fixation if other limiting factors do not impede the process.

Most heterotrophic N_2-fixing bacteria in non-rhizosphere soil depend on available carbon materials for their field energy source. Organic materials in soil generally are not readily decomposable in nature as humic acid. Thus, organic materials incorporated freshly into paddies as dead plant tissues or root tissues left in soil would be a good source of energy in N_2 fixation by the heterotrophic bacteria in natural paddy soil environment. Assuming that the total dry weight of rice plants is 10 t ha^{-1} (25% as roots), organic matter residues that remain in a paddy field would be about 2.5 t ha^{-1}. Rice straw much more favorably stimulates N_2 fixation by heterotrophic bacteria in paddy fields. Fixation measured by the N-15 technique was equivalent to 42–45 kg N_2 ha^{-1} in a soil at field capacity and 13–150 kg N_2 ha^{-1} in flooded soil when the soil was amended with less than 1% straw [90]. Rates of fixation were as high as 500–1,000 kg N_2 ha^{-1} in the soils amended with 5–20% straw and incubated under flooded conditions in this study. However, that is not the usual case in the paddy field since 1% straw would be equivalent in amount to 20 t ha^{-1} assuming that 1 ha contains 2 million kilograms soil.

3.3.7. Interaction of N_2-fixing bacteria with other soil microorganisms. N_2-fixing bacteria thrive in the presence of other microorganisms that behave as antagonists or as synergists. Most of the studies related to such interactions have been performed *in vitro* [33, 54, 56, 64]. However some recent observations or experiments concern processes occurring in the rhizosphere of plants growing in non-sterile soils.

Some data suggested the antagonistic effect of actinomycetes on *Azotobacter* [21, 79, 105] or *Beijerinckia* [26]. Roussos *et al.* [97] pointed out that the low percentage of nitrogenase-positive isolates in the endorhizosphere of rice cultivar Moroberekan grown in Senegalese *sol gris* might be attributed to antagonistic actinomycetes.

Actually the effect of actinomycetes on root colonization by N_2-fixing bacteria appears to be more complex than expected. Thus by inoculating rice growing in a non-sterile soil with two actinomycete strains, Rinaudo *et al.* [96] observed a 25% decrease in the rhizosphere ARA (Experiment 1, in Table 6). But when a mixed inoculum comprising the same actinomycete strains plus an N_2-fixing bacterium was used, the rhizosphere ARA was 75% higher than when inoculation was made with the N_2-fixing strain alone (Experiment 2, in Table 6). The decrease in ARA observed in Experiment 1 was explained by possible antagonism between the introduced actinomycete strains and by the native N_2-fixing microflora. In Experiment 2, the increase in ARA resulting from the inoculation with the mixed culture was attributed to a synergistic interaction between the introduced N_2-fixing strain and the actinomycetes.

According to Remacle and Rouatt [86], in the early stages of barley, pectinolytic bacteria contribute to the decomposition of roots and seed reserves, and liberate available carbohydrates, thus stimulating the multiplication of *Azotobacter*.

Table 6. Inoculation of the rhizosphere of rice cv Moroberekan grown in a non-sterile soil (*sol gris*) by an N_2-fixing bacterium (F4) and two actinomycetes (Al and A15) [96]

Inoculum	ARA (nmol C_2H_4 h^{-1} ± S.E.)	
	Plant^{-1}	g (dry root)$^{-1}$
Experiment 1		
Control	173± 23	4640± 900
Al + A15	130± 83	3100±1950
Experiment 2		
F4	132± 34	3100± 950
F4 + Al + A15	234±176	5690±3850

S.E. = standard error of mean

Such a stimulatory effect was observed in the rhizosphere of rice grown in sterilized soil inoculated with a mixture of N_2-fixing bacteria and pectinolytic strains [24]. Interestingly, in a survey dealing with 34 *Azospirillum* strains isolated from the rice rhizosphere in Senegal, Diem *et al.* [24] found that 53% of the strains studied could use pectine as a substrate to fix N_2.

Other soil organisms may be involved in antagonistic processes. The results obtained by Rinaudo and Germani [92] show that nematodes of the genus *Hirschmanniella* (specific parasites of flooded rice) could be held responsible for the limitation of non-symbiotic N_2 fixation in submerged rice soils (Fig. 3).

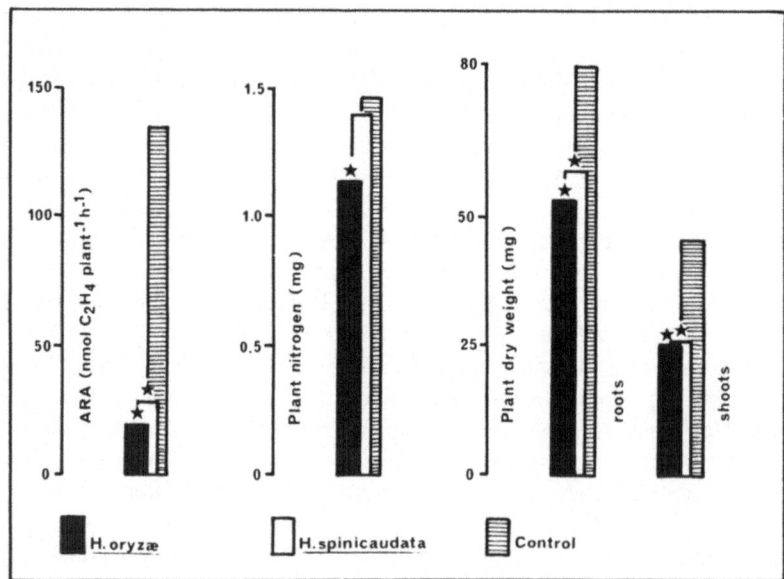

★ : Significantly different from control at P:0,01

Fig. 3. Effect of the infestation of the soil with *Hirschmanniella oryzae* and *Hirschmanniella spinicaudata* on the ARA and the plant growth of 3-week old rice seedlings (cv Moroberekan) [92].

3.4. Conclusion

Comparing the symbiotic N_2-fixing system with the rice rhizosphere N_2-fixing system shows differences which are summarized in Table 7. In the rice rhizosphere, N_2 fixers appear to find a favorable niche, but there is no tight relationship between plant growth and activity of the N_2-fixing microorganisms probably because they remain in competition with the other components of the microflora. The looseness of the relationship between the plant and the N_2 fixers explains that this system is very sensitive to the effect of environmental factors and, thus is most unstable. Therefore a recent international symposium on N_2 fixation aptly recommended that this relationship, often termed as an 'associative symbiosis', should be referred to as 'biocoenosis' (National Academy of Sciences, 1979).

4. Managing N_2 fixation in paddy soils

The yield potential of a rice crop cannot fully be realized without the application of fertilizer nitrogen. However, in many rice-growing countries the use of fertilizers is very much limited, and yield production virtually depends on the natural fertility of paddy soil. Two major aspects concern the nitrogen nutrition to promote rice growth: (1) the enrichment of nitrogen element in paddy fields, and (2) the efficiency of nitrogen in rice production.

Rice growing countries with sufficient supply of chemical fertilizers, on the other hand, are currently very much interested in exploring a possible use of biological source of nitrogen for agriculture because of their environmental problems and shortage of energy sources.

To enrich the nitrogen element we should consider improving the environmental conditions for increasing N_2 fixation. The various factors related to the free-living or associative N_2 fixation already discussed and to the introduction of a symbiotic N_2-fixing system should be considered. The efficient use of fixed nitrogen in the field management by preventing nitrogen losses through leaching or denitrification, and the promotion of associative N_2 fixation in the rice rhizosphere by seed inoculation of N_2-fixing bacteria or screening of rice varieties for higher N_2 fixation are to be investigated in the years to come.

4.1. Organic matter application

The production of straw in Japan is 19 million and its direct application to paddy fields was recently widely adapted by Japanese farmers [107]. An investigation in 1972 indicated that the average production in Japanese paddy was 4.93 t ha^{-1} for grain and 4.89 t ha^{-1} for straw. Straw materials added to rice paddies should be sources of a good energy for N_2 fixers. Carbon dioxide released into air from the decaying straw material could be used by the crop for photosynthesis [125], and subsequently may affect the N_2 fixation in the rice rhizosphere.

Table 7. Comparison of legume and rhizosphere N_2-fixing systems

Nitrogen-fixing system	Bacterial strains	Specificity	Bacterial infection	Amount of energy (expressed in moles of glucose) required to fix 1 mole of nitrogen	Protection against oxygen	Fate of fixed N_2
Legume-*Rhizobium* symbiosis	*Rhizobium*	Certain	Intracellular	1	By plant cell and pigments such as leghaemoglobin	Directly used by the legume as NH_4^+-N
Rhizosphere N_2-fixing system	Many belonging to different genera, often associated with other microorganisms	Probably no specificity	Apparently limited to rhizoplane, moribund or dead root cortical cells, and to root residues	3–4 (aerobic fixers) 8–10 (anaerobic fixers). Some environmental factors could presumably increase efficiency two- or three-fold	Protection by O_2 consumption of root respiration / Partially protection by waterlogging	Indirectly and partly directly used by the plant

However, in many Southeast Asian countries, the rice straw are traditionally burned to prevent rice disease problems. Williams *et al.* [122] reported no measurable difference in 5-year rice yield averages between burning and incorporation at any level of nitrogen fertilizers even when the rate of straw incorporation was increased to 30 t ha^{-1}.

The application of organic materials such as straw or cellulose stimulates biological N$_2$ fixation in flooded soils [14, 20, 67, 68, 71, 72, 73, 84, 89, 124]. However, there are a few reports on the actual measurement of N$_2$-fixing activity in rice field into which organic materials were incorporated during a growth period of rice plant [72, 114]. A field experiment by Matsuguchi [72] clearly indicated that the application of rice straw stimulated the N$_2$-fixing activity of paddy soils even with the addition of nitrogen fertilizer at 50 kg N ha^{-1} (Fig. 4).

Ponnamperuma [81] reported that straw incorporation, increased the total N content in paddy soil to 39–57 kg N ha^{-1} per season.

4.2. Promotion of associative N$_2$ fixation

The nitrogen nutrition of the rice plant is very much related to its photosynthetic activity and carbon metabolism. Carbon supply to the rice rhizosphere during the reproductive stages should be a very important factor in promoting the N$_2$ fixation by the associative N$_2$-fixing bacteria. It has been suggested that appreciable energy for heterotrophic diazotrophs may be supplied in a number of aquatic systems [55] and cereal crops [70]. Any method to stimulate rice plants to provide energy should

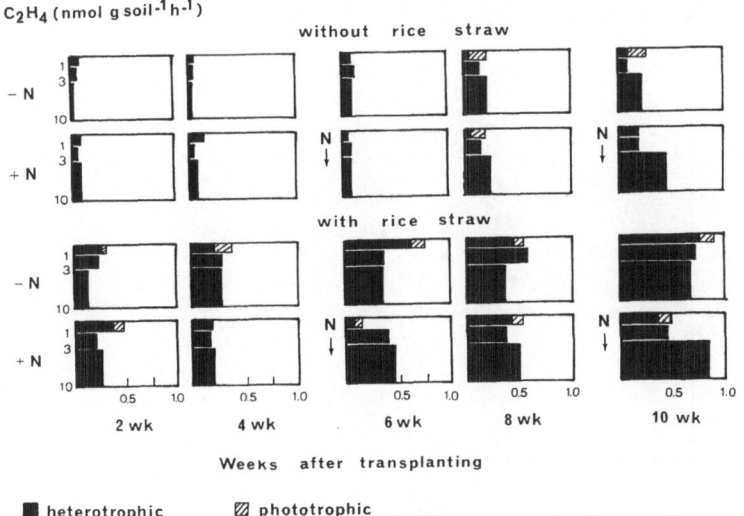

Fig. 4. Effect of N application on heterotrophic and phototrophic N$_2$-fixing activities in paddy plow layers (Tochigi Ando soil) with and without rice straw [72].

96

be considered significant and investigated. Hale and Moore [43] reviewed extensively the works on the various factors affecting root exudation such as plant, environment, and chemicals.

The associative N_2-fixation in the rhizosphere can also be promoted by the application of organic materials to the paddy field in addition to the N_2 fixation of free-living heterotrophs.

The screening of available rice cultivars for higher N_2-fixing activities in the rice rhizosphere should be worthwhile. Varieties seem to differ in N_2-fixing activity and the correlation between root weight and nitrogen fixing activity is high (Fig. 5). Further systematic screening for rice varieties in accordance with photosynthetic activity, color intensity of leaves, or root-oxidizing activity should be explored.

At the reproductive stage, the N_2-fixing bacteria are the majority of microflora

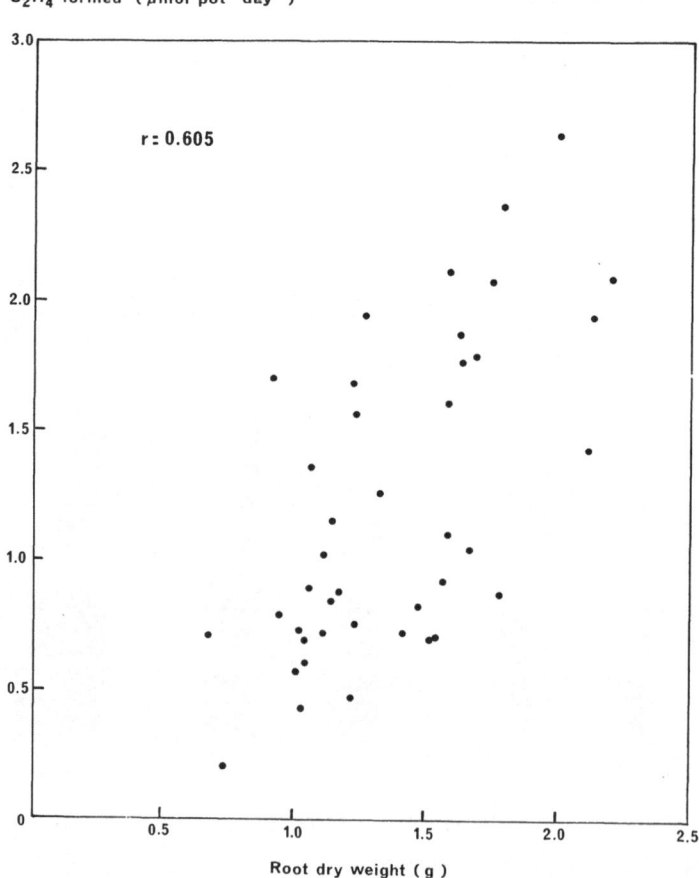

Fig. 5. Acetylene-reducing activity and root weight of 41 rice varieties (each value is the mean of 5 replications) [61].

inhabiting the rice rhizosphere [121]. The N_2-fixing bacteria inhabiting both the inside root tissue and surface root are closely associated with the rice plant. However, there is no report on specific N_2-fixing bacteria associated closely with rice roots. Artificial inoculation of some grasses with bacteria known to fix N_2 in the rhizosphere such as *Azospirillum lipoferum* was reported to increase the yields of grasses [104]. Successful inoculations of maize or wheat with *Azospirillum brasilense* have also been recently reported [46, 87, 88]. For rice, few studies have been carried out except by Gauthier and Rinaudo [39]. Using 32 different strains of *Azospirillum*, these authors showed that the rhizosphere ARA of 3-week old rice seedlings grown in test tubes could be significantly altered by inoculation: rhizosphere ARA could be either increased (+ 60%) or decreased (− 80%) in a non-sterile Senegalese *sol gris*. Such responses varied widely with the soil type. Another experiment carried out using similar experimental devices showed that inoculation with *Azospirillum brasilense* Sp7, significantly increased the plant growth, but this increase was not related to the ARA (Table 8). Thus it was hypothesized that growth-stimulating substances produced by *Azospirillum* Sp7 were involved. Up to now, this beneficial effect of inoculation observed in the case of young seedlings, has not been observed in larger-scale experiments. The lack of response in the latter experiments was attributed to the fact that introduced *Azospirillum* strains had been eliminated by competition, a process which did not occur in small devices, where the inoculum was much larger compared to the native microflora.

4.3. Field management

Maintaining soil in flooded conditions seems to be the key to the nitrogen enrichment in paddy soils [80]. Flood fallowing increased total N content by an average of 50 kg ha^{-1} per season in Maahas clay soil [81]. Rice plants readily translocated the nitrogen fixed in the rhizosphere into plant parts [52, 133]. The rhizosphere of flooded rice at the reproductive stage would have played a role in the nitrogen nutrition of rice by N_2 fixation, thereby contributing to rice yield.

The grain yield of rice correlates with the amount of nitrogen at the reproductive stages of rice plants, which determines the number of spikelets per unit area and percentage of ripened grains. In Japan rice plants showing high grain yield generally absorb 30–40% of the total plant N during the reproductive stages after panicle initiation. Japanese farmers now topdress nitrogen fertilizers to paddy rice at the reproductive stage. The nitrogen application between panicle initiation to heading stage is called 'Hogoe' and that between heading to maturity is called 'Migoe' in Japan.

However, application of fertilizer nitrogen appears to suppress the activities of N_2-fixing *Azospirillum* in the rice rhizosphere [76]. Trolldenier [109] reported that in a long-term fertility trial, the application of N, P, or K to the paddy field significantly stimulated the nitrogenase activity in the rice rhizosphere. Inoculation of N_2-fixing bacteria known as nitrogenase-depressed mutant strains, which are able to

Table 8. Influence of soil inoculation with *Azospirillum brasilense* Sp7, upon ARA and growth of 3-week old rice seedlings cv Moroberekan (Gauthier and Rinaudo, unpublished)

Soil	Inoculum	ARA (nmol C_2H_2 h^{-1} $plant^{-1}$ ± S.E.)	Plant growth	
			Aerial parts (mg $plant^{-1}$ ± S.E.)	Roots (mg $plant^{-1}$ ± S.E.)
Sol gris	0	60 ± 56	48 ± 18	68 ± 11
	+	50 ± 21	136 ± 13	85 ± 12
Bel Air	0	200 ± 33	75 ± 14	45 ± 5
	+	322 ± 146	106 ± 17	47 ± 3
Boundoum	0	61 ± 16	62 ± 15	48 ± 7
	+	47 ± 31	125 ± 18	84 ± 9

S.E. = standard error of mean

produce nitrogenase even in the presence of ammonium and to export large quantities of fixed nitrogen [99], may be significant in promoting N_2 fixation in paddy fields that received nitrogen fertilizers.

The chemical fertilizers in rice production should be applied without prohibiting the biological N_2-fixing activities. Similarly, some pesticides have a depressive effect on the heterotrophic N_2-fixing bacteria in soils [103]. Caution should be observed in their application in rice paddies. Symbiotic N_2 fixation by legume is still considered the most important biological process in agricultural industry since it greatly affects plant production of legumes and soil nitrogen fertility. Intercropping or mix-cropping of legumes with the rice crop may promote nitrogen enrichment in rice paddies. Legumes that are able to grow in flooded soil conditions and enrich soil with nitrogen should be worth studying [36].

4.4. Conclusion

Recent advancements in biological N_2-fixation studies are revealing the mechanisms involved in the natural environment of the nitrogen element in rice fields. However, they only answer the question why conventional rice farmers were able to produce rice for the last several hundred years without chemical fertilizers.

To produce more rice to feed the increasing populations under the circumstances of energy shortage we, in the rice-growing countries, should be aware of field management problems in promoting N_2 fixation in paddy soils. Unfortunately, there is yet no feasible N_2-fixation technique and it is hoped that a revolutional technology would soon solve this problem. At the same time it would be a very important consideration to manage field practice such as the application of fertilizers and other agriculture chemicals or water controls in a way to avoid inhibiting N_2-fixing activities by soil bacteria.

5. Epilogue

Only recently scientists have realized why paddy fields can maintain their nitrogen fertility for a long time. The rapid progress of N_2-fixation studies in this chapter, proves the importance of heterotrophic N_2 fixation in maintaining the nitrogen fertility in paddy soils under natural environments. However, the exact contribution of phototrophic N_2 fixers and heterotrophic N_2 fixers to the nitrogen nutrition of the rice crop is still unclear. It appears to depend on the environmental conditions of the paddy, plant age, or some other factors.

The major limiting factors in heterotrophic N_2 fixation in flooded rice paddies are the supply of organic materials as energy source and perhaps the supply of N_2 in paddy. The presence of the rice plant is apparently the most important factor. The maximum N_2-fixing activity of rice rhizosphere is reached at the plant's flowering stage, as in the case of nodule activity. At this stage, the rice plant has developed its root system and attained maximum values in dry weight, surface area, and air-transporting system. Flooding soil creates a unique ecological site, a so-called aerobic-anaerobic interface in both the soil surface and the rhizosphere of the rice plant. The ecological site seems important in the heterotrophic N_2 fixation in paddy.

Microbial studies of N_2 fixation in the rice rhizosphere should be promoted to identify the N_2-fixing bacteria that contribute most to the fixing of N_2. It is unlikely that a specific bacterial species is associated with rice roots as in the rhizobium—legume relationship.

In the coming years, varietal screening for higher N_2 fixation, inoculation of seed with an associative N_2-fixing bacterium, or management of paddy fields to make conditions favorable for heterotrophic N_2 fixation, would be further explored to promote biological N_2 fixation in rice paddy.

References

1. Alexander, M. 1961 Introduction to soil microbiology. John Wiley and Sons.
2. Alimagno, B.V. and Yoshida, T. 1975 Growth and yield of rice in Maahas soil inoculated with nitrogen-fixing blue-green algae. Philipp. Agric. 59, 80—90.
3. Alimagno, B.V. and Yoshida, T. 1977 *In situ* acetylene-reduction assay of biological nitrogen fixation in lowland rice soils. Plant and Soil 47, 239—244.
4. App, A.A., Watanabe, I., Alexander, M., Daez, C., Santiago, T. and DeDatta, S.K. 1980 Non symbiotic nitrogen fixation associated with the rice plant in flooded soils. Soil Sci. 130, 283—289.
5. Armstrong, W. 1969 Rhizosphere oxidation in rice; an analysis of inter-varietal differences in oxygen flux from the roots. Physiol. Plant. 22, 296—303.
6. Asanuma, S., Tanaka, H. and Yatazawa, M. 1979 Rhizoplane microorganisms of rice seedlings as examined by scanning electron microscopy. Soil Sci. Pl. Nutr. 25, 539—551.
7. Ayers, W.A., and Thornton, R.H. 1968 Exudation of amino acids by intact and damaged roots of wheat and peas. Plant and Soil 28, 193—207.
8. Balandreau, J., and Knowles, R. 1978 The rhizosphere. In: Interactions between non-pathogenic soil microorganisms and plants, pp. 243—268. Dommergues, Y.R. and Krupa, S.V. (eds.), Elsevier, Amsterdam.
9. Balandreau, J., Rinaudo, G., Hamad-Fares, I. and Dommergues, Y. 1975 Nitrogen fixation in the rhizosphere of rice plants. In: Nitrogen Fixation by Free-living Microorganisms, pp. 57—70. Stewart, W.D.P. (ed.), Cambridge Univ. Press.
10. Balandreau, J., Rinaudo, G., Oumarov, M., and Dommergues, Y. 1976 Asymbiotic N_2 fixation in paddy soils. In: Proceedings International Symposium on N_2 fixation, pp. 611—628. Washington State Univ. Press, Pullman.
11. Balandreau, J., Millier, C.R. and Dommergues, Y. 1974 Diurnal variations of nitrogenase activity in the field. Appl. Microbiol. 27, 662—665.
12. Balandreau, J., Ducerf, P., Hamad-Fares, I., Weinhard, P., Rinaudo, G., Millier, C. and Dommergues, Y. 1978 Limiting factors in grass nitrogen fixation. In: Limitations and Potentials for Biological Nitrogen Fixation in

the Tropics, pp. 275–302. Döbereiner, J., Burris, R.H. and Hollaender, A. (eds.), Plenum Press, New York and London.

13. Barber, D.A. and Martin, J.K. 1976 The release of organic substances by cereal roots in soil. New Phytol. 76, 69–80.

14. Barrow, N.J., and Jenkinson, D.S. 1962 The effect of waterlogging on fixation of nitrogen by soil incubated with straw. Plant and Soil 16, 258–262.

15. Becking, J.H. 1961 Studies on nitrogen-fixing bacteria of the genus *Beijerinckia*. I. Geographical and ecological distribution in soils. Plant and Soil 14, 49–81.

16. Becking, J.H. 1978 *Beijerinckia* in irrigated rice soils. Environmental Role of Nitrogen-fixing Blue-green Algae and Asymbiotic Bacteria. Ecol. Bull. (Stockholm) 26, 116–129.

17. Boddey, R.M., Quilt, P., and Ahmad, N. 1978 Acetylene reduction in the rhizosphere of rice: Method of assay. Plant and Soil 50, 567–574.

18. Boddey, R.H., and Ahmad, N. 1979 Seasonal variations in nitrogenase activity of various rice varieties measured with an *in situ* acetylene reduction technique in the field. In: Proc. Int. Workshop on Associative N_2-fixation, CENA, Piracicaba, July 22, 1979.

19. Bont, J.A.M. De, Lee, K.K. and Bouldin, D. 1978 Bacterial oxidation of methane in rice paddy. Bull. Ecol. Res. Commun. 26, 91–96.

20. Brouzes, R., Lasik, J. and Knowles, R. 1969 The effect of organic amendment, water content and oxygen on the incorporation of $^{15}N_2$ by some agricultural and forest soils. Can. J. Microbiol. 15, 899–905.

21. Brown, M.E. 1974 Seed and root bacterization. Ann. Rev. Phytopathol. 12, 181–197.

22. Burris, R.H. 1974 Methodology. In: The biology of nitrogen fixation, pp. 9–23. Quipsel, A. (ed.), North Holland.

23. Diem, H.G. and Dommergues, Y.R. 1979 Significance and improvement of rhizospheric nitrogen fixation. In: Recent advances in biological nitrogen fixation, pp. 190–226. Subba Rao, N.S. (ed.), Oxford and IBH Publishing Co., New Delhi.

24. Diem, H.G., Gauthier, D. and Rinaudo, G. 1981 Utilisation de la pectine par *Azospirillum*. Cah. ORSTOM, ser. Biol. (in press).

25. Diem, H.G., Rougier, R., Hamad-Fares, I., Balandreau, J., and Dommergues, Y.R. 1978 Colonization of rice roots by diazotroph bacteria. Environmental role of nitrogen-fixing blue-green algae and asymbiotic bacteria. Ecol. Bull. (Stockholm) 26, 305–311.

26. Diem, H.G., Schmidt, E.L. and Dommergues, Y.R. 1978 Use of the fluorescent-antibody technique to study the behaviour of a *Beijerinckia* isolate in the rhizosphere and spermosphere of rice. Environmental role of nitrogen-fixing blue-green algae and asymbiotic bacteria. Ecol. Bull. (Stockholm) 26, 312–318.

27. Döbereiner, J. and Campelo, A.B. 1971 Non-symbiotic nitrogen-fixing bacteria in tropical soils. In: Nitrogen fixation in natural and agricultural habitats. Plant and Soil, special volume, pp. 457–470.

28. Döbereiner, J. and Day, J.M. 1976 Associative symbiosis and free-living systems. In: Proc. 1st international symposium on nitrogen fixation, pp. 513–538. Newton, W.E. and Nyman, C.J. (eds), Washington State University Press.

29. Döbereiner, J. and De-Polli, H. 1979 Diazotrophic rhizocoenoses. In: Associative N_2 fixation. Vose B. and Ruschel A.P. (eds.) Vol 1, CRC Press, Boca Raton, Florida.

102

30. Döbereiner, J. and Ruschel, A.P. 1964 To the study of rhizosphere effects on soil microorganisms with special reference to *Beijerinckia*. Biologie du sol 1, 6–9.
31. Döbereiner, J., Nery, M. and Marriel, I.E. 1976 Ecological distribution of *Spirillum lipoferum*, Beijerinck. Can. J. Microbiol. 22, 1464–1473.
32. Dommergues, Y.R. 1978 The plant-microorganism system. In: Interactions between non pathogenic soil microorganisms and plants, pp. 1–36. Dommergues, Y.R. and Krupa, S.V. (eds.), Elsevier, Amsterdam.
33. Dommergues, Y.R. and Muftaftschiev, S. 1965 Fixation synergique de l'azote atmospherique dans les sols tropicaux. Ann. Inst. Pasteur 109, 112–120.
34. Dommergues, Y., Balandreau, J., Rinaudo, G. and Weinhard, P. 1973 Nonsymbiotic nitrogen fixation in the rhizosphere of rice, maize and different tropical grasses. Soil Biol. Biochem. 5, 83–89.
35. Dommergues, Y.R. and Rinaudo, G. 1979 Factors affecting N_2 fixation in the rice rhizosphere. In: Nitrogen and Rice, pp. 241–260. International Rice Research Institute, Los Banös, Philippines.
36. Dreyfus, B.C. and Dommergues, Y.R. 1981 Nitrogen-fixing nodules induced by Rhizobium on the stem of the tropical legume *Sesbania rostrata*. FEMS Microbiol. Letters 10, 313–317.
37. Eskew, D.L., Eaglesham, A.R.J. and App, A.A. 1981 Heterotrophic $^{15}N_2$ fixation and distribution of newly fixed N in a rice-flooded soil system. Plant. Physiol. 68, 48–52.
38. Garcia, J.L., Raimbault, M., Jacq, V., Rinaudo, G. and Roger, P. 1974 Activités microbiennes dans les sols de rizieres du Senégal: relation avec les caractéristiques physicochimiques et influence de la rhizosphere. Rev. Ecol. Biol. Sol. 11, 169–185.
39. Gauthier, D. and Rinaudo, G. 1981 Etude de quelques souches d'*Azospirillum* isolées au Sénégal: effet de l'inoculation par ces souches d'un système sol-riz non stérile. Cah. ORSTOM, sér. Biol. 43, 27–31
40. Gilmour, J.T., Gilmour, C.M. and Johnston, T.H. 1978 Nitrogenase activity of rice plant root systems. Soil Biol. Biochem. 10, 261–264.
41. Grist, D.H. 1965 Rice. 4th edition, pp. 221–223. Longmans and Green, London.
42. Habte, M. and Alexander, M. 1980 Effect of rice plants on nitrogenase activity of flooded soils. Appl. Environ. Microbiol. 40, 507–510.
43. Hale, M.G. and Moore, L.D. 1979 Factors affecting root exudation II. 1970–1978. Advances in Agronomy 31, 93–124.
44. Hamad-Fares, I. 1976 La fixation de l'azote dans la rhizosphere du riz. Thèse, Université de Nancy I, France.
45. Hardy, R.W.F., Holsten, R.H., Jackson, E.K. and Burns, R.C. 1968 The acetylene-ethylene assay for N_2 fixation: Laboratory and field evaluation. Plant Physiol. 43, 1185–1207.
46. Hegazi, N.A., Monib, M. and Vlassak, K. 1979 Effect of inoculation with Spirilla and Azotobacter on nitrogenase activity on roots of maize grown under subtropical condition. Appl. Environ. Microbiol. 38, 621–625.
47. Hirota, Y., Fujii, T., Sano, Y. and Iyama, S. 1978 Nitrogen fixation in the rhizosphere of rice. Nature, London 276, 416–417.
48. IRRI. 1973 Modern varieties without fertilizer. IRRI Reporter 3/73. The International Rice Research Institute, Los Baños, Philippines.
49. IRRI. 1978 Annual Report for 1977., Los Baños, Laguna, Philippines, p. 7.
50. Ishizawa, S., Suzuki, T. and Araragi, M. 1975 Ecological study of free-living nitrogen fixers in paddy soil. In: Nitrogen fixation and nitrogen cycle. JIBF synthesis, pp. 41–50. Takahashi, H. (ed.), Tokyo University Press.

51. Ishizawa, S. and Toyota, H. 1960 Studies on the microflora of Japanese soil (in Japanese, English summary). Bull. Natl. Inst. Agric. Sci. Jpn. Ser. B. 14, 204—284.
52. Ito, O., Cabrera, D. and Watanabe, I. 1980 Fixation of dinitrogen-15 associated with rice plants. Appl. Environ. Microbiol. 39, 554—558.
53. Jensen, H.L. 1965 Non-symbiotic nitrogen fixation. In: Soil nitrogen. Bartholomew, W.V. and Clark, F.E. (eds.), Amer. Soc. Agron. Series. 10, 436—480.
54. Jensen, V. and Holm, E. 1975 Associative growth of nitrogen fixing bacteria with other microorganisms. In: Nitrogen fixation by free-living microorganisms, pp. 101—119. Stewart, W.D.P. (ed.), Cambridge University Press.
55. Knowles, R. 1977 The significance of asymbiotic dinitrogen fixation by bacteria. In: A Treatise on Dinitrogen Fixation IV: Agronomy and Ecology, pp. 33—83. Hardy, R.W.F. and Gibson, A.H. (eds.), John Wiley & Sons.
56. Kobayashi, M., Katayama, T. and Okuda, A. 1965 Nitrogen-fixing microorganisms in paddy soils. 15. Nitrogen fixation in a mixed culture of photosynthetic bacteria (*Rhodopseudomonas capsulatus*) with other heterotrophic bacteria (5). *Azotobacter agilis*. Soil Sci. Plant Nutr. 11, 1—5.
57. Konishi, C. and Seino, K. 1961 Studies on the maintenance mechanism of paddy soil fertility in nature. Bull. Hokuriku Agric. Expt. Sta. 2, 41—136.
58. Koyama, T. and App, A. 1979 Nitrogen balance in flooded rice soils. In: Nitrogen and Rice, pp. 95—104. International Rice Research Institute, Los Baños, Philippines.
59. Lakshmi Kumari, M., Kavimandan, S.K. and Subba Rao, N.S. 1976 Occurrence of nitrogen-fixing *Spirillum* in roots of rice, sorghum, maize, and other plants. Indian J. Exp. Biol. 14, 638—649.
60. Lee, K.K. and Watanabe, I. 1977 Problems of the acetylene reduction technique applied to water-saturated paddy soils. Appl. Environ. Microbiol. 34, 654—660.
61. Lee, K.K. and Yoshida, T., 1977 An assay technique of measurement of nitrogenase activity in root zone of rice for varietal screening by the acetylene reduction method. Plant and Soil 46, 127—134.
62. Lee, K.K., Alimagno, B.V. and Yoshida, T. 1977 Field technique using acetylene reduction method to assay nitrogenase activity and its association with the rice rhizosphere. Plant and Soil 47, 519—526.
63. Lee, K.K., Castro, T.F. and Yoshida, T. 1977 Nitrogen fixation throughout growth, and varietal differences in nitrogen fixation by the rhizosphere of rice. Plant and Soil 48, 613—619.
64. Line, M.A. and Loutit, M.W. 1973 Nitrogen fixation by mixed cultures of aerobic and anaerobic microorganisms in an aerobic environment. J. Gen. Microbiol. 74, 179—180.
65. Luxmoore, R.J., Stolzy, L.H. and Letey, J. 1970. Oxygen diffusion in the soil-plant system. IV Oxygen concentration profiles, respiration rates and radial losses predicted for rice roots. Agrom. J. 62, 329—332.
66. MacRae, I.C. 1975 Effect of applied nitrogen upon acetylene reduction in the rice rhizosphere. Soil Biol. Biochem. 7, 337—338.
67. MacRae, I.C. and Castro, T.F. 1967 Nitrogen fixation in some tropical rice soils. Soil Sci. 103, 277—280.
68. Magdoff, F.R. and Bouldin, D.R. 1970 Nitrogen fixation in submerged soil-sand-energy material media and the aerobic-anaerobic interface. Plant and Soil 33, 49—61.
69. Mahmoud, S.A.Z., El-Sawy, M., Ishac, Y.Z. and El-Safty, M.M. 1978 The

effects of salinity and alkalinity on the distribution and capacity of N_2 fixation by *Azotobacter* in Egyptian soils. In: Environmental Role of Nitrogen-fixing Blue-green Algae and Asymbiotic Bacteria. Ecol. Bull. (Stockholm) 26, 99–109.

70. Martin, J.K. 1977 Factors influencing the loss of inorganic carbon from wheat roots. Soil Biol. Biochem. 9, 1–7.

71. Matsuguchi, T. 1977 Nitrogen transfer and the role of biological nitrogen fixation in paddy fields under intensive rice cultivation. JARQ 11, 204–210.

72. Matsuguchi, T. 1979 Factors affecting heterotrophic nitrogen fixation in submerged rice soils. In: Nitrogen and rice, pp. 207–222. International Rice Research Institute, Los Baños, Philippines.

73. Matsuguchi, T., and Yoo, I.D. 1979 Stimulation of phototrophic N_2 fixation in paddy fields through rice straw application. Workshop on Nitrogen Cycling in South East Wet Monsoonal Ecosystems. University of Chieng Mai, Thailand, Nov. 5–10, 1979.

74. Matsuguchi, T., Tangcham, B., and Patiyuth, S. 1975 Free-living nitrogen fixers and acetylene reduction in tropical rice field. Jpn. Agric. Res. Q. 8, 253–256.

75. Matsuo, Y., and Takahashi, E. 1977 Statistical analysis of soil science data. Part 2. Quantitative estimate of natural supply of nitrogen in paddy field over Japan (in Japanese). Abstr. 1977 Meeting, Soil Sci. Soil Manure, Japan 23, 88.

76. Nayak, D.N., and Rao, V.R. 1977 Nitrogen fixation by *Spirillum* sp. from rice roots. Arch. Microbiol. 115, 359–360.

77. Okuda, A. and Yamaguchi, M. 1956 Nitrogen fixing microorganisms in paddy soils. Distribution of non-sulphur purple bacteria in paddy soils. Soil Plant Food 2, 131–133.

78. Old, K.M., and Nicolson, T.H. 1975 Electron microscopial studies of the microflora of roots in sand dune grasses. New Phytol. 74, 51–58.

79. Patel, J.J., and Brown, M.E. 1969 Interactions of *Azotobacter* with rhizosphere and root surface microflora. Plant and Soil 31, 273–281.

80. Ponnamperuma, F.N. 1972 The chemistry of submerged soils. Adv. Agron. 24, 29–96.

81. Ponnamperuma, F.N. 1980 The nitrogen supply in tropical wetland rice soils. A paper presented at a Special Workshop on Nitrogen Fixation and Utilization in Rice Fields. April 28–30, 1980. IRRI, Los Baños, Laguna, Philippines.

82. Racho, V.V., and De Datta, S.K. 1968 Nitrogen economy of cropped and uncropped flooded rice soils under field conditions. Soil Sci. 105, 419–427.

83. Raalte, M.H. Van. 1941 On the oxygen supply of rice roots. Ann. Bot. Gardens, Buitenzorg 51, 43–58.

84. Rao, V.R. 1976 Nitrogen fixation as influenced by moisture content, ammonium sulfate and organic sources in a paddy soil. Soil Biol. Biochem. 8, 445–448.

85. Rao, V.R., Kalininskaya, T.A., and Miller, Yu M. 1973 The activity of non-symbiotic nitrogen fixation in soils of rice fields studied with [15]N. Microbiologiya 42, 729–734.

86. Remacle, J., and Rouatt, J.W. 1968 Culture mixte d'*Azotobacter chroococcum* et de germes pectinolytiques dans la rhizosphere de l'orge. Ann. Inst. Pasteur 115, 745–754.

87. Rennie, R.J. 1980 ^{15}N-isotope dilution as a measure of dinitrogen fixation by *Azospirillum brazilense* associated with maize. Can. J. Bot. 58, 21–24.

88. Rennie, R.J., and Larson, R.I. 1979 Dinitrogen fixation associated with disomic chromosome substitution lines of spring wheat. Can. J. Bot. 57, 2771–2775.

89. Rice, W.A., and Paul, E.A. 1972 The organisms and biological process involved in asymbiotic nitrogen fixation in waterlogged soil amended with straw. Can. J. Microbiol. 18, 715–723.

90. Rice, W.A., Paul, E.A., and Wetter, L.R. 1967 The role of anaerobiosis in symbiotic nitrogen fixation. Can. J. Microbiol. 13, 829–836.

91. Rinaudo, G. 1977 La fixation d'azote dans la rhizosphere du riz: importance du type varietal. Cah. ORSTOM, ser. Biol. 12, 117–119.

92. Rinaudo, G., and Germani, G. 1981 Effect of the nematodes *Hirschmanniella oryzae* and *Hirschmanniella spinicaudata* on the N_2 fixation in the rice rhizosphere. Rev. Nematol. 4, 171–172.

93. Rinaudo, G., Balandreau, J., and Dommergues, Y. 1971 Algal and bacterial non-symbiotic nitrogen fixation in paddy soils. In: Biological nitrogen fixation in natural and agricultural habitats. Plant and Soil, special volume, pp. 471–479.

94. Rinaudo, G., Hamad-Fares, I., and Dommergues, Y. 1977 N_2 fixation in the rice rhizosphere: methods of measurements; practices suggested to enhance the process. In: Biological N_2 fixation in Farming Systems of the Tropics, pp. 313–324. Ayanaba, A. and Dart, P.J. (eds.), John Wiley and Sons, New York.

95. Rinaudo, G., Aufeuvre, M.A., and Boureau, M. 1978 Dispositif pour l'estimation de la fixation d'azote rhizospherique en sol drainé ou engorgé. Cah. ORSTOM, sér. Biol. 13, 165–170.

96. Rinaudo, G., Gauthier, D., and Dommergues, Y. 1979 Enhancement of associative N_2 fixation through manipulation of the rhizosphere microflora. In: Associative N_2 fixation. Vose B. and Ruschel A.P. (eds.) Vol 1, CRC Press, Boca Raton, Florida.

97. Roussos, S., Garcia, J.L., Rinaudo, G., and Gauthier, D. 1980 Distribution de la microflore hétérotrohpe aérobie et en particulier des bactéries dénitrifiantes et fixatrices d'azote libres dans la rhizophère du riz. Ann. Microbiol. 131A, 197–207.

98. Rouquerol, T. 1962 Sur les phénomène de fixation de l'azote dans les rizières de Camargue. Ann. Agron. 13, 325–346.

99. Shanmugam, K.T., O'Gara, F., Andersen, K., and Valentine, R.C. 1978 Biological nitrogen fixation. Ann. Rev. Plant Physiol. 29, 263–276.

100. Shiga, H. 1971 Forty one year's three element experiment for rice. (in Japanese). Hort. and Agr. 26, 523–525.

101. Shiga, H., and Ventura, W. 1979 Nitrogen-supplying ability of paddy soils under field conditions in the Philippines. Soil Sci. and Plant Nutr. 22, 387–399.

102. Silva, M.F.S. Da, and Döbereiner, J. 1978 Occurrence of *Azospirillum* sp. in soils and roots. In: Limitations and potentials for biological nitrogen fixation in the tropics, p. 372. Döbereiner, J., Burris, R.H. and Hollaender, A. (eds.), Plenum Press, New York and London.

103. Simon-Sylvestre, G., and Fournier, J.C. 1979 Effect of pesticides on the soil microflora. Advances in Agronomy 31, 1–92.

104. Smith, R.L., Bouton, J.H., Schank, S.C., Quesenberry, K.H., Tyler, M.E., Milam, J.R., Gaskins, M.H., and Littell, R.C. 1976 Nitrogen fixation in grasses inoculated with *Spirillum lipoferum*. Science 193, 1003–1005.

105. Strzelczyk, E. 1961 Studies on the interaction of plants and free-living

nitrogen-fixing microorganisms. II. Development of antagonists of *Azotobacter* in the rhizosphere of plants at different stages of growth in two soils. Can. J. Microbiol. 7, 507–513.

106. Takai, Y., Koyama, T., and Kamura, T. 1956 Microbial metabolisms in reduction process of paddy soil (Part I). Soil Sci. Plant Nutr. 2, 63–66.

107. Tanaka, A. 1973 Methods of handling rice straw in various countries. International Rice Commission Newsletter 22(2), 1–20.

108. Tien, T.M., Diem, H.G., Gaskins, M.H. and Hubbell, D.H. 1981 Production of polygalacturonie acid trans-eliminase by *Azospirillum* species. Can. J. Microbiol. 27, 426–431.

109. Trolldenier, G. 1975 Influence of fertilization on atmospheric nitrogen fixation in rice fields. In: Proceedings of the 11th Colloquium of the International Potash Institute held in Bornholm, Denmark, pp. 287–292.

110. Trolldenier, G. 1977 Mineral nutrition and reduction processes in the rhizosphere of rice. Plant and Soil 47, 193–202.

111. Trolldenier, G. 1977 Influence of some environmental factors on nitrogen fixation in the rhizosphere of rice. Plant and Soil 47, 203–217.

112. Umali-Garcia, M., Hubbell, D.H., and Gaskins, M.H. 1978 Process of infection of *Panicum maximum* by *Spirillum lipoferum*. In: Environmental role of nitrogen-fixing blue-green algae and asymbiotic bacteria. Ecol. Bull. (Stockholm) 26, 373–379.

113. Venkataraman, G.S. 1979 Algal inoculation in rice fields. In: Nitrogen and Rice, pp. 311–321. International Rice Research Institute, Los Baños, Philippines.

114. Wada, H., Panichsakpatana, S., Kimura, M., and Takai, Y. 1978 Nitrogen fixation in paddy soils. I. Factors affecting N_2 fixation. Soil Sci. Plant Nutr. 24, 357–365.

115. Wada, H., Panichsakpatana, S., Kimura, M., and Takai, Y. 1979 Organic debris as micro-site for nitrogen fixation. Soil Sci. Plant Nutr. 25, 453–456.

116. Watanabe, I. and Barraquio, W.L. 1979 Low levels of fixed nitrogen required for isolation of free-living N_2-fixing organisms from rice roots. Nature 277, 565–566.

117. Watanabe, I. and Cholitkul, W. 1979 Field studies on nitrogen fixation in paddy soils. In: Nitrogen and Rice, pp. 223–239. International Rice Research Institute, Los Baños, Philippines.

118. Watanabe, I., Lee, K.K., Alimagno, B.V., Sato, M., Del Rosario, D.C., and De Guzman, M.R. 1977 Biological nitrogen fixation studies by *in situ* acetylene-reduction assays. IRRI Research Paper Series No. 3, International Rice Research Institute, Los Baños, Philippines.

119. Watanabe, I., Lee, K.K., and Alimagno, B.V. 1978 Seasonal change of N_2-fixing rate of rice field assayed by *in situ* acetylene reduction technique. I. Experiments in long-term fertility plots. Soil Sci. Pl. Nutr. 24, 1–13.

120. Watanabe, I., Lee, K.K., and De Guzman, M. 1978 Seasonal change of N_2-fixing rate in lowland technique. II. Estimate of nitrogen fixation associated with rice plants. Soil Sci. Plant Nutr. 24, 465–471.

121. Watanabe, I., Barraquio, W.L., De Guzman, M.R., and Cabrera, D.A. 1979 Nitrogen-fixing (acetylene-reduction) activity and population of aerobic heterotrophic nitrogen-fixing bacteria associated with wetland rice. Appl. Environ. Microbiol. 37, 813–819.

122. Williams, W.A., Morse, M.D., and Ruckman, J.E. 1972 Burning vs incorporation of rice crop residues. Agron. 64, 467–468.

123. Yamaguchi, M. 1979 Biological nitrogen fixation in flooded rice field. In:

Nitrogen and Rice, pp. 193—204. International Rice Research Institute, Los Baños, Philippines.

124. Yoneyama, T., Lee, K.K., and Yoshida, T. 1977 Decomposition of rice straw residues in tropical soils. IV. The effect of rice straw on nitrogen fixation by heterotrophic bacteria in some Philippine soils. Soil Sci. Plant Nutr. 23, 289—295.

125. Yoshida, S., Coronel, V., Parao, F.T., De Los Reyes, E. 1974 Soil carbon dioxide flux and rice photosynthesis. Soil Sci. Plant Nutr. 20, 381—386.

126. Yoshida, T. 1975 Microbial metabolism of flooded soils. In: Soil biochemistry, Vol. 3. pp. 83—122. Paul, E.A. and McLaren A.D. (eds.).

127. Yoshida, T. 1978 Microbial metabolism in rice soils. In: Soils and Rice, pp. 445—463. International Rice Research Institute, Los Baños, Philippines.

128. Yoshida, T. and Ancajas, R.R. 1973 Nitrogen-fixing activity in upland and flooded rice fields. Soil Sci. Soc. Amer. Proc. 37, 45—46.

129. Yoshida, T. and Ancajas, R.R. 1971 Nitrogen fixation by bacteria in the root zone of rice. Soil Sci. Soc. Amer. Proc. 35, 156—158.

130. Yoshida, T., Ancajas, R.R. and Bautista, E.M. 1973 Atmospheric nitrogen fixation by photosynthetic microorganisms in a submerged Philippine soil. Soil Sci. Plant Nutr. 19, 117—123.

131. Yoshida, T. and Broadbent, F.E. 1975 Movement of atmospheric nitrogen in rice plants. Soil Sci. 120, 288—291.

132. Yoshida, T., Takai, Y. and Del Rosario, D. 1975 Molecular nitrogen content in a submerged rice field. Plant and Soil 42, 633—660.

133. Yoshida, T. and Yoneyama, T. 1980 Atmospheric dinitrogen fixation in the flooded rice rhizosphere determined by the N-15 isotope technique. Soil Sci. Plant Nutr. 26, 551—559.

4. N_2-fixing tropical non-legumes

J.H. BECKING

1. Introduction

Non-leguminous plants with root nodules possessing N_2-fixing capacity are present in a large number of phylogenetically unrelated families and genera of dicotyledonous angiosperms. These plants occur in a wide variation of habitats and show a large range of morphological forms. Some of them are small prostrate herbs (e.g. *Dryas* spp.), others shrubs (e.g. *Ceanothus* spp. and *Colletia* spp.), while others are stout tree-like woody species (e.g. *Alnus* spp. and *Casuarina* spp.).

The only species of horticultural/agricultural significance is *Rubus ellipticus*, because it is a raspberry species producing a soft, edible fruit. All the other non-leguminous nodulated species are of no agricultural importance for crop production. The woody species are, however, important in forestry for reforestation and wood production and all of them play a prominent role in plant succession of natural ecosystems by covering bare soil of disturbed areas or sites. The section dealing with management and prospects for use in the tropics will discuss some of these properties in more detail.

2. Nodulated species

The older literature dealing with non-leguminous N_2-fixing dicotyledonous angiosperms can be found in the reviews of Becking [20, 21, 23, 26, 28] and Bond [40]. The present communication will cover more recent contributions in the field, with emphasis on the tropical species. Table 1 presents a complete enumeration of the non-leguminous dicotyledonous taxons possessing root nodules, including the more recent discoveries. As evident from this table, these nodulated plants comprise 8 orders, 9 families, 18 genera, and about 175 species of dicotyledonous plants.

With respect to root nodulation, two types of root-nodule symbiosis can be distinguished. The majority of these symbiosis are actinorhizal, i.e. caused by microorganisms belonging to the actinomycetes of the genus *Frankia* of the family Frankiaceae [22, 25]. In one genus of dicotyledons, however, i.e. the genus *Parasponia* of the Ulmaceae, a true bacterium (Eubacteriales) is involved in the symbiosis. This bacterium belongs to the genus *Rhizobium* of the Rhizobiaceae. It is to a certain degree promiscuous, as it can also produce root nodulation and effective N_2 fixation in a number of tropical legumes such as cowpea (*Vigna* spp.) and other leguminous species.

Y.R. Dommergues and H.G. Diem (eds.), Microbiology of Tropical Soils and Plant Productivity. ISBN 978-94-009-7531-6.
© 1982 Martinus Nijhoff/Dr W. Junk Publishers, The Hague/Boston/London.

Table 1. Classification of the non-leguminous dinitrogen-fixing Dicotyledons with *Frankia* symbioses

Order	Family	Tribe	Genus	Number of nodulated species (in parentheses total number of species)[a]	
Casuarinales	Casuarinaceae	–	*Casuarina*	25	(45)
Myricales	Myricaceae	–	*Myrica*	26	(35)
			Comptonia	1	(1)
Fagales	Betulaceae	Betuleae	*Alnus*	33	(35)
			Elaeagnus	17	(45)
	Elaeagnaceae	–	*Hippophae*	1	(3)
Rhamnales			*Shepherdia*	3	(3)
		Rhamneae	*Ceanothus*	31	(55)
	Rhamnaceae		*Discaria*	6	(10)
		Colletieae	*Colletia*	3	(17)
			Trevoa	1	(6)
Coriariales	Coriariaceae	–	*Coriaria*	14	(15)
		Rubieae	*Rubus*	1	(250) (429)[b]
Rosales	Rosaceae	Dryadeae	*Dryas*	3	(4)
		Cercocarpeae	*Purshia*	2	(2)
			Cercocarpus	4	(20)
Cucurbitales	Datiscaceae		*Datisca*	2	(2)

[a] Taxonomic estimates mainly based on Willis [136]
[b] According to Focke [58] 429 *Rubus* species occur worldwide, but Willis [136] gives as an estimate 250 *Rubus* species

2.1. Actinorhizal symbioses

The root nodulation in the genus *Casuarina* (single genus) of the Casuarinaceae is well known in the tropics. This species occurs spontaneously in Southeast Asia, including the Southwest Pacific and Australia. Representatives of this large genus (45 species) have, however, also been introduced in recent times in parts of Africa such as North Africa, Tunisia and Morocco, Dakar, and the Cape Verde Islands off the coast of Dakar; in tropical and subtropical American areas such as Florida [56, 109, 110]; and in some localities in Asia [7].

The genus *Alnus* of the Betulaceae, with 35 species, occurs mainly in temperate regions, but important representatives also occur in the tropics. For example, *Alnus jorullensis* occurs in South America and *A. nepalensis*, is found at the higher elevations in Southeast Asia (Nepal), while species like *A. japonica* and *A. maritima* are often introduced and thrive well at higher elevations in tropical Asia (Indonesia, Philippines).

The genus *Elaeagnus* of the Elaeagnaceae has many species present in natural ecosystems in Southeast Asia, including *Elaeagnus latifolia* and *E. conferta* on Java,

Indonesia [31], and *E. philippensis* in the Philippines [6]. Of the representatives of the genus *Myrica* (Myricaceae), several montane species such as *M. javanica* naturally inhabit the higher regions of Indonesia and the Philippines [8, 20, 23].

Of the order Coriariales, which has a single family and a single genus with 15 representatives, some species occur in Asia at the higher elevations, e.g. *Coriaria japonica* and *C. nepalenis.* For both species, nodulation and N_2 fixation has been established [45, 75].

In the Rosaceae, root nodulation has already been reported in three North-American species of the genus *Cercocarpus,* i.e. *C. betuloides, C. montanus* and *C. paucidentatus.* Recently, nodulation has also been observed in *Cercocarpus ledifolius* growing as pioneer species in *Pinus flexilis* stands in California in the United States [88]. The latter species probably also has a montane neotropic distribution. Of the tribe Dryadeae of the Rosaceae, root nodulation is reported in the genera *Dryas* (Fig. 1) and *Purshia.* Representatives of the genus *Dryas* have a northern temperate and subarctic distribution, but this plant genus occurs also at the higher elevations (montane zone) in North America, Europe (Alps), and some mountains in Asia such as the Himalayas. The genus *Purshia* of the tribe Dryadeae has a solely North-American distribution, like most members of the above-mentioned genus *Cercocarpus* (tribe Cercocarpeae of the Rosaceae). In the tribe Rubieae of Rosaceae, however, one member of this extensive group, i.e. *Rubus ellipticus* (Fig. 2), has been observed and confirmed to possess root nodulation and N_2 fixation [31, 40]. The observation of root nodulation in *Rubus* initially caused some surprise and even disbelief, because although nodulation was known in the tribes Dryadeae and Cercocarpeae of the Rosaceae, the genus *Rubus* is relatively unrelated to them. The nodulated species *Rubus ellipticus* occurs naturally on continental Asia, and on some islands: e.g., Sri Lanka and Luzon in the Philippines. So far nodulation and N_2 fixation (acetylene reduction) has only been reported for *Rubus ellipticus* specimens grown in Java, Indonesia. Root nodulation in one species of the large genus *Rubus* (tribe Rubieae) with about 250 species [136] or 429 species [58], is quite notable. Moreover in all non-leguminous nodulated plants investigated so far, nodulation is a generic character. Nodulation, however, is apparently far from a generic character in the genus *Rubus,* though perhaps other nodulating species will eventually be found in this very large genus.

In the Rhamnaceae, root nodulation in a representative of the genus *Colletia* of the order Rhamnales has been reported. The *Colletia* species has a principally neotropic distribution. Root nodulation in *Colletia* was first observed by Bond in a specimen of *C. paradoxa* (syn. *C. cruciata*) growing in the Glasgow Botanical Gardens, Scotland. This nodulation was reported at the final I.B.P. conference, Section PP-N at Edinburgh in 1973 [40]. Subsequently, Medan and Tortosa [89] mentioned nodulation in *Colletia paradoxa* and *C. spinosissima* plants growing in the Botanical Gardens of Buenos Aires, Argentina. They also reported nodulation occurring locally in natural vegetations in four species of the genus *Discaria* also belonging to the Rhamnaceae, i.e. *Discaria americana, D. serratifolia, D. trinervis* and *D. nana.* At present, nodulation in *Discaria* has only been reported in the

112

Fig. 1. Dryas sp. (tribe Dryadeae of Rosaceae). Although having a mainly temperate and sub-arctic distribution, some *Dryas* species occur at high elevations on Euro-Asiatic mountains, where they are prominent as pioneer plants on bare rocky soils. (a) *Dryas* sp., plant habitus in natural environment; (b) root nodules of *Dryas drummondii* Richardson; bar scale = 1 cm.

temperate species *D. toumatou* endemic to New Zealand [93] and in an unclassified species growing in the Botanical Gardens of Edinburgh, raised from seeds from neotropic origin, i.e. from wild plants in Chile [40]. *Colletia* species introduced in Asia also have been observed to bear root nodules, and N_2-fixing activity was established (acetylene reduction test), i.e. in *Colletia paradoxa* (syn. *C. cruciata*)

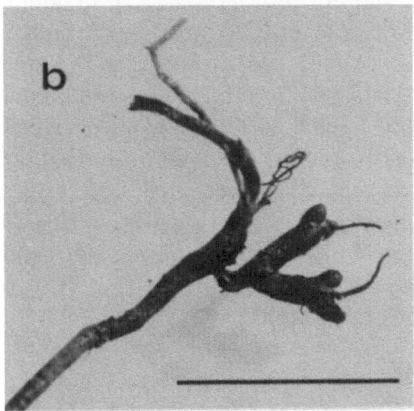

Fig. 2. Rubus ellipticus J.E. Smith. (a) Twig with leaves and flower buds; bar scale $= 4$ cm; (b) root nodules; bar scale $= 1$ cm.

and in *C. armata* (syn. *C. spinosa*) growing in the Cibodas Mountain Gardens (altitude 1450 m) at Mt. Pangrango-Gedeh, W. Java, Indonesia [31]. Moreover, nodulation was observed in *C. paradoxa* and *C. armata* plants growing in various Botanical Gardens in Europe (Amsterdam, Cologne, Munich and Nantes)(Becking, unpublished).

Recently, also in another genus of the Rhamnaceae of neotropic and subtropic distribution, i.e. *Trevoa*, which is closely related to *Colletia*, root nodulation has been observed. The species *Trevoa trinervis* has been found to bear root nodules and subsequent acetylene reduction tests have revealed N_2 fixation in natural habitat [104] comparable to that of *Ceanothus* and *Cercocarpus* measured under similar conditions [55, 67, 77]. *Trevoa trinervis* is an important matorral shrub in Chile occurring primarily on disturbed sites, more or less in the same way *Ceanothus* and *Cercocarpus* species occur as xerophytic chaparral shrubs in California.

Unlike the above-mentioned Californian species, however, the leaves of *Trevoa trinervis* are drought deciduous. Therefore, with the onset of drought stress, although the *Trevoa* rapidly sheds its leaves, photosynthesis is maintained by chlorophyllous tissue of young stems and spines. With regard to the latter phenomenon, *Trevoa* closely resembles certain *Colletia* species (Fig. 3), which also shed their leaves during the unfavourable season, but in which phyllocladioid stems and spines continue with photosynthetic activity and serve the nodules with indispensable carbohydrates.

A recent discovery of this type of symbiosis has been the observation of root nodulation in the genus *Datisca* of the Datiscaceae of the order Cucurbitales. Up to now, this order was not considered to contain nodulated species as there is no affinity of this order to other nodulated taxons. Nodulation and N_2 fixation (acetylene reduction) was observed in *Datisca* in the two representatives of this genus, i.e. *Datisca cannabina* (see Fig. 4), with a Mediterranean distribution over the Indo-Arabian region to the Himalayas and Central Asia, and *D. glomerata*, found in southwestern, and northwestern Mexico. In both species, root nodulation was observed and N_2 fixation was confirmed by Chaudhary [47, 48, 137]. Root nodulation in *Datisca cannabina* had already been reported, however, by Severini [108], more than 55 years ago. Severini had in addition already conducted some growth experiments with nodulated plants in nitrogen-deficient medium and had demonstrated their N_2-fixing ability. In spite of the fact that this reference of root nodulation of *Datisca cannabina* had been cited in Metcalfe and Chalk's book [90], Severini's observation escaped notice until recently. This omission is probably due to the fact that the observed root nodulation was not connected with possible N_2 fixation.

2.2. Rhizobium symbioses

This type of symbiosis in non-leguminous plants has only recently been discovered in members of the genus *Parasponia* of the Ulmaceae of the Urticales. Evidence of root nodulation in the Ulmaceae by a *Rhizobium* species was first reported by Trinick [125], who mentioned the nodulation and N_2 fixation of *Trema aspera* occurring in the Pangia District of Papua (New Guinea). Later this species was classified as *Trema cannabina* var. *scabra* [126], but recently it has been reclassified as a *Parasponia* species [3, 4, 30]. The confusion between *Trema* and *Parasponia* is understandable, since representatives of both genera are morphologically very similar. In the past, specimens of both genera have been regularly confused as one can see from the many reidentifications and name changes found on labels of herbarium specimens present in the Herbarium Bogoriensis, Bogor, Indonesia and Rijksherbarium, Leiden University, The Netherlands. Members of both genera can be distinguished only by minor morphological characteristics. When these characters become known, however, it is relatively easy to discriminate between the two genera. *Parasponia* can be distinguished from *Trema* by its imbricate

Fig. 3. Colletia paradoxa (Spreng.) Escalante (syn. *C. cruciata* Gillies ex Hook.). (a) Twig of a mature plant showing phyllocladioid stems with spines and flowers; (b) root nodules; bar scale = 2 cm.

116

Fig. 4. Datisca cannabina L. (Datiscaceae, Cucurbitales). (a) Twig of mature plant with flowers; bar scale = 3 cm; (b) young plant; bar scale = 6 cm; (c) root nodules; bar scale = 1 cm. (Root nodules are abnormal because they showed no acetylene reduction activity.)

perianth lobes of the male flowers and by the presence of intrapetiolar connate stipules enclosing the terminal bud [11, 113].

Just as the presence of root nodules in *Datisca* was recorded much earlier in the literature, the presence of root nodules in *Parasponia* was recognized at a very early date. In a symposium on 'Green manure in Indonesia', Ham [65] in 1909 called attention to the fact that the non-leguminous tree called 'anggrung' in the Javanese language (central and eastern Java) or 'kuraj' in the Sundanese language (western Java) bore root nodules and that these root nodules probably had the capacity to 'collect nitrogen' as do so many leguminous plant species. The vernacular names 'anggrung' and 'kuraj' are sometimes also used indiscriminately for *Trema* and *Parasponia* species on Java.

Subsequently, Backer and Van Sloten [12] were able to confirm root nodulation in *Trema* in one case, but not in other cases. At that time, however, no rigid discrimination had been made between members of the genera *Trema* and *Parasponia*. It is therefore likely that the species were confused and that the tree bearing root nodules was in fact a *Parasponia* species. This situation was definitely resolved by Clason [49], who demonstrated clearly the pioneer habit of *Parasponia* on the volcanic ash soils of Mt. Kelut (eastern Java, Indonesia) in an extensive study on the regeneration of the flora in that region. He also observed the profuse root nodulation in this species. In addition, he stated very plainly that 'nitrogenous food is possibly obtained in this way' [49]. Dried herbarium specimens of root nodules of *Parasponia* collected at that time by Clason on Mt. Kelut are still present in the herbarium material of the Herbarium Bogoriensis at Bogor, Indonesia, but Clason was not the first who called attention to the pioneer properties of *Parasponia*. Junghuhn [74] had much earlier described *Parasponia* as a pioneer vegetation on Mt. Merapi (central Java, Indonesia) on places cleared by volcanic activity like solidified lava flows. He remarked that it was an active colonizer of bare, virgin soils on this mountain at altitudes between 1500 and 1800 m. Both Junghuhn and Clason attributed the plant species to *Parasponia parviflora* Miq., but herbarium material from these localities has recently been classified as *Parasponia rugosa* Bl [113]. This is the same species found in Papua by Trinick [125] and having a distribution from central Java over the Lesser Sunda Islands, some Greater Sunda Island (Sulawesi), and the Philippines as far as Papua.

Root nodulation in *Parasponia* has now been recorded in three of the five species within the genus, i.e. *P. rugosa* [49, 125], *P. parviflora* [3, 4, 30] (see Fig. 5a–c) and *P. andersonii* [127]. It is likely then that the two other unexamined *Parasponia* species confined to New Guinea are also nodulated.

It is noteworthy that the *Trema* species have a more western distribution than *Parasponia*. It is found in the western parts of the Malaysian archipelago (western Indonesia and Malay Peninsula), Africa, and the American continent, while the representatives of the genus *Parasponia* have a typical Asiatic distribution and are common in the eastern parts of the Malaysian archipelago (i.e. central and eastern Indonesia) as far as New Guinea (Papua).

Rhizobial root nodulation has also been reported in some xerophylic desert

Fig. 5. Rhizobial root nodules of *Parasponia* (Ulmaceae, Urticales). (a) Twig of mature plant with flowers of *Parasponia parviflora* Miq.; bar scale = 3 cm; (b) young nodulated plant of *Parasponia parviflora* Miq.; bar scale = 1 cm; (c) and (d) root nodules; in both bar scale = 1 cm.

plants belonging to the Zygophyllaceae. Sabet [105] observed root nodulation occurring in a number of species of the genera *Zygophyllum* (i.e. *Z. coccineum*, *Z. album*, *Z. decumbens* and *Z. simplex*), *Fagonia* (*F. arabica*), and *Tribulus* (*T. alatus*) found in the poor sandy soils of the Egyptian deserts. Mostafa and Mahmoud [94] reported the isolation of *Rhizobium*-like strains from these root nodules. These strains could be cross-inoculated to produce effective root nodules with some legumes such as *Trifolium alexandrinum* and *Arachis hypogaea*. They reported in addition, however, that some of these zygophyllaceous species apparently had their own nodule bacterium. Both the Sabet and Mostafa and Mahmoud publications noted that when these zygophyllaceous plants were cultivated in sterilized and unsterilized soil, the nodulated plants in the unsterilized soil (or in the sterilized soil after innoculation) grew more vigorously compared to non-nodulated plants, which in due course showed nitrogen deficiency symptoms. The researchers concluded from these experiments that the root nodules supplied the host plant with nitrogen. Athar and Mahmood [9] confirmed the presence of root nodules on zygophyllaceous plants for *Zygophyllum simplex*, *Fagonia cretica*, and *Tribulus terrestris* growing in dry sandy soil low in nutritional elements near Karachi in South Pakistan. From morphological observations and staining techniques, they concluded that the endophyte within the root nodules was a *Rhizobium* species. The author of this chapter was able to examine *Zygophyllum coccineum* plants in Egyptian desert soil alongside the road between Cairo and Suez. Nodular structures were observed along the roots (Fig. 6b). Subsequent acetylene reduction tests carried out in the field *in situ* with this material, however, revealed no nitrogenase activity of these root nodules (Becking, unpublished).

3. Biology of the symbioses

3.1. Actinorhizal symbioses

About 175 plant species covering 17 genera, 8 families, and 7 orders of worldwide distribution have been reported to bear actinomycete-induced root nodules with N_2-fixing capacity (Table 1). The actinomycete involved has been classified to belong to the genus *Frankia* of the family Frankiaceae [22, 25].

Actinorhizal root nodules are morphologically and anatomically distinct from legume root nodules. Two main types can be distinguished (Fig. 7a—d). In the *Alnus* type, a coralloid root nodule, usually lacking nodule roots, is formed by dichotomous branching. The nodular lobes originate from lateral roots with inhibited or very slowly growing apical meristems. In the *Myrica/Casuarina* type of root nodule, the apex of each nodule lobe, in contrast, produces a normal but negative geotropic root. In this way the root nodule becomes clothed with upward growing rootlets. The distinction between these two root nodule types is, however, not clearcut. Thus, *Alnus rubra* inoculated with the *Alnus glutinosa* endophyte produces root nodules in which the nodular lobes give rise to negative-geotropic

Fig. 6. Zygophyllum coccineum L. (Zygophyllaceae, Malpighiales). (a) Young plant; (b) root nodules; bar scale = 1 cm.

terminal rootlets (see Fig. 7d), and [20, 21, 23] and *Casuarina equisetifolia* growing in the field often exhibits coralloid root nodules reminiscent of the *Alnus* type.

As pointed out earlier by Becking [20, 21, 23, 26], the very initial root-nodule development of these actinomycetous root nodules is different from that of leguminous plants. In actinorhizal symbioses, a pre-nodule is first formed, with a morphological structure apparent only as a slight thickening of the main root. Pre-nodules appear in the longitudinal section to consist of only a few host-cell layers, and only a restricted number of cortical cells are invaded by the *Frankia* species coming from the root hair. From this pre-nodule stage, the true root nodule is formed by a lateral root primordium developing from meristematic proliferation of pericycle and mid-cortical cells of the main root. Not until a later stage of root-nodule development does the *Frankia* infection progress into the lateral root initiated within the pre-nodule. In the pre-nodule stage, the endophyte is already partially present in its vesicular form within a cortical parenchyma cell layer of only 3–4 cells thick[21, 26]. These pre-nodules, however, do not fix N_2. Glass-slide techniques have shown that pre-nodule initiation is accompanied by an extensive

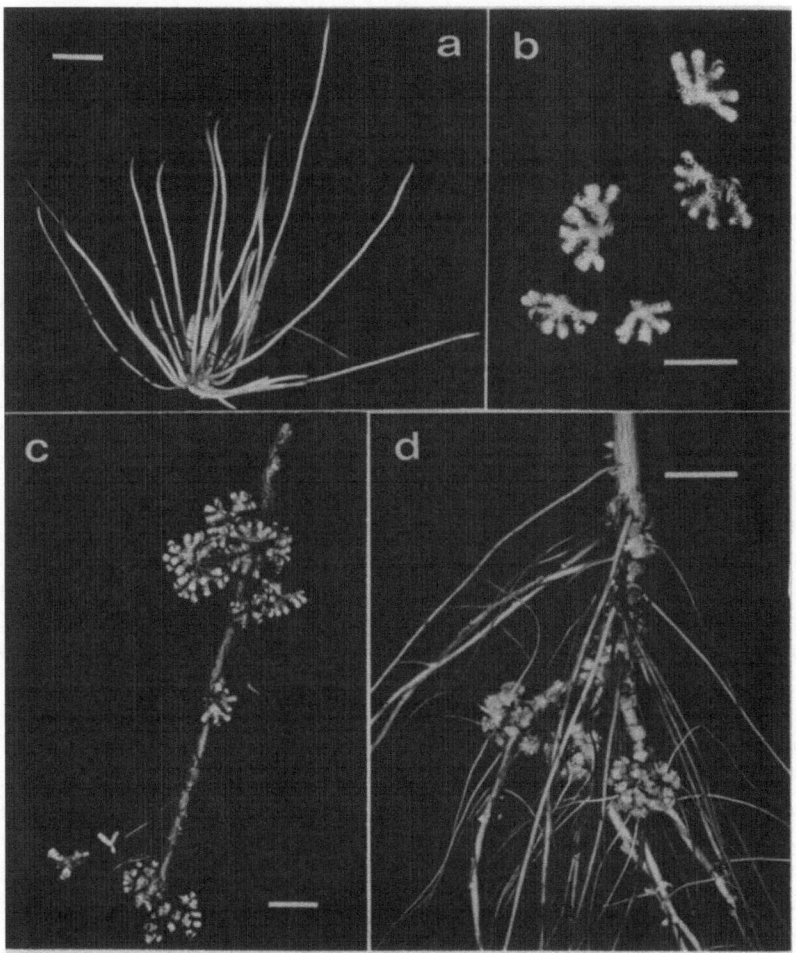

Fig. 7. Alnus and *Myrica/Casuarina* type of root nodules. (a) *Casuarina equisetifolia* L. root nodules showing that the apex of each nodule lobe gives rise to a negative geotropic root; (b) detached and divided root nodules of *Alnus glutinosa* L. (Gaertn.), showing dichotomous branching of nodule lobes; (c) coralloid root nodule of *Alnus glutinosa* L. (Gaertn.); (d) root nodules of *Alnus rubra* produced by an *Alnus glutinosa* inoculum. The nodular lobes produce some negatively geotropic root like those in *Casuarina* spp. and *Myrica* spp. In all figures bar scale = 1 cm.

curling of the initially straight root hairs at the infection site [21, 23, 26]. This stimulus is probably due to the production of auxin or related growth substances by the endophyte. At least one root hair becomes infected with the endophyte at the infection site. Torrey [122], Callaham and Torrey [43], and Torrey and Callaham [124] stated that only one or very few root-hair infections give rise to pre-nodule formation, a situation apparently different from that of leguminous root nodules. In the case of the latter, root-hair infections and abortive infections are generally associated with the root-nodule formation. Penetration of the actino-mycetes into the root hair comes at the root-hair tip, and subsequent events observed in the root hair are very reminiscent of those observed in leguminous root nodules. A pocket is formed at the root-hair tip by the invagination of the host-cell wall, and at this point an infection thread containing actinomycete cells develops inside the host cell. During its growth, the thread is accompanied by the host cell nucleus [21, 26]. When the infection thread reaches the main root, it is still sheathed by cell-wall material made up presumably of cellulose, pectins, and polysaccharides produced by the host. While this encapsulation material surrounds all the endophyte filaments throughout its entire life as endophyte in living host cells, it is absent when the host cell is dead or has been killed by the action of the endophyte. In the latter case, another form of the endophyte develops, characterized by the presence of spores (formerly called granulated bodies) liberated from sporangia [32, 130, 131]. When the *Frankia* endophyte invades new cortical parenchyma cells, penetration is probably the result of dissolution of the host-cell wall on its path from cell to cell as seen in micrographs (Fig. 8). Sometimes the endophyte does not pass the host cells intra-cellularly, but it continues its path for some distance intercellularly between host cells before re-entering a new host cell. When the endophyte proceeds from one cortical parenchyma cell to another, it apparently stimulates hypertrophy and cell division in the host-cell tissue. The enlarged host cells and the large nuclei they possess suggest that they are polyploids, but definite proof is still lacking. Moreover, it is worth noting that the endophyte does not penetrate a zone of cortical parenchyma cells adjacent to the epidermis nor those cells close to the endodermis surrounding the stele [21, 26]. This restriction in growth is probably the result of a physiological interaction, presumably hormonal, between the plant and the endophyte.

The study of the fine structure of the endophyte has revealed the presence of hyphae, vesicle structures, and spore-like particles seen through transmission and scanning electron microscopy mainly in the *Alnus* endophyte [24, 25, 26, 27, 32]. An *Alnus glutinosa* host cell containing vesicular endophyte structures is shown in the scanning electron micrograph of Fig. 9, while mature spore-like endophytic structures seen through transmission electron microscopy are presented in Fig. 10. Transmission of these spores to adjacent host cells is shown in transmission electron micrograph (Fig. 11). Gardner [61] studied the fine structure of other non-leguminous root nodules, such as those of *Hippophaë rhamnoides, Myrica gale* and *M. cerifera, Ceanothus velutinus,* and *Casuarina cunninghamiana.* Strand and Laetsch [115, 116] studied the cell and endophytic structures of the root nodules of *Ceanothus integerrimus* extensively. Finally, Lalonde and Knowles [81, 82], Lalonde and Devoe [80], and Lalonde *et al.* [84] contributed more detailed cytological studies on the sheathed material around the endophyte and on the origin of the membrane envelopes. They used cytological staining techniques in combination with transmission electron microscopy and freeze-etching micro-scopy, all in the *Alnus crispa* var. *mollis* root nodules.

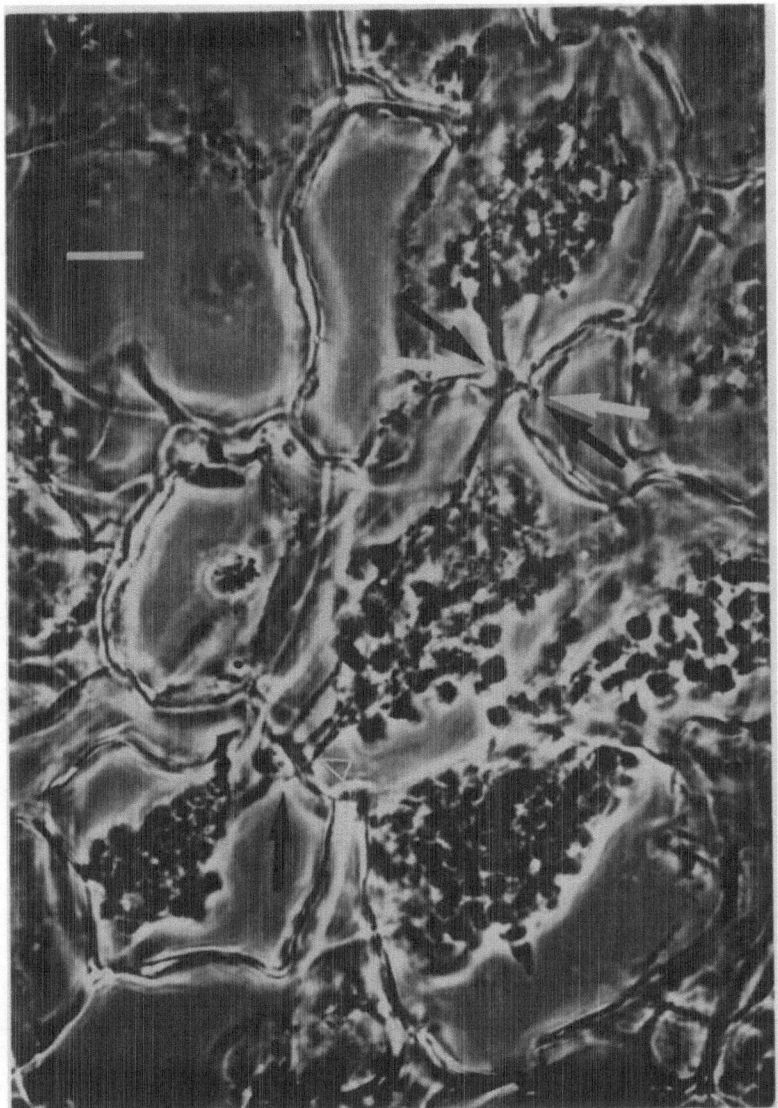

Fig. 8. Paraffin-wax section through a root nodule of *Alnus glutinosa* L. (Gaertn.) showing the perforation of the host-cell wall by the hyphae of the *Frankia* endophyte penetrating the successive host cells. In the center of the host cell, vesicular endophytic structures are visible; bar scale = 10 µm.

Root nodule initiation and root nodule growth of the *Myrica/Casuarina* root nodules have been studied extensively by Torrey and his group [42, 43, 95, 122, 123).

In the *Myrica/Comptonia* symbiosis, as in the *Alnus* symbiosis, a pre-nodule is

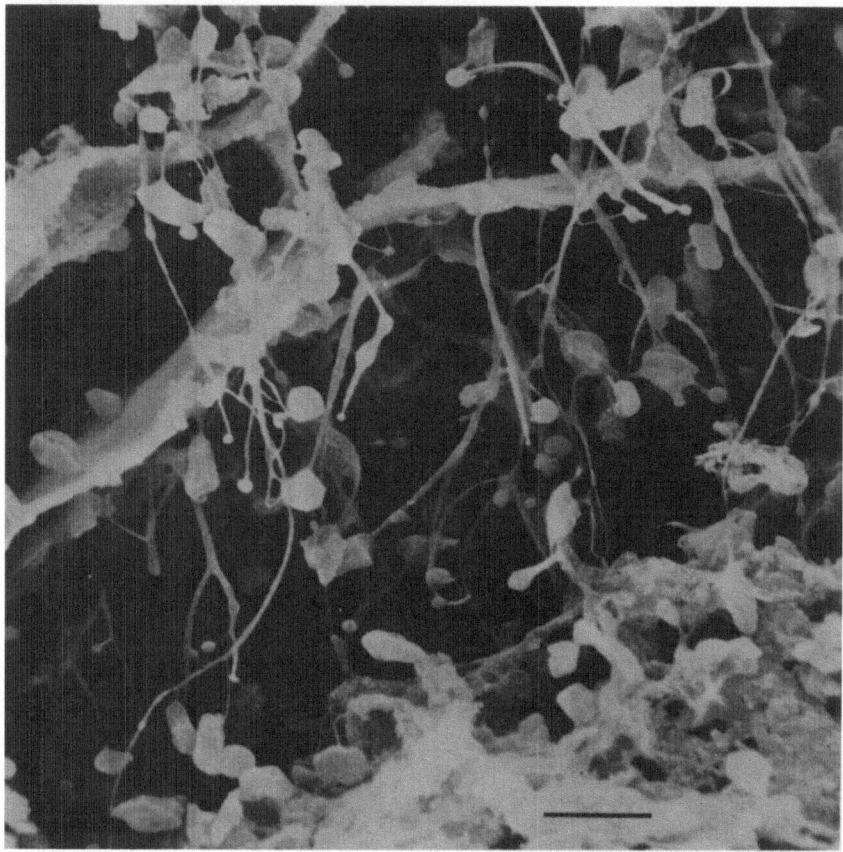

Fig. 9. Scanning electron micrograph of an opened host cell of *Alnus glutinosa* L. (Gaertn.) showing the vesicular structures of the *Frankia* endophyte at the tip of the hyphae; bar scale = 4 μm.

formed subsequent to root-hair infection. This stage is succeeded by three different phases of root-nodule development: nodule-lobe formation, a transitional or arrested stage of variable duration, and nodule-root primordium originating endogen-nodules, the primary nodule lobe is a lateral root primordium originating endogen-ously within the pre-nodule with a formation involving pericycle, endodermis, and cortical cell derivates. The nodule lobe develops slowly as the cortical parenchyma cells are invaded by the actinorhizal endophyte. After a period of arrest of variable duration, from a few days to several weeks, the nodule-lobe meristem begins altered developments, forming the elongate nodule root, which undergoes slow but continuous growth to about 3 to 4 cm in final length. New nodule-lobe primordia are initiated endogenously at the base of the existing nodule lobes, ultimately forming a cluster of nodule roots. Each nodule root has a terminal apical meristem with reduced root-cap formation and a modified root structure possessing an

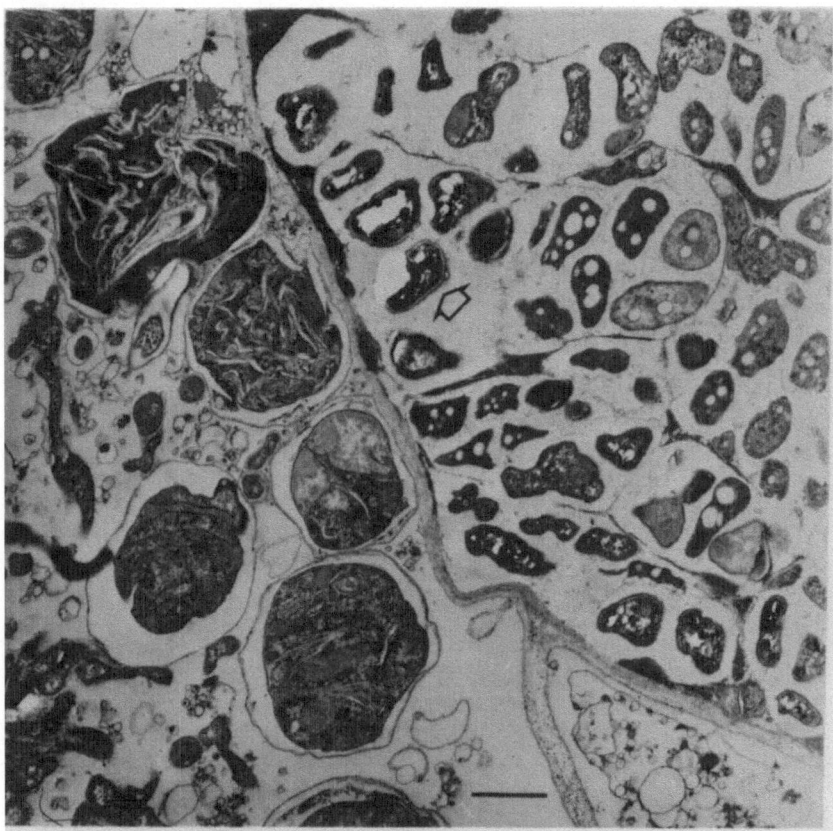

Fig. 10. Transmission electron micrograph of vesicles and hyphae (to the left) and mature spores (to the right) of the *Frankia* endophyte within the host cell. Note the dead host cell devoided of cytoplasmic content in the plant cell containing spores. The cell wall of the spore is very thick (see arrow) and these spore structures probably function as 'Dauerspores' for the survival and transmission of the endophyte; bar scale = 1 μm.

elaborate cortical intercellular space system and a reduced central cylinder. The endophyte is restricted to cortical cells of the nodule lobe and is totally absent from tissues of the nodule root. A probable function of the nodule roots in *Myrica/ Comptonia* is to facilitate gas diffusion to the N_2-fixing endophyte site in the nodule lobe when nodules occur under conditions of low oxygen tension.

An anatomical study of nodule formation in *Casuarina* showed essentially the same sequence of events as the above-described observation for *Myrica/Comptonia* root nodules. Anatomical analysis of nodule formation in *Casuarina* revealed that the large number of nodule lobes formed are the result of repeated endogenous lateral root initiations, one placed upon another in a complexly branched and truncated root system. The endophyte-infected cortical tissues derived from successive root primordia form the swollen nodular mass [122].

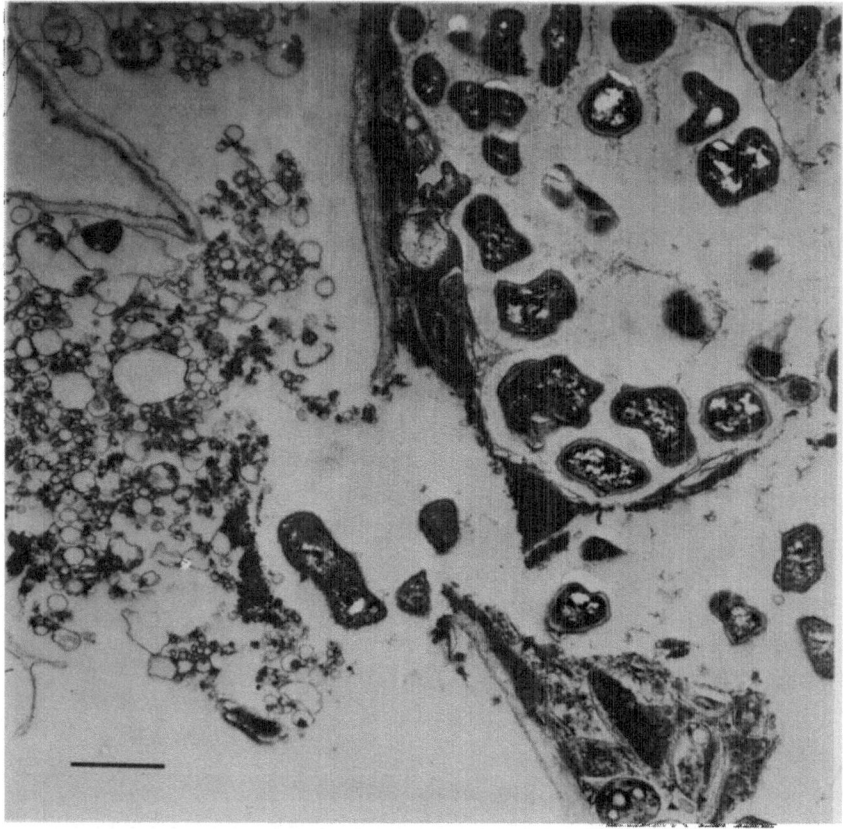

Fig. 11. Transmission electron micrograph of host cells of *Alnus glutinosa*, showing the transmission of spores of the *Frankia* endophyte from a dead host cell to an adjacent living host cell by the rupture of the host-cell wall; bar scale = 1 μm.

3.2. Rhizobium symbioses

The internal structure of the root nodule of *Parasponia rugosa* (formerly classified as *Trema aspera* or *T. cannabina* var. *scabra*) has been described by Trinick [125, 126] and Trinick and Galbraith [128]. Initially the nodule structure was described as bearing some resemblance to that of leguminous plants, but later, more affinities with non-leguminous root nodules were found. In transection the root nodules show a central vascular bundle with bacteroid infected host-cell tissue forming a horseshoe-shaped zone around it. The central vascular stele is a typical feature of non-leguminous root nodules. As in other non-leguminous root nodules, *Parasponia* nodules possess an apical meristematic zone which provides for the continuous elongation of the nodules. The infection thread has been observed to persist in this form and to penetrate the newly formed host cells immediately behind the apical meristem. The infection thread can either pass directly through the host-cell wall to infect other cells, or it can enter young host cells from the intercellular spaces between the host cells. As in legumes and the actinorhizal non-legumes nodulated by the *Frankia* endophyte, the host cells invaded by the *Rhizobium* endophyte become hypertrophied. However, in contrast to the infection threads in leguminous root nodules, the *Rhizobium* cells in *Parasponia* species are very rarely released from the infection threads. Usually the infection thread continues to grow until it has completely filled the host cell. The wall of the infection thread was found by the above-mentioned investigators [126, 128] to be continuous with the host-cell wall, and therefore probably to consist of cellulose and pectins. The persistence of infection threads within host cells appears to be a peculiar feature of *Parasponia* root nodules. Because more than two thirds of the infected host cells throughout the nodule contained these structures, it was presumed that N_2 fixation occurred in the bacteria within the thread structures. Thus, bacteria inside the infection threads may have properties similar to bacteroids enclosed in a membrane envelope found in leguminous host cells.

A study of the internal structure of *Parasponia andersonii* root nodules showed that the infection-thread structures were even more dominant in this species than in the preceding species, since *Rhizobium* cells had never been found to be released from the infection threads into the host cytoplasm [127]. The persistence of the thread structures was confirmed by an examination of serial sections of host cells containing either little or very extensive infection. The absence of 'released' rhizobia in the light microscope sections may be due to either the improved resolution obtained in the Araldite-embedding material in comparison with wax used in the study with *Parasponia rugosa* root nodules [128] or to a difference between the species. Transmission electron micrographs showed that the walls of the threads varied greatly in thickness, and thread structures were often observed without a rigid wall and enclosed only by a cytoplasmic membrane. In view of the observed differences between the two *Parasponia* species, it was surmised that the two species probably represented different evolutionary stages of the *Rhizobium* infection [127].

A preliminary study of *Parasponia parviflora* root nodules from Java, Indonesia by Becking [30] showed that these nodules were of the coralloid type. They differ from the *Alnus* root-nodule type of *Frankia* symbioses, however, in that nodule branching is more irregular and not strictly dichotomous. Moreover, the larger root nodules of the *Alnus* type are usually sessile on the thicker roots, whereas the *Parasponia* root nodules often shown thin bases (Fig. 5c and d) at the point of attachment with the main root. Becking found bacteroids to be regularly released into the host cytoplasm. When released, however, these *Rhizobium* cells maintain their normal straight shape (Fig. 12) and do not become distorted as is usually the case in bacteroids of leguminous plants. Moreover, a large proportion of the rhizobial cells are found in thin-walled membrane envelopes consisting of a double-layered cytoplasmic membrane (Fig. 13). These membrane structures are long and thin, and within the host cell, the longitudinal axis of the envelopes is often orientated

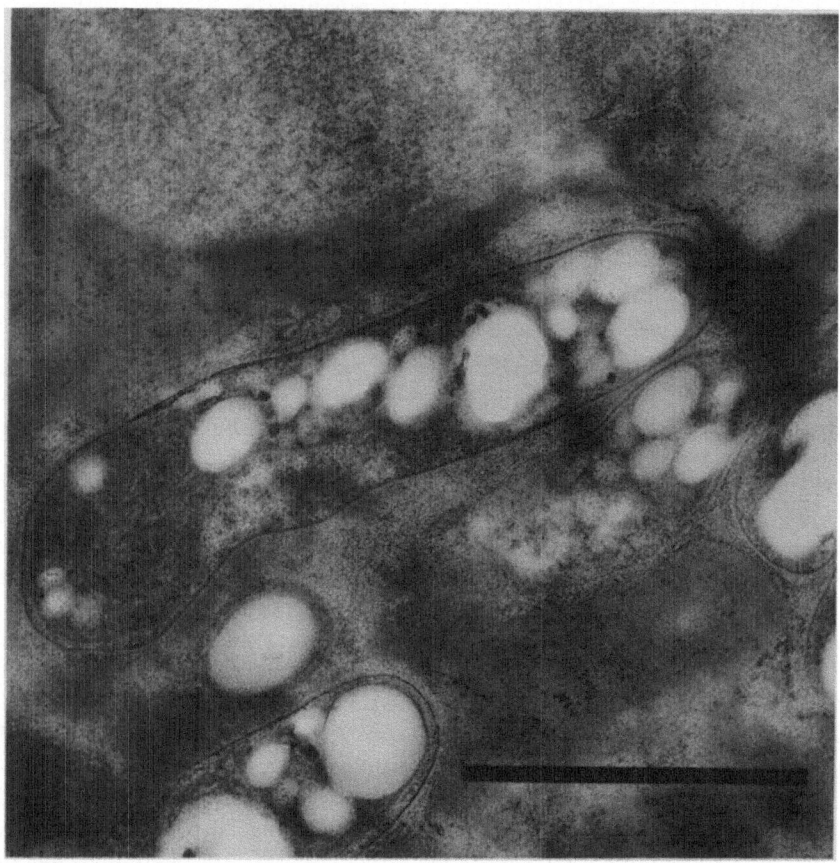

Fig. 12. Transmission electron micrograph of a free *Rhizobium* bacteroid in the cytoplasm of *Parasponia parviflora* Miq. root nodule cell. The reserve material inside the bacterium is poly-β-hydroxybutyrate; bar scale = 1 μm.

in the same direction as adjacent membrane envelopes. In extreme cases, thread-like envelope structures may develop (Fig. 14). In these elongated membrane structures, about 5–25 bacteria can be generally counted in cross-section [30], but sometimes there is only one. The total number of *Rhizobium* bacteroids within these membrane sacks or stretched envelopes in one direction is certainly a multiple of that observed in cross-section. In the membrane envelopes the individual *Rhizobium* cells are usually orientated with the long axis parallel to the long axis of the membrane structure or thread. The thread-like structures are usually very uneven and variable in length and width, so one has the impression that the thicker threads are either composed of multi-layered threads or that they become thicker through some internal growth process (Fig. 14).

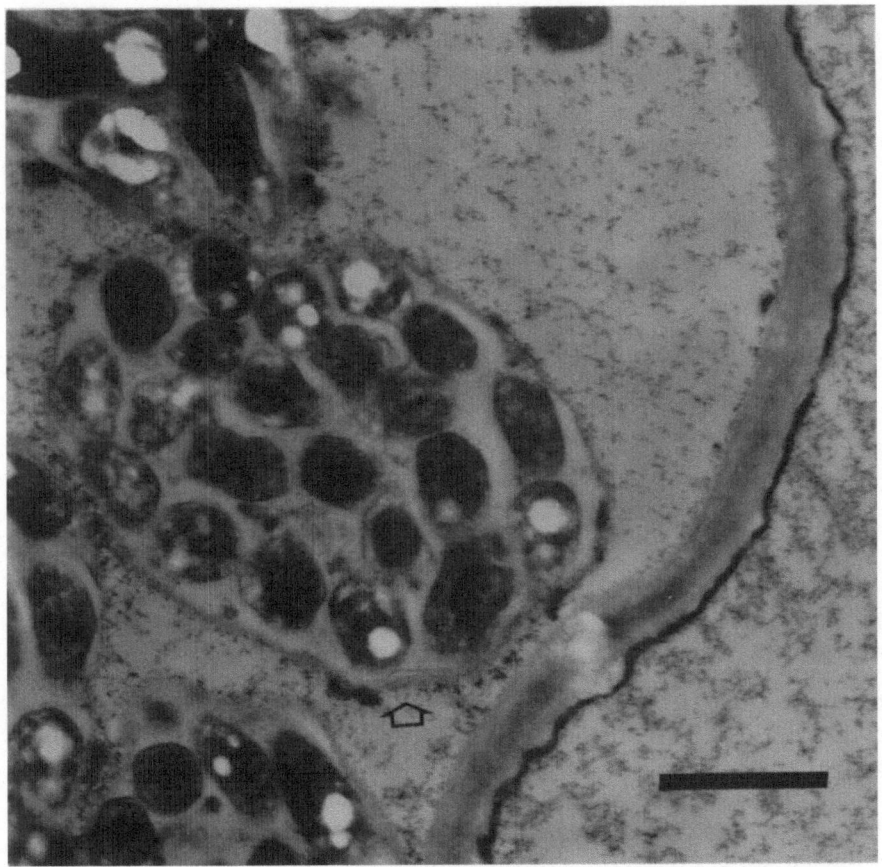

Fig. 13. Transmission electron micrograph. Transverse section through the membrane envelope enclosing *Rhizobium* cells in a *Parasponia parviflora* Miq. host cell. Note the living cytoplasm of the host cell and the double-layered cytoplasmic structure of the envelopes (see arrow); bar scale = 1 μm.

4. Isolation of the endophyte

4.1. Actinorhizal symbioses

Most of the earlier literature concerning isolation of the endophyte is summarized by Becking [21, 23, 26, 27, 28]. One study particularly worth mentioning was conducted by Pommer [97] who claimed that he had isolated the root-nodule endophyte of *Alnus glutinosa*. The isolate was in general morphologically very similar to the symbiotic endophyte (*Frankia* sp.) observed within the root-nodule cells, and moreover, it could induce root nodulation in aseptically grown *Alnus glutinosa* seedlings within 4–6 weeks. The isolate grows, although very slowly, on agar plates

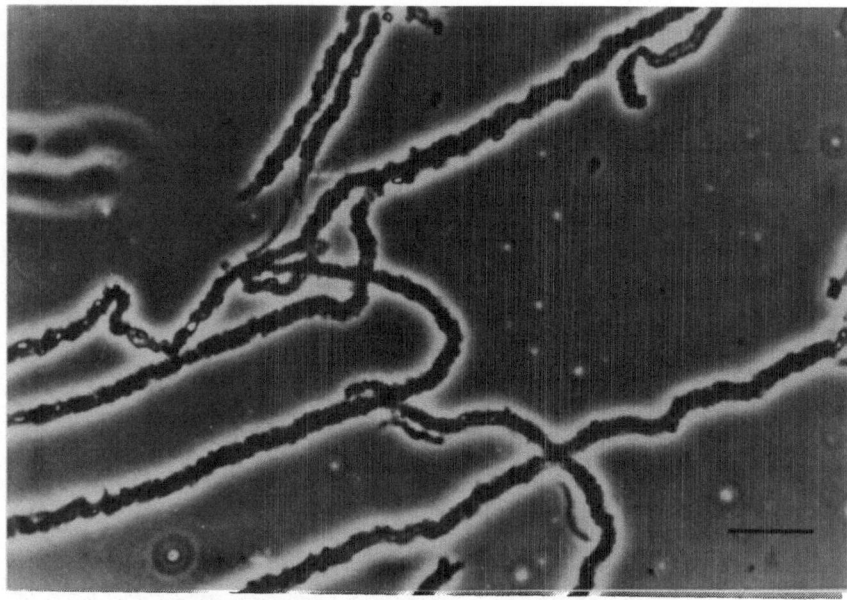

Fig. 14. Thread-like structures in *Parasponia parviflora* Miq. host cells enclosing *Rhizobium* bacteria. Note the variable thickness of the threads; phase contrast micrograph; bar scale = 1 μm.

containing a simple glucose-asparagin agar supplied with *Alnus* root-nodule extract and exposed to normal atmospheric conditions. Although it is likely that Pommer really isolated the endophyte (unfortunately the isolate was subsequently lost, Pommer personal communication), this experiment could not be repeated by others, e.g. Quispel [98] and Becking (personal observation).

By employing a complex nutrient medium used for the *in vitro* cultivation of plant tissue, Becking [19, 21, 23, 26] isolated the root-nodule endophyte of *Alnus glutinosa*. The actinomycetous endophyte thus obtained did not produce growth outside the plant tissue on this medium. The endophyte only invaded new-formed callus tissue very slowly, and because of this insufficient transmission, the endophyte was often lost in the subsequent callus explants. The isolated endophyte in this monoxenic culture produced root nodulation in axenic *Alnus glutinosa* seedlings, but the resulting root nodules contained only the hyphal form of the *Frankia* endophyte and no vesicles. Most likely as a result, these root nodules were ineffective with respect to nitrogenase activity [19].

Lalonde *et al.* [83] reported the isolation of the *Alnus crispa* var. *mollis* endophyte on a rather complex medium used for growth of plant tissues [66]. The isolate lost its resemblance to the nodular endophyte after several subcultures, and it failed to produce nodulation in aseptic *Alnus crispa* var. *mollis* seedlings. Extensive immuno-labelling tests were carried out in order to demonstrate that the isolate was an ineffective strain of the nodular endophyte. Since the substances involved in these immuno-labelling reaction are unknown and may be also present in other actinomycete species (as previously shown by these authors in the same publication), definite proof of the strains identity with the nodular endophyte cannot be given. In fact, it is rather unlikely that this strain is the endophyte.

Using micro-dissection techniques and enzyme degradation, Callaham *et al.* [44] were able to isolate a slow growing actinomycete from the root nodules of *Comptonia peregrina* in a medium containing sucrose, mineral salts, amino acids, and vitamins, or in a yeast extract mannitol/sucrose medium supplemented with growth substances (thiamine-HCl, nicotine acid, pyridoxin-HCl, etc.) used in standing liquid medium in petri dishes. The isolate grew very slowly in axenic culture in the medium under these conditions, but somewhat better growth was obtained in yeast extract liquid medium (Difco Yeast Extract, 0.5% w/v) in unshaken test tubes filled only with 5–6 cm medium. The isolate appeared to be micro-aerophilic, since it grew best at the bottom of tubes of liquid medium, but it failed to grow under complete anaerobiosis. The isolate was able to produce root nodulation in *Comptonia* seedlings kept in sand culture in the greenhouse. Using the same isolation procedure as for the initial isolation, Callaham *et al.* were able to reisolate the endophyte from the nodules. According to the authors, the Koch's postulate was in this way fulfilled. It is reasonable to assume that the real endophyte was involved. However since the experiment was not conducted in a monoxenic culture with axenic *Comptonia* seedlings, the conclusion that Koch's postulate was fulfilled is disputable. Another factor which weakens that conclusion is the appearance in one of the experiments of a number of nodules in the uninoculated controls caused by contamination. The independent evaluation tests performed by Lalonde [78] suffer from the above-mentioned challengeable immuno-labelling reactions and therefore are inconclusive. Moreover, although surface-sterilized *Comptonia* seeds were used at the start of the experiment, no tests have been reported to verify how long the monoxenic conditions of the plant cultures were maintained.

New isolation techniques or modifications of the earlier method of Callaham *et al.* [44] have led to the isolation of several *Frankia* strains from root nodules of different non-leguminous plant species. Using a technique of sucrose-density sedimentation the researchers were able to separate the actinomycete from the root nodule and in this way isolate *Frankia* species from the root nodules of *Elaeagnus umbellata* (Elaeagnaceae) and *Alnus viridis* ssp. *crispa* [14, 15]. In these experiments, the sucrose-density fractions were poured on agar plates with a nutrient medium containing yeast extract, 0.5%; dextrose, 1.0%; casamino-acids, 0.5%; H_3BO_3, 1.5 mg l^{-1}; $ZnSO_4 \cdot 7H_2O$, 1.5 mg l^{-1}; $MnSO_4 \cdot 2H_2O$, 4.5 mg l^{-1}; $NaMoO_4 \cdot 2H_2O$, 0.25 mg l^{-1}; $CuSO_4 \cdot 5H_2O$, 0.04 mg l^{-1}; vitamin B_{12}, 1.6 mg l^{-1}; and Difco agar, 0.8%. The plates were sealed with Parafilm and incubated at 28 °C. Colonies of a finely filamentous actinomycete appeared on the plates after approximately 3–5 weeks.

Quispel and Tak [100, 101] stressed the use of an extract with petrol ether-soluble substances from Alder roots (or root nodules) to the nutrient medium for isolating and growing the *Frankia* species from *Alnus glutinosa* root nodules.

They succeeded in isolating what they called the spore⁻ type of endophyte, but they failed to grow the *Alnus glutinosa* spore⁺ type of endophyte under these conditions.

Gauthier *et al.* [62] were able to isolate two *Frankia* strains from the root nodules of *Casuarina equisetifolia* by plating 1 ml of a 10^{-3} suspension of young nodules on a so-called QMOD medium, i.e. a medium containing per liter deionized water: yeast extract (BBL), 500 mg; Bacto-Peptone (Difco), 5 mg; glucose, 10 g; KH_2PO_4, 300 mg; NaH_2PO_4, 200 mg; $MgSO_4 \cdot 7H_2O$, 200 mg, KCl 200 mg; ferric citrate (citric acid and ferric citrate, 1% sol.), 1 ml; some minor salts; lipid supplement, agar 15 g [79]. Small colonies (ca. 0.1 mm in diameter) appeared on this medium in 3–4 weeks and the microscope had to be used to observe and isolate the strains. The *Frankia* colonies could be easily discriminated from those of other bacteria (contaminants) by their typical 'starfish-like' appearance. These strains of *Frankia* were unable to nodulate *Casuarina equisetifolia* but effectively nodulated *Hoppohaë rhamnoides* (Gauthier, personal communication). The absence of nodulation with *Casuarina equisetifolia* is not yet understood.

5. Taxonomy and morphology

The endophyte of actinomycetous root nodules have been attributed to the genus *Frankia* of the family Frankiaceae by Becking [22, 25]. The taxonomic division in species was based on the morphology of the endophyte within the host tissue, such as the structure and dimensions of the hyphae, spherical vesicular bodies or club-shaped structures at the tip of the hyphae, and the presence or absence of spore-like endophytic cells, sometimes also called 'bacteria-like' cells, 'bacteroids', or 'granulae'. Moreover, analogous to the subdivision of the genus *Rhizobium*, use was made of cross-inoculation groups and an attempt was made to determine cell-wall constituents of the endophyte, because the classification of the actinomycetes is based on major cell-wall types [86].

Originally, the endophyte was described to be an obligate symbiont [22] because of the researcher's failure to isolate the species. Later, however, it was observed that there was considerable growth of the endophyte as a hyphae in monoxenic cultures in the rhizosphere of *Alnus glutinosa* seedlings [26]. In addition it was noted that there must be a free stage or viable form of the endophyte in soil, because it could induce root nodulation in axenic *Alnus* seedlings. For this reason, the term 'obligate' in respect to the association has purposefully been omitted in the description of this group in Bergey's Manual [37]. The above-mentioned description distinctly states that there is a free stage of the *Frankia* endophyte in soil and that the endophyte is moreover probably micro-aerophilic [25]. The latter property was proven to be the case after its final isolation [44].

The vesicular or club-shaped structures at the tip of the hyphae in *Frankia* species are probably degenerated sporangia modified under the influence of the host cytoplasm and very probably associated with N_2 fixation. This conclusion is based on the observation that nodule slices containing the endophyte predominantly in vesicular form have more N_2 fixation (acetylene reduction) than similar slices containing the endophyte as hyphae or spore-like structures, but lacking vesicles [25, 29]. Such has been proven in *Frankia* sp. CpI1 strain cultured in pure culture in a defined nutrient medium, where the acetylene reduction activity of the culture appeared simultaneously with vesicle formation [121]. N_2 fixation (acetylene reduction) in pure culture was also established in *Frankia* sp. strain isolated from *Casuarina equisetifolia* root nodules. In these strains, the highest specific activity of acetylene reduction was observed when p_{O_2} was 10 kPa [62].

Cell-wall analyses performed in search for diaminopimilic acid (DAP) using paper chromatography and chemical methods revealed the presence of meso-DAP (but not LL-DAP), arabinose, galactose and glycine in the *Frankia* species of *Alnus glutinosa* root nodules. The cell-wall preparations were obtained by a filtration technique of nodule-tissue homogenates using filter cloth of different mesh and in which isolated hyphal and vesicular endophytic clusters were sedimented on cloth of 10 μm mesh. These endophyte clusters were carefully freed from adhering host-cell components by repeated washings with phosphate buffer. The endophytic cells were subsequently ruptured by ultra-sonic treatment and the sub-cellular fragments thoroughly washed with buffer in order to remove cytoplasmic constituents prior to hydrolysis and cell-wall analyses [25, 26, 27]. Although we could find DAP in the *Alnus* endophyte, we did not detect it in the *Casuarina* endophyte, and Quispel [99] reported its absence in the *Frankia* species of *Myrica gale* root nodules.

The division of the *Frankia* species in cross-inoculation groups has been subjected to criticism, since in a number of cases cross-inoculation was observed between representatives of different groups, e.g. the effective nodulation and N_2 fixation in the *Frankia* sp. CpI1 strain isolated from root nodules of *Comptonia peregrina* in *Alnus* species. Such compatibility is, however, not a reason to reject the general division, because incompatibility barriers are also not strict for certain *Rhizobium* strains and have apparently not devaluated the existing classification of this genus according to cross-inoculation groups. Rodriguez-Barrueco and Bond [103] observed that an *Alnus glutinosa* inoculum (crushed nodules) produced root nodulation in *Myrica gale*, but that the reciprocal combination of *Myrica gale* inoculum tested on *Alnus glutinosa* plants did not give nodulation. Recently, however, Miguel *et al.* [91] mentioned that they could produce root nodulation with the reverse combination. These authors also observed that a *Hippophaë rhamnoides* inoculum produced root nodulation in *Myrica gale*, *Coriaria myrtifolia* and *Elaeagnus angustifolia*, but that this inoculum did not nodulate *Alnus glutinosa*. Most of these experiments were, however, performed in water cultures under non-axenic conditions using an inoculum of crushed root nodules of water- or field-grown plants. Therefore an unambiguous proof cannot be presented because the involvement of other actinomycetes can not be excluded.

Further, a distinction between normal and abnormal combinations should probably be made, since in the combination of the *Alnus glutinosa* endophyte in the *Myrica gale* host, the appearance of the endophyte within the host is abnormal and more like that of the *Alnus* endophyte (Rodriguez-Barrueco, personal communication). Moreover, the N_2-fixing capacity of this combination is much lower than usual. In addition, the normal endophytes of *Alnus* and *Myrica* cannot be identical, since the *Alnus* endophyte contains DAP and the *Myrica* endophyte does not, and this presence or absence of DAP in the cell walls of actinomycetes has diagnostic value [17, 18, 85, 87].

Analyses of the cell-wall components of the free-living actinomycete isolates from root nodules growing in axenic culture [14, 15, 16, 44, 62] will definitively give an insight into the taxonomy and interspecific relationships of *Frankia* species.

5.1. Rhizobium symbioses

The isolation of the rhizobial endophyte of non-leguminous root nodules (*Parasponia* spp.) does not offer any problems. Homogenates of surface-sterilized root nodules treated with ethanol 70% (v/v), and subsequently with hydrogen peroxide 6% (v/v) plated on agar medium containing mannite and yeast extract produced good rhizobial growth. This medium has the following composition: distilled water, 1000 ml; yeast extract, 1.0 g; mannite, 10 g; K_2HPO_4, 0.5 g; $MgSO_4 \cdot 7H_2O$, 0.5 g; NaCl, 0.1 g; $CaCO_3$, 3.0 g; trace-element solution of Allen and Arnon [5], 1 ml; agar, 12 g (pH 6.8—7.0)(Becking, unpublished). The *Rhizobium* strains often appear on these plates as a sole organism; in other cases (with less severe surface disinfection), they are at least the dominant organism.

6. Biochemical progress

6.1. Nitrogenase activity

N_2 fixation by nodule breis of non-leguminous plants using $^{15}N_2$ and C_2H_2 was first reported by Sloger and Silver [111, 112] and Sloger [110]. The nitrogenase activity of the nodule breis was found to be much lower than that of intact nodules. The presence of a reducing agent (sodium dithionite) and the absence of O_2 was essential during homogenization, but O_2 was required, presumably for the production of ATP, during exposure of the homogenate.

Van Straten *et al.* [132] and Akkermans *et al.* [2] reported reasonable nitrogenase activity in nodule homogenates with the acetylene reduction assay, as long as 0.3 M sucrose and 100 mM dithionite ($Na_2S_2O_4$) were present during anaerobic homogenization. The addition of Na-dithionite was essential in this experiment, because the nodule material contains large amounts of phenolic compounds which after oxidation inhibit nitrogenase activity. The reaction was found to be ATP-dependent, but strangely enough, the addition of an ATP-generating system (creatine phosphate (Cr $\sim$ P)/creatine phosphokinase or phospho (enol) pyruvate

(PEP)/pyruvate kinase (PK)) decreased the nitrogenase activity during short-term experiments. Apparently, $Cr \sim P$ and PEP inhibited acetylene reduction, but the nature of this inhibition could not be elucidated. The observed nitrogenase activity concerned mainly the vesicle clusters, as 60% of the homogenate activity was recovered by the resuspended residue from a $10 \mu m$ filter. The $10 \mu m$ filtrate consisting of hyphal fragments and disrupted vesicle clusters contained only 29% of the activity of the homogenate. The latter activity was all particle bound. The remaining 11% of the activity was lost during the procedure. This finding suggests again that the sites of N_2 fixation are the vesicles confirming earlier experiments of strong tetrazolium-reducing activity of the vesicles [1] and that tissue slices of nodules containing host cells filled with vesicles showed higher nitrogenase (C_2H_2 reduction) activity than nodule slices containing the other forms of the endophyte [25, 29].

Cell-free nitrogenase activity in actinorhizal root nodules was obtained by Benson et al. [36] with *Alnus glutinosa* root nodules by disruption of the root nodules in liquid nitrogen to release the actinomycetal endophyte. The powdered root nodules were suspended and centrifuged four times in $100 \, mM$ potassium phosphate buffer (pH 7.4) containing $10 \, mM$ dithionite and then repeatedly washed in tris-buffer ($20 \, mM$ tris-HCl, pH 7.2) with $20 \, mM$ ascorbate and $2 \, mM$ dithionite to free the homogenate from most of the inhibitory phenolic compounds released during homogenization. Nitrogenase (C_2H_2 reduction) activity was stable even when degassed buffer alone was used in the later washings, indicating that dithionite is not needed in late washing to retain activity. Cell-free preparations were made by mixing the sedimented material with solid polyvinyl-polypyrrolidone (PVP) and sonication for 1.5–2.0 min. Prolonged sonication rapidly inactivated the C_2H_2-reducing activity. The sonicated material was centrifuged down and the supernatant containing the cell-free extract had an initial activity of $0.5–1.0 \mu mole$ of C_2H_2 reduced per g of nodules per h, or about half the activity of the particulate homogenates. The cell-free enzyme is somewhat unstable after isolation as shown by the non-linear time course in 60 min. The cell-free nitrogenase required ATP, Mg^{2+}, and $Na_2S_2O_4$ for acetylene reduction.

Recently, N_2 fixation (acetylene reduction) has been established in *Frankia* species growing *in vitro* in pure culture in a defined medium [62, 121]. These free-living *Frankia* species were the *Frankia* sp. CpI1 strain isolated from *Comptonia peregrina* root nodules and two *Frankia* strains (str. D11 and G2) isolated from *Casuarina equisetifolia* root nodules. In *Frankia* sp. CpI1, inoculum size had an effect on growth (and vesicle formation) and acetylene reduction activity; both increased at higher inoculum density. In the *Frankia* sp. CpI1 strain, acetylene reduction was greater in stationary culture than in shaken culture. In the *Frankia* sp. strains isolated from *Casuarina* root nodules, the oxygen requirements for nitrogenase (C_2H_2) reduction) activity were studied in more detail.

6.2. Hydrogenase activity

Nitrogenase-dependent hydrogen evolution has been observed in detached legume and non-legume root nodules and in reaction mixtures containing cell-free nitrogenase.

An evaluation of the magnitude of hydrogen evolution seen as energy loss in terms of the efficiency of electron transfer to N_2 via nitrogenase, suggests that hydrogen production may severely reduce N_2 fixation in some plants. For most leguminous plants, it has been shown that 40–60% of the energy of the electron flow was lost through hydrogen evolution, but in some tropical legumes (i.e. *Vigna sinensis* and *Vigna radiata*) and in some non-leguminous plants (i.e. *Alnus rubra, Purshia tridentata, Elaeagnus angustifolia, Ceanothus velutinus* and *Myrica californica*), the relative efficiency is much higher, at 70–90%. The latter plants apparently have evolved a mechanism of minimizing net hydrogen production by recycling the produced hydrogen [107].

Regarding the above-mentioned observation, Benson *et al.* [36] studied the hydrogen production in crude homogenates and cell-free extracts of *Alnus glutinosa* root nodules. The H_2 evolution was ATP dependent and occurred at a rate comparable to the rate of C_2H_2 reduction. Since intact actinorhizal root nodules evolve little hydrogen, there must be a highly active or tightly coupled uptake hydrogenase in the system. As explained above, the relation between the uptake hydrogenase and ATP-dependent H_2 production by nitrogenase is of special interest, because part of the ATP expended for H_2 production can be recovered by coupled hydrogenase-catalyzed reoxidation of H_2. This process increases the efficiency of the system.

6.3. Other investigations

Using translocation studies with ^{14}C-labelled photosynthate to the root nodules of first-year *Alnus glutinosa* plants growing under natural illumination, but at constant temperature, Wheeler [135] showed that a maximum influx of new photosynthates occurred at the time of the midday peak in N_2 fixation. An analysis of fluctuations in the levels of the main free sugars present in the nodules suggested that a substantial part of the nodule carbohydrate was unavailable for N_2 fixation, and that maximal rates of fixation are attained only when new photosynthates are entering the nodules. Similar diurnal changes in the N_2 fixation rate were observed by Bond and Mackintosh [41] in *Casuarina cunninghamiana*, using the N-15 method.

Studies of hormones in non-leguminous root nodules have revealed the presence of cytokinin-like substances in *Alnus glutinosa* and *Myrica gale* root nodules [38, 102]. Analyses of the cytokinin extracts of different plant parts showed that a zeatin-9-glucoside-like substance was the prominent cytokinin in nodules and leaves of *Alnus*, but that a zeatin riboside-like substance was the major cytokinin present in the roots and root pressure sap [68, 69, 70, 71]. Cytokinin levels determined by bioassay were also observed in other non-leguminous root nodules such as *Purshia tridentata*, *Myrica gale*, *Hippohaë rhamnoides* and *Colletia paradoxa* [72]. In addition to the above-mentioned hormones, gibberellin-like (GA-like) substances were detected in various parts of young nodulated plants as estimated by the lettuce hypocotyl bioassay [73].

7. Management and prospects for use in the tropics

Up till now, the only nodulated non-legume found to be of agricultural importance for food production is *Rubus ellipticus* (Rosaceae). Under normal conditions (in the field), this plant species can fix N_2 at a rate comparable with that achieved by other non-legume root nodules in similar conditions. This raspberry species is taxonomically a representative of the sub-genus *Idaeobatus* [58] along with the common raspberry *Rubus idaeus* and some other species. The fruit of *Rubus ellipticus* is similar in size to the common raspberry, is yellow in colour, and has good culinary quality. According to Bailey [13], the species, at least for some period, was grown as a crop plant in Florida and California under the name of Golden Evergreen Raspberry. It is possible that by crossing *R. ellipticus* with another raspberry or with soft fruits belonging to the genus *Rubus* (blackberries), hybrids can be obtained in which the nodulating habit is combined with other useful agricultural features of these *Rubus* species. Hybridization within the genus *Rubus* is common, including hybridization between raspberries and blackberries. Many natural hybrids of wild *Rubus* species have also been reported [34, 53, 64].

The economic importance of the other non-leguminous nodulated species is found mainly in their capacities to increase the nitrogen status of soils by their N_2-fixation ability and to reclaim land very poor in nitrogen. In certain soils deficient also in other elements, supplemental amelioration procedures are often necessary in addition to the use of N_2-fixing non-legumes. Thus, generally speaking, the cultivation of these nodulated plants will reduce the requirement for manufactured nitrogenous fertilizer and well increase the productivity of sites now deficient in nitrogen.

In temperate regions, extensive use has been made of Alder species (*Alnus glutinosa*, *A. viridis crispa*, etc.) to reforest distressed soils in the revegetation of mine spoils, rubbish heaps, and eroded land [46, 76].

For a long time both in Europe and in the United States, N_2-fixing non-leguminous plant species have been employed for nitrogen accretion in forest ecosystems. The species have been used either as the only or major component of the final harvest (Red Alder) or as a rotation crop, as in the case of alternating stands of Alder and needle-leaved trees (Black Alder — Spruce or Red Alder — Douglas Fir), or in unbroken continuity as in the case of Alder-Poplar mixtures [10, 51, 52, 106, 114, 117, 118, 120, 129, 133]. Nitrogen accretion and redistribution in the ecosystem by litter fall and organic matter decomposition is well known for Alder stands [39, 54, 60, 92, 96, 117, 119, 134, 139]. In silvicultural systems, other combinations are sometimes employed, such as Autumn-Olive, *Elaeagnus umbellata*, and Black Walnut, *Juglans nigra* [59], Pitch Pine, *Pinus rigida*, or Black Pine, *P. thunbergii*, grown in combination with Bayberry, *Myrica pensylvanica* [120]. Many of these non-leguminous plants, furthermore, can grow on waste or mine-land soils under rather adverse conditions of heavy metal content and other toxic agents [46, 57]. Some woody N_2-fixing trees like *Alnus rubra* are in present-day forestry highly valued for wood production, as they produce a high biomass yield per unit of land area [39, 63, 117, 118, 138]. Alder wood is valued either for timber, fiber, or fuel, depending on its source and quality [33, 63].

Similar utilization of nodulated non-legumes is certainly possible in the humid and more arid tropics. As a counterpart of the commercially important European Black Alder, *Alnus glutinosa*, and the North-American Red Alder, *Alnus rubra*, *Alnus jorullensis* of South America is equally valuable. Up till now, relatively little has been done with this tree species in its native country Colombia. Selection and breeding of improved genotypes (as now have been obtained for *Alnus glutinosa* and *A. rubra*) will certainly increase the productivity and economic value of this species in the tropics. Moreover, certain temperate species of woody N_2 fixers may be used in the higher regions of tropical countries. Good growth and wood production by these non-legumes in these new areas may be possible. For instance the Green Alder, *Alnus viridis* has been imported in New Zealand for revegetation of eroded mountain slopes [35].

Moreover, some woody species typically found in the tropics are of considerable importance for reforestation and wood production in these areas. The littoral *Casuarina equisetifolia* is particularly valuable for the revegetation of dune sands and bare sandy soils near the coast, while the montane species *Casuarina junghuhniana* (syn. *C. montana*) is a very good colonizer of bare soil or eroded areas in mountainous regions of the tropics (e.g. Java, Sumatra in Indonesia). The Southeast Asiatic/Australian *Casuarina* species are widely imported in other tropical countries (Africa, South America and southern USA) for silvicultural purposes. *Casuarina* has a great silvicultural potential, and it would be profitable to make more extensive silvicultural investigations on the use of the various *Casuarina* species [c. 45 spp] in the tropics; these species are especially valuable because they exhibit relatively rapid growth combined with a good N_2 fixation capacity and show special potential to grow in hot, dry climates. In a quantitative estimate of the N increase in a sand-dune soil of the Cap Vert peninsula North of Dakar, West Africa, Dommerques [56] found that the initial bare soil contained 80 kg N ha^{-1}, and after afforestation for 13 years, the soil contained 309 kg N ha^{-1}, i.e. 137 kg N in the A_{00} horizon, 28 kg N in the A_0 horizon, and 144 kg N in the mineral horizon of 0–10 cm. Thus, the afforestation of this area produced an increase in soil N of 229 kg N ha^{-1}. In addition, the standing timber of the *Casuarina* forest was estimated to contain 531 kg N ha^{-1}, signifying an average net N_2 fixation of 58.5 kg N annum^{-1}, reminiscent of the N_2 fixation by temperate Alder species [1, 50, 96].

The potential for mixed plantations of commercial woody species and actinorhizal plants should also be studied in the tropics, such as in the case of the interplanting of *Casuarina* species or of *Myrica* species (e.g. *M. pillulifera* or *M. javanica*) in *Pinus merkusii* plantations for tropical stands of conifers.

Finally, the value of a thorough investigation of natural ecosystems for finding economically valuable species should not be underestimated. In natural ecosystems, nodulated non-legumes often play a prominent role in plant succession covering bare soil of disturbed areas. Such sites may be caused by soil erosion, land-slides, fire, volcanic activity, or human interference in the form, for example, of road-building activities or savanah fires set for agricultural purposes or forest clearings. Many nodulated non-legumes are pioneer plants of forest denuded of timber or colonizers of bare soil forming a necessary chain in natural plant succession. Because of their pioneer character, nodulated non-legumes, especially the smaller species, are relatively short-lived. Through their nitrogen input they assist the revegetation of the area by other more aggressive plant species. After the latter have been established, the nodulated non-legumes are soon overgrown. For this reason, these non-leguminous nodulated plants are rarely found in climax vegetations, and when they do persist in such situations, it is merely due to the presence of extreme environmental conditions, e.g. *Myrica javanica* as a regular component of crater vegetation (fumarole gases) on Java and Sumatra (Indonesia) and *Parasponia rugosa* on recently erupted mud (lahar) streams produced by volcanic activity, e.g. at Mt. Kelut on Java, Indonesia. Besides the *Parasponia* species, *Myrica javanica* is also an active colonizer of bare, rocky soil produced by solidified lava streams. *Parasponia rugosa* is in addition a prominent colonizer of open soils after removal of the climax vegetation, such as in tea plantations, where it grows as weed in between the tea plant rows (Pangia district, Papua). The West-Javanese species *Parasponia parviflora* is a prominent pioneer plant of denuded primary forest and occurs therefore frequently in secondary forest. A study of non-leguminous nodulated plants in their natural habitat will certainly result in a selection of species applicable in current agricultural systems.

This book was in press when Diem *et al.* (C.R. Acad. Sc. Paris, Ser. D, 1982) reported the isolation of an infective and effective strain of *Frankia* from *Casuarina* nodules.

References

1. Akkermans, A.D.L. 1971. Nitrogen fixation and nodulation of *Alnus* and *Hippophaë* under natural conditions. Ph.D. Thesis, Univ. Leiden, The Netherlands, 85 pp.
2. Akkermans, A.D.L., van Straten, J. and Roelofsen, W. 1977 Nitrogenase activity of nodule homogenates of *Alnus glutinosa*: A comparison with the *Rhizobium*—pea system. In: Recent Developments in Nitrogen Fixation, pp. 591—603. Newton, N., Postgate, J.R. and Rodriguez-Barrueco, C., (eds.), Proceedings 2nd International Symposium on Nitrogen Fixation, Salamanca, 1976, Academic Press, London.
3. Akkermans, A.D.L., Abdulkadir, S. and Trinick, M.J. 1978 N_2-fixing root

nodules in Ulmaceae: *Parasponia* or (and) *Trema* spp.? Plant and Soil 49, 711–715.

4. Akkermans, A.D.L., Abdulkadir, S. and Trinick, M.J. 1978 Nitrogen fixing root nodules in Ulmaceaea. Nature, London, 274, 190.

5. Allen, M.B. and Arnon, D.I. 1955 Studies on nitrogen-fixing blue-green algae I. Growth and nitrogen fixation by *Anabaena cylindrica* Lemm. Pl. Physiol. 30, 366–372.

6. Aspiras, R.B., Miranda Jr, H.C. and Pabale, R.C. 1980 Dinitrogen fixation in the root nodules of *Elaeagnus philippensis*. Kalikasan, Philipp. J. Biol. 9 (1): 104–107.

7. Aspiras, R.B., Pabale, R.C. and Miranda Jr, H.C. 1980 Dinitrogen fixation in the root nodules of *Casuarina* spp. Kalikasan, Philipp. J. Biol. 9 (2–3): 317–320.

8. Aspiras, R.B., Paris, C.B. and de la Cruz, A.R. 1981 Nitrogen fixation by nodulated nonlegumes found in high elevations in the Philippines. Kalikasan, Philipp. J. Biol. 10 (1): (in press).

9. Athar, M. and Mahmood, A. 1972 Root nodules in some members of Zygophyllaceae growing at Karachi University Campus. Pakistan J. Bot., London, 4, 209–210.

10. Atkinson, W.A., Bormann, B.T. and De Bell, D.S. 1979 Crop rotation of Douglas-Fir and Red Alder: A preliminary biological and economic assessment. Bot. Gaz. 140 (suppl.), S102–S107.

11. Backer, C.A. and Bakhuizen van den Brink, R.C. 1965 Flora of Java (*Spermatophytes* only), vol. 2, *Angiospermae*, Families 111–160. P. Noordhoff, Groningen, The Netherlands.

12. Backer, C.A. and van Slooten, D.F. 1924 Geïllustreerd Handboek der Javaansche Theeonkruiden en hunne beteekenis voor de cultuur. Algemeen Proefstation voor Thee. Drukkerijen Ruygrok & Co., Batavia, Indonesia.

13. Bailey, L.H. 1927 The Standard Cyclopedia of Horticulture, vol. 3, Macmillan Co. New York and London.

14. Baker, D. and Torrey, J.G. 1979. The isolation and cultivation of actinomycetous root nodule endophytes. In: Symbiotic Nitrogen Fixation in the Management of Temperate Forest, pp. 38–56. Gordon, J.C., Wheeler, C.T. and Perry, D.A. (eds.), Forest Research Laboratory, Oregon State University, Corvallis, U.S.A.

15. Baker, D., Kidd, G.H. and Torrey, J.G. 1979 Separation of actinomycete nodule endophytes from crushed nodule suspensions by sephadex fractionation. Bot. Gaz. 140 (suppl.), S49–S51.

16. Baker, D., Newcomb, W., and Torrey, J.G. 1980 Characterization of an ineffective actinorhizal microsymbiont, *Frankia* sp. EuI1 (Actinomycetales). Can. J. Microbiol. 26, 1072–1089.

17. Becker, B., Lechevalier, M.P., Gordon, R.E. and Lechevalier, H.A. 1964 Rapid differentiation between *Nocardia* and *Streptomyces* by paper chromatography of whole-cell hydrolysates. Appl. Microbiol. 12, 421–423.

18. Becker, B., Lechevalier, M.P. and Lechevalier, H.A. 1965 Chemical composition of cell-wall preparations from strains of various genera of aerobic actinomycetes. App. Microbiol. 13, 236–243.

19. Becking, J.H. 1965 *In vitro* cultivation of Alder root-nodule tissue containing the endophyte. Nature, London, 207, 885–887.

20. Becking, J.H. 1966 Interactions nutritionelles plantes-actinomycetes. Rapport Général. Annls Inst. Pasteur, Paris, Suppl. 111, 211–246.

21. Becking, J.H. 1968 Nitrogen fixation by non-leguminous plants. Symposium Nitrogen in Soil, Groningen, May 17–19, 1967. Stikstof, Dutch Nitrogenous Fertilizer Review 12, 47–74.

22. Becking, J.H. 1970 *Frankiaceae* fam. nov. (*Actinomycetales*) with one new

combination and six new species of the genus *Frankia* Brunchorst 1886, 174. Int. J. Syst. Bacteriol. 20, 201–220.

23. Becking, J.H. 1970 Plant-endophyte symbiosis in non-leguminous plants. 2nd Conference on Global Impacts of Applied Microbiology, Addis Abeba, Ethiopia, November 5–11, 1967. Plant and Soil 32, 611–654.

24. Becking, J.H. 1973 Biological fixation of atmospheric nitrogen: Other systems. 4th Conference on Global Impact of Applied Microbiology, São Paulo, Brazil, July 23–28, 1973, pp. 421–460. Publ. Brasilian Microbiological Society.

25. Becking, J.H. 1974 Key of Family Frankiaceae and Genus *Frankia*. In: Bergey's Manual of Determinative Bacteriology, pp. 702–706, 871–872. Buchanan, R.E. and Gibbons, N.E. (eds.), Publ. Williams & Wilkins, Baltimore, U.S.A.

26. Becking, J.H. 1975 Root nodules in Non-legumes. In: The Development and Function of Roots, Third Cabot Symposium, Harvard Univ., Mass., U.S.A., April 8–12, 1974, pp. 507–566. Torrey, J.C. and Clarkson, D.T. Academic Press, London.

27. Becking, J.H. 1976 Actinomycete symbioses in non-legumes. Proceedings 1st International Symposium on Nitrogen Fixation, June 3–7, 1974. Pullman, Washington, pp. 581–591. Newton, W.E. and Nyman, C.J. (eds.), Washington State University Press, U.S.A.

28. Becking, J.H. 1977 Nitrogen-fixing associations in higher plants other than legumes. In: *Dinitrogen* (N_2) fixation, vol. 2, Chapter 6, pp. 185–275. Hardy, R.W.F. and Silver, W.S. (eds.), John Wiley & Sons, Inc. Publ., New York, U.S.A.

29. Becking, J.H. 1977 Endophyte and association establishment in non-leguminous nitrogen-fixing plant. 2nd Intern. Symposium on N_2 Fixation, September 13–17, 1976. Salamanca, Spain. In: Recent Developments in Nitrogen Fixation, pp. 551–567. Newton, W., Postgate, J.R. and Rodriguez-Barrueco, C. (eds.) Academic Press, London.

30. Becking, J.H. 1979 Root nodule symbiosis between *Rhizobium* and *Parasponia* (Ulmaceae). Plant and Soil 51, 289–296.

31. Becking, J.H. 1979 Nitrogen fixation by *Rubus ellipticus* J.E. Smith. Plant and Soil 53, 541–545.

32. Becking, J.H., de Boer, W.E. and Houwink, A.L. 1964 Electron microscopy of the endophyte of *Alnus glutinosa*. Antonie van Leeuwenhoek, Journal of Microbiology and Serology 30, 343–376.

33. Becking Sr, J.H. 1972 Enige gegevens en kanttekeningen over de betekenis van de zwarte els voor de houtteelt in Nederland. Nederl. Bosbouw Tijdschr. 44 (5): 128–131.

34. Beijerinck, W. 1956 *Rubi neerlandici*, Bramen en Frambozen in Nederland; hun bouw, leefwijze, verwantschap, verspreiding en gebruik (with a summary in English). Verhandelingen der Koninkl. Nederl. Akad. van Wetensch. 2e reeks, 51, No. 1, 1–156, 82 plates.

35. Benecke, U. 1969 Symbionts of Alder nodules in New Zealand. Plant and Soil 30, 145–149.

36. Benson, D.R., Arp, D.P. and Burris, R.H. 1979 Cell-free nitrogenase and hydrogenase from actinorhizal root nodules. Science N.Y. 205, 688–689.

37. Bergey's Manual of Determinative Bacteriology, 8th ed. 1974 Buchanan, R.E. and Gibbons, N.E. (eds.), Williams & Wilkins Company, Baltimore, U.S.A.

38. Bermudez de Castro, F., Canizo, A., Costa, A., Miguel, C. and Rodriguez-Barrueco, C. 1977 Cytokinins and nodulation of the non-legumes *Alnus glutinosa* and *Myrica gale*. 2nd Intern. Symposium on N_2 Fixation, September

142

13–17, 1976, Salamanca, Spain. In: Recent Developments in Nitrogen Fixation, pp. 539–550. Newton, W., Postgate, J.R. and Rodriguez-Barrueco, C. (eds.), Academic Press, London.

39. Bollen, W.B., Chen, C.-S., Lu, K.C. and Tarrant, R.F. 1967 Influence of Red Alder on fertility of a forest soil. Microbial and chemical effects. Res. Bull. 12, Forest Res. Laboratory, School of Forestry, Oregon State Univ., Corvallis, 61 pp.

40. Bond, G. 1976 The results of the IBP survey of roof nodule formation in non-leguminous angiosperms. In: Symbiotic Nitrogen Fixation in Plants, pp. 443–474, Nutman, P.S. (ed.), Cambridge University Press, Cambridge, England.

41. Bond, G. and Mackintosh, A.H. 1975 Diurnal changes in nitrogen fixation in the root nodules of *Casuarina*. Proc. R. Soc. London B, 192, 1–12.

42. Bowes, B., Callaham, D. and Torrey, J.G. 1977 Time-lapse photographic observations of morphogenesis in root nodules of *Comptonia peregrina* (Myricaceae). Am. J. Bot. 64: 516–525.

43. Callaham, D. and Torrey, J.G. 1977 Prenodule formation and primary nodule development in roots of *Comptonia* (Myricaceae). Can. J. Bot. 55: 2306–2318.

44. Callaham, D., Del Tredici, P. and Torrey, J.G. 1978 Isolation and cultivation *in vitro* of the actinomycete causing root nodulation in *Comptonia*. Science, N.Y. 199: 899–902.

45. Canizo, A. and Rodriguez-Barrueco, C. 1958 Nitrogen fixation by *Coriaria nepalensis* Wall. Rev. Ecol. Biol. Sol 15, 453–458.

46. Carpenter, Ph.L. and Hensley, D.L. 1979 Utilizing N_2-fixing woody plant species for distressed soils and the effect of lime on survival. Bot. Gaz. 140 (Suppl.), S76–S81.

47. Chaudhary, A.H. 1978 The discovery of root nodules in new species of non-leguminous angiosperms from Pakistan and their significance. In: Limitations and Potentials for Biological Nitrogen Fixation in the Tropics, p. 359. Döbereiner, J., Burris, R.H. and Hollaender, A. (eds.), Plenum Press, New York.

48. Chaudhary, A.H. 1979 Nitrogen-fixing root nodules in *Datisca cannabina* L. Plant and Soil 51, 163–165.

49. Clason, E.W. 1935 The vegetation of the Upper-Badak region of Mount Kelut (East Java). Bulletin Jardin Botanique, Buitenzorg (Bogor) 13, 509–518.

50. Crocker, R.L. and Major, J. 1955 Soil development in relation to vegetation and surface age at Glacier Bay, Alaska. J. Ecol. 43, 427–448.

51. De Bell, D.S. 1975 Short-rotation culture of hardwoods in the Pacific Northwest. Iowa State J. Res. 49, 345–352.

52. De Bell, D.S. and Radwan, M.A. 1979 Growth and nitrogen relations of coppiced black cottonwood and Red Alder in pure and mixed plantings. Bot. Gaz. 140 (Suppl.), S97–S101.

53. de Jongh, S.E. 1971 Overzicht der Nederlandse bramen. I. Inleiding en Homalacanthi. Rijksherbarium, Leiden, The Netherlands, 182 pp.

54. Delver, P. and Post, A. 1968 Influence of alder hedges on the nitrogen nutrient of apple trees. Plant and Soil 28, 325–336.

55. Delwiche, C.C., Zinke, P.J. and Johnson, C.M. 1965 Nitrogen fixation by *Ceanothus*. Pl. Physiol. 40, 1045–1047.

56. Dommergues, Y. 1963 Evaluation du taux de fixation de l'azote dans un sol dunaire reboisé en filao (*Casuarina equisetifolia*). Agrochimica 7, 335–340.

57. Fessenden, R.J. and Sutherland, B.J. 1979 The effect of excess soil copper on the growth of black spruce and green alder seedlings. Bot. Gaz. 140 (Suppl.), S82–S87

143

58. Focke, W.O. 1894 Die Natürlichen Pflanzenfamilien, pp. 1–61. Engler, A. and Prantl, P. (eds.), Teil III, 3. Abteilung *Rosaceae*, Verlag W. Engelmann, Leipzig.
59. Funk, D.T., Schlesinger, R.C. and Ponder jr, F. 1979 Autumn-olive as a nurse plant for black walnut. Bot. Gaz. 140 (Suppl.), S110–S114.
60. Gants, G.V. 1940 The silvicultural importance of the grey alder as an accumulator of nitrogen. Mitt. Kirov. forsttech. Akad. No. 58, 178–189.
61. Gardner, I.C. 1976 Ultrastructural studies of non-leguminous root nodules. In: Symbiotic Nitrogen Fixation in Plants, pp. 485–496. Nutman, P.S. (ed.), Cambridge University Press, Cambridge, England.
62. Gauthier, D., Diem, H.G. and Dommergues, Y. 1981 *In vitro* nitrogen fixation by two Actinomycete strains isolated from *Casuarina* nodules. App. Environ. Microbiol. 41, 306–308.
63. Gordon, J.C. and Dawson, J.O. 1979 Potential uses of nitrogen-fixing trees and shrubs in commercial forestry. Bot. Gaz. 140 (Suppl.), S88–S90.
64. Gustafsson, Å. 1943 The genesis of the European blackberry flora. Lunds Univ. Årsskrift. N.F. Adv. 2, 39, Nr. 6, 1–199.
65. Ham, S.P. 1909 Discussion on 'Groene Bemesting' ('Green Manure'), Handelingen 10de Congres Ned. Indisch Landbouw Syndicaat II (2) 26, *loc. cit.* pp. 25–27.
66. Harvey, A.E. 1967 Tissue culture of *Pinus monticola* on a chemically defined medium. Can. J. Bot. 45, 1783–1787.
67. Hellmers, H. and Kelleher, J.M. 1959 *Ceanothus leucodermis* and soil nitrogen in southern California mountains. Forest Sci. 5, 275–278.
68. Henson, I.E. and Wheeler, C.T. 1977 Hormones in plants bearing nitrogen-fixing root nodules: distribution and seasonal changes in levels of cytokinins in *Alnus glutinosa* (L.) Gaertn. J. Exp. Bot. 28, 205–214.
69. Henson, I.E. and Wheeler, C.T. 1977 Hormones in plants bearing nitrogen-fixing root nodules: partial characterization of cytokinins from root nodules of *Alnus glutinosa* (L.) Gaertn. J. Exp. Bot. 28, 1076–1086.
70. Henson, I.E. and Wheeler, C.T. 1977 Hormones in plants bearing nitrogen-fixing nodules: metabolism of 8-^{14}C-zeatin in root nodules of *Alnus glutinosa* (L.) Gaertn. J. Exp. Bot. 28, 1087–1098.
71. Henson, I.E. and Wheeler, C.T. 1977 Hormones in plants bearing nitrogen-fixing root nodules: cytokinin transport from the root nodules of *Alnus glutinosa* (L.) Gaertn. J. Exp. Bot. 28, 1099–1110.
72. Henson, I.E. and Wheeler, C.T. 1977 Hormones in plants bearing nitrogen-fixing root nodules: cytokinin levels in roots and root nodules of some non-leguminous plants. Z. Pflanzenphysiol. 84, 179–182.
73. Henson, I.E. and Wheeler, C.T. 1977 Hormones in plants bearing nitrogen-fixing root nodules: gibberellin-like substances in *Alnus glutinosa* (L.) Gaertn. New Phytol. 78, 373–381.
74. Junghuhn, F. 1853 Java, zijne gedaante, zijn plantentooi en zijn inwendige bouw, vol. 1, Publ. C.W. Mieling, 's-Gravenhage, The Netherlands.
75. Kataoka, T. 1930 On the significance of the root-nodules of *Coriaria japonica* A. Gr. in the nitrogen nutrition of the plant. Japan J. Botany 5, 209–218.
76. Kohnke, H. 1941 The black alder as a pioneer tree on sand dunes and eroded land. J. Forestry 39, 333–334.
77. Kummerow, J., Alexander, J.V., Neel, J.W. and Fishbeck, K. 1978 Symbiotic nitrogen fixation in *Ceanothus* roots. Am. J. Bot. 65, 63–69.
78. Lalonde, M. 1978 Confirmation of the infectivity of a free-living actinomycete isolated from *Comptonia peregrina* root nodules by immunological and ultrastructural studies. Can. J. Bot. 56, 2621–2635.
79. Lalonde, M. and Calvert, H.E. 1979 Production of *Frankia* hyphae and spores

144

as an infective inoculant for *Alnus* species. In: Symbiotic Nitrogen Fixation in the Management of Temperate Forests, pp. 95–110. Gordon, J.C., Wheeler, C.T. and Perry, D.A. (eds.), Forest Res. Laboratory, Oregon State University, Corvallis, U.S.A.

80. Lalonde, M. and Devoe, I.W. 1976 Origin of the membrane enclosing the *Alnus crispa* var. *mollis* Fern. root nodule endophyte as revealed by freeze-etching microscopy. Physiol. Plant Pathol. 8, 123–129.

81. Lalonde, M. and Knowles, R. 1975 Ultrastructure of the *Alnus crispa* var. *mollis* Fern. root nodule endophyte. Can. J. Microbiol. 21, 1058–1080.

82. Lalonde, M. and Knowles, R. 1975 Ultrastructure, composition, and biogenesis of the encapsulation material surrounding the endophyte in *Alnus crispa* var. *mollis*. Can. J. Bot. 53, 1951–1971.

83. Lalonde, M., Knowles, R. and Fortin, J.-A. 1975 Demonstration of the isolation of non-infective *Alnus crispa* var. *mollis* Fern. nodule endophyte by morphological immunolabelling and whole cell composition studies. Can. J. Microbiol. 21, 1901–1920.

84. Lalonde, M., Knowles, R. and Devoe, I.W. 1976 Absence of 'Void area' in freeze-etched vesicles of the *Alnus crispa* var. *mollis* Fern. root nodule endophyte. Arch. Microbiol. 107, 263–267.

85. Lechevalier, M.P. and Fekete, E. 1971 Chemical Methods as Criteria for Separation of Actinomycetes into Genera. Workshop Subcommittee on Actinomycetes of the American Society for Microbiology, Rutgers University, New Brunswick, New Jersey, 23 pp.

86. Lechevalier, M.P. and Lechevalier, H.A. 1970 Chemical composition as criterion in the classification of aerobic actinomycetes. Int. J. syst. Bacteriol. 20, 435–443.

87. Lechevalier, H.A., Lechevalier, M.P. and Becker, B. 1966 Comparison of the chemical composition of cell walls of Nocardiae with that of other aerobic actinomycetes. Int. J. syst. Bacteriol. 16, 151–160.

88. Lepper, M.G. and Fleschner, M. 1977 Nitrogen fixation by *Cercocarpus ledifolius* (Rosaceae) in pioneer habitats. Oecologia (Berlin) 27, 333–338.

89. Medan, D. and Tortosa, R.D. 1976 Nodulos radicales en *Discaria* y *Colletia* (Ramnaceas). Boletín de Sociedad Argentina de Botánica 17, 323–336.

90. Metcalfe, R.C. and Chalk, L. 1957 Anatomy of the Dicotyledons, vol. 1, 2nd ed. Clarendon Press, Oxford. *loc. cit.* p. 697.

91. Miguel, C., Cañizo, A., Costa, A. and Rodriguez-Barrueco, C. 1978 Some aspects of the *Alnus*-type root nodule symbiosis. In: Limitations and Potentials for Biological Nitrogen Fixation in the Tropics, pp. 121–133. Döbereiner, J., Burris, R.H. and Hollaender, A. (eds.), Plenum Press, New York.

92. Mikola, P. 1958 Liberation of nitrogen from alder leaf litter. Acta Forestalia Fennica 67, 1–10.

93. Morrison, T.M. and Harris, G.P. 1958 Root nodules in *Discaria toumatou* Raoul Choix. Nature, London, 182, 1746–1747.

94. Mostafa, M.A. and Mahmoud, M.Z. 1951 Bacterial isolates from root nodules of Zygophyllaceae. Nature, London, 167, 446–447.

95. Newcomb, W., Peterson, R.L., Callaham, D. and Torrey, J.G. 1978 Structure and host-actinomycete interaction in developing root nodules of *Comptonia peregrina*. Can. J. Bot. 56, 502–531.

96. Ovington, J.D. 1956 Studies of the development of woodland conditions under different trees. IV. The ignition loss, water, carbon and nitrogen content of the mineral soil. J. Ecology 44, 171–179.

97. Pommer, E.H. 1959 Über die Isolierung des Endophyten aus den Wurzelknöllchen von *Alnus glutinosa* Gaertn. und über erfolgreiche Re-Infektionsversuche. Ber. Deuts. Bot. Ges. 72: 138–150.

98. Quispel, A. 1960 Symbiotic nitrogen fixation in non-leguminous plants. V. The growth requirements of the endophyte of *Alnus glutinosa*. Acta Botan. Neerl. 9, 380—396.
99. Quispel, A. 1974 The endophytes of the root nodules in non-leguminous plants. In: The Biology of Nitrogen Fixation, pp. 499—520. Quispel, A. (ed.), North-Holland Publishing Company, Amsterdam.
100. Quispel, A. and Tak, T. 1978 Ineffective cultures of the endophyte of *Alnus glutinosa*. Proc. Steenbock-Kettering Intern. Symposium on Nitrogen Fixation, Univ. of Wisconsin, Madison, U.S.A. Abstract C-54.
101. Quispel, A. and Tak, T. 1978 Studies on the growth of the endophyte of *Alnus glutinosa* (L.) Vill. in nutrient solutions. New Phytol. 81, 587—600.
102. Rodriguez-Barrueco, C. and Bermudez de Castro, F. 1973 Cytokinin-induced pseudonodules on *Alnus glutinosa*. Physiologia Plantarum 29, 277—280.
103. Rodriguez-Barrueco, C. and Bond, G. 1976 A discussion of the results of cross-inoculation trials between *Alnus glutinosa* and *Myrica gale*. In: Symbiotic Nitrogen Fixation in Plants, pp. 561—565. Nutman, P.S. (ed.), International Biological Programme, vol. 7. Cambridge University Press, Cambridge, England.
104. Rundel, P.W. and Neel, J.W. 1978 Nitrogen fixation by *Trevoa trinervis* (*Rhamnaceae*) in the Chilean Matorral. Flora 167, 127—132.
105. Sabet, Y.S. 1946 Bacterial root nodules in the Zygophyllaceae. Nature, London, 157, 656—657.
106. Scamoni, A. 1960 Waldgesselschaften und Waldstandorte. Akademie Verlag, Berlin, 326 pp.
107. Schubert, K.R. and Evans, H.J. 1976 Hydrogen evolution: A major factor affecting the efficiency of nitrogen fixation in nodulated symbionts. Proc. Natn. Acad. of Sci. U.S.A. 73, 1207—1211.
108. Severini, C. 1922 Sui tubercoli radicali di *Datisca cannabina*. Annali di Bontanica 15, 29—51.
109. Silver, W.S. and Mague, T. 1970 Assessment of nitrogen fixation in terrestrial environments in field conditions. Nature, London 227, 378—379.
110. Sloger, C. 1968 Nitrogen fixation by tissues of leguminous and non-leguminous plants. Ph.D. thesis, Univ. Florida, U.S.A., 96 pp.
111. Sloger, C. and Silver, W.S. 1965 Note on nitrogen fixation by excised root nodules and nodular homogenates of *Myrica cerifera* L. In: Non-heme Iron Proteins: Role in Energy Conversion, pp. 299—302. San Pietro, A. (ed.), Antioch Press, Yellow Springs, Ohio, U.S.A.
112. Sloger, C. and Silver, W.S. 1966 Nitrogen fixation by excised root nodules and nodular homogenates of *Myrica cerifera* L. Abstracts 9th International Congress Microbiology, Moscow, p. 285.
113. Soepadmo, E. 1977 Ulmaceae. In: *Flora Malesiana*, Ser. I, vol. 8, pp. 31—76. Van Steenis, C.G.G.J. (ed.), Noordhoff-Kolf N.V., Jakarta.
114. Stone, E.L. 1955 Observations on forest fertilization in Europe. Proc. Natl. Joint Comm. Fert. Applications 31, 81—87.
115. Strand, R. and Laetsch, W.M. 1977 Cell and endophyte structure of the nitrogen fixing root nodules of *Ceanothus integerrimus*. I. Fine structure of the nodule and its endosymbiont. Protoplasma 93, 165—178.
116. Strand, R. and Laetsch, W.M. 1977 Cell and endophyte structure of nitrogen fixing root nodules of *Ceanothus integerrimus*. II. Progress of the endophyte into young cells of the growing nodule. Protplasma 93, 179—190.
117. Tarrant, R.F. 1961 Stand development and soil fertility in a Douglas-fir — Red Alder plantation. Forest Sci. 7, 238—246.
118. Tarrant, R.F. and Trappe, J.M. 1971 The role of *Alnus* in improving the forest environment. Plant and Soil (Special Volume), 335—348.
119. Tarrant, R.F., Isaac, L.A. and Chandler, R.F. 1951 Observations on litter fall

146

and utilization. U.S. Department of Agriculture, Forest Service, Miscellaneous Publication No. 881, 44 pp.

120. Tiffney jr, W.N. and Barrera, J.F. 1979 Comparative growth of pitch and Japanese black pine in clumps of the N_2-fixing shrub, bayberry. Bot. Gaz. 140 (Suppl.), S108–S109.

121. Tjepkema, J.D., Ormerod, W. and Torrey, J.G. 1980 Vesicle formation and acetylene reduction activity in *Frankia* sp. CP11 cultured in defined nutrient media. Nature, London 287, 633–635.

122. Torrey, J.G. 1976 Initiation and development of root nodules of *Casuarina* (Casuarinaceae). Am. J. Bot. 63, 335–344.

123. Torrey, J.G. and Callaham, D. 1978 Determinate development of nodule roots in actinomycete-induced root nodules of *Myrica gale* L. Can. J. Bot. 56, 1357–1364.

124. Torrey, J.C. and Callaham, D. 1979 Early nodule development in *Myrica gale*. Bot. Gaz. 140 (Suppl.), S10–S14.

125. Trinick, M.J. 1973 Symbiosis between *Rhizobium* and the non-legume *Trema aspera*. Nature, London, 244, 459–460.

126. Trinick, M.J. 1976 *Rhizobium* symbiosis with a non-legume. Proceedings 1st International Symposium on Nitrogen Fixation, pp. 507–517. Newton, W.E. and Nyman, G.J. (eds.), Washington State University Press, U.S.A.

127. Trinick, M.J. 1979 Structure of nitrogen-fixing nodules formed by *Rhizobium* on roots of *Parasponia andersonii Planch*. Can. J. Microbiol. 25, 565–578.

128. Trinick, M.J. and Galbraith, J. 1976 Structure of root nodules formed by *Rhizobium* on the non-legume *Trema cannabina* var. *scabra*. Arch. Microbiol. 108, 159–166.

129. Van der Meiden, H.A. 1960 Handboek voor de Populierenteelt ('Handbook for poplar culture'). Publ. Koninkl. Nederl. Heidemaatschappij, Arnhem, Holland, 3rd ed., 291 pp.

130. Van Dijk, C. 1978 Spore formation and endophyte diversity in root nodules of *Alnus glutinosa* (L.) Vill. New Phytol. 81, 601–615.

131. Van Dijk, C. and Merkus, E. 1976 A microscopial study of the development of a spore-like stage in the life cycle of the root nodule endophyte of *Alnus glutinosa* (L.) Gaertn. New Phytol. 77, 73–91.

132. Van Straten, J., Akkermans, A.D.L. and Roelofsen, W. 1977 Nitrogenase activity of endophyte suspensions derived from root nodules of *Alnus*, *Hippophaë*, *Shepherdia* and *Myrica* spp. Nature, London, 266, 257–258.

133. Virtanen, A.I. 1957 Investigations on nitrogen fixation by the alder. II. Associated culture of spruce and inoculated alder without combined nitrogen. Physiol. Plant. 10, 164–169.

134. Voigt, G.K. and Steucek, G.L. 1969 Nitrogen distribution and accretion in an Alder ecosystem. Proc. Sci. Soc. Amer. 33, 946–949.

135. Wheeler, C.T. 1971 The causation of the diurnal changes in nitrogen fixation in the nodules of *Alnus glutinosa*. New Phytol. 70, 487–495.

136. Willis, J.C. 1973 A Dictionary of the Flowering Plants and Ferns, 8th ed., revised by H.K. Airy Shaw. Cambridge University Press, Cambridge, England.

137. Winship, L.J. and Chaudhary, A.H. 1979 Nitrogen fixation by *Datisca glomerata*: a new addition to the list of actinorhizal diazotrophic plants. In: Symbiotic Nitrogen Fixation in the Managements of temperate forests, p. 485, poster.

138. Worthington, N., Ruth, R.H. and Matson, E.E. 1962 Red Alder, its management and foliage nutrient content of some Pacific northwest tree species. J. Forestry 49, 914–915.

139. Zavitkovski, J. and Newton, M. 1971 Litter fall and litter accumulation in Red Alder stands in western Oregon. Plant and Soil 35, 257–268.

5. Free-living blue-green algae in tropical soils

P.A. ROGER and P.A. REYNAUD

1. Introduction

One result of the fertilizer price increase during the last decade is a renewed interest in biological N_2 fixation as a means of reducing the use of N fertilizer. However, biological N_2 fixation requires energy generally obtained by the catabolism of photosynthetically fixed carbon (photosynthate). Among the N_2-fixing microorganisms, only blue-green algae (BGA) are able to generate their own photosynthate from CO_2 and water. This trophic independence makes BGA especially attractive as a biofertilizer. The agronomic potential of BGA was recognized in 1939 by De [14], who attributed the natural fertility of tropical paddy fields to N_2-fixing BGA. Since rice forms the staple diet of a high proportion of the world's population in areas where N fertilizer is rarely available, research on BGA in tropical soils has been focussed mainly on the paddy field ecosystem. Relative to the amount known about paddy soils, very little is known about other soils.

2. Occurrence of BGA in tropical soils

Blue-green algae (BGA) are photosynthetic prokaryotic microorganisms, some of which are capable of N_2 fixation. This resulting trophic independence which has already been stressed, and a great adaptability to environmental factors should enable BGA to be ubiquitous. However Watanabe [81] and Watanabe and Yamamoto [84] found that N_2-fixing BGA are not present in every environment: of 911 samples only 46 (5%) harbored N_2-fixing species. This surprisingly low value is probably due to unsuitable methodology and to the small size of the samples. Watanabe's results suggested that N_2-fixing BGA grow more abundantly in tropical and subtropical regions than in temperate and subtemperate regions.

2.1. Upland soils

Upland soils in arid climates are probably very inhospitable to many microorganisms because the temperatures are high and water is severely limited. BGA are especially resistant to such adverse conditions; thus they are the dominant components of the microflora in many cases [17]. A study of the savanna soils in the Congo indicated that the flora was almost entirely composed of BGA [10]; N_2-

Y.R. Dommergues and H.G. Diem (eds.), Microbiology of Tropical Soils
and Plant Productivity. ISBN 978-94-009-7531-6.
© 1982 Martinus Nijhoff/Dr W. Junk Publishers, The Hague/Boston/London.

fixing BGA have been reported to develop profusely in sugar cane and maize fields in India [63]. Large populations of *Calothrix* sp. in pearl millet fields and of *Gloeotrichia* sp. in sugarcane fields were found by the authors in Senegal.

A qualitative study of the algal flora of dried soil samples from experimental upland fields (pH 7.8–8.3) at IARI, New Delhi, indicated that BGA were dominant in all the soil samples, Chlorophyceae were poorly represented and Xantophyceae were absent. Among the BGA, numerous N_2-fixing forms were observed [16]. Soil algae from regions around the Gulf of Mexico and areas in Ecuador and Colombia were studied by Durrell [15]. In 120 samples he found 62 species of algae; 46 of these species belonged to BGA. About half of the samples contained N_2-fixing species; *Nostoc muscorum* was observed in 21% of the samples and *Nostoc paludosum* in 13%. Other N_2-fixing species were observed in less than 4% of the samples.

2.2. Paddy soils (submerged soils)

The paddy field ecosystem provides a favourable environment for the growth of BGA with respect to their requirements for light, water, high temperature and nutrient availability. This could be the reason BGA grow in higher abundance in paddy soils than in upland soils [84] as reported in the widely different climatic conditions of India [35] and Japan [36]. By pooling data obtained in Senegal it appears that N_2-fixing BGA were recorded in 86 out of 89 paddy soils [48]. However, Venkataraman [78] pointed out that 'contrary to general belief, N_2-fixing BGA are not invariably present in tropical rice soils, and that an all India survey showed that out of 2213 soil samples from rice fields, only about 33% harboured N_2-fixing forms'. The heterogeneous and sometimes limited distribution of N_2-fixing BGA is still not well understood because no systematic analysis has correlated the presence or absence of BGA with environmental factors [30].

2.3. Quantitative evaluations

The lack of satisfactory methods for estimating biomasses of the different algal groups [17] has certainly limited ecological studies of soil algae. Plating techniques, most frequently used, are advantageous in providing qualitative and quantitative results simultaneously; however the accuracy of the count depends on the reliability of the particular dilution method. Filamentous forms are hard to separate into individual cells whereas moniliform filaments, which are easily separated, may give inflated figures of abundance [17].

Algal enumerations are often limited by an inadequate sampling methodology. Most of the results are expressed as numbers of algae per gram of soil, which do not take into account algae present in the floodwater of submerged soils and do not permit any extrapolation at the field level (what is the dry weight of soil colonized by algae in one hectare of a paddy field?). A more satisfactory way to evaluate algal population is to determine the number of algae per cm^2 by using core samples with a well-defined diameter, each core sample including the first centimeter of soil and the corresponding floodwater column [52]. Such a procedure allows comparisons and extrapolations at the field level. Roger and Reynaud [55] found that the distribution of soil algae is log-normal (logarithms of numbers are normally distributed) and that many samples are required to obtain a significant evaluation. For example, the mean value of *Anabaena* sp. biomass based upon 40 samples (each sample being obtained by mixing 10 sub-samples) taken in a 0.25 ha paddy field still exhibited a confidence interval of $+ 32\%$ and $- 27\%$ of the mean. One fact at least is well established: in paddy soils (see Roger and Kulasooriya [58]), BGA numbers vary within large limits, from a few to 10^7 units g^{-1} dry soil. In upland soils lower values ranging from a few to 10^6 units g^{-1} dry soil have been generally reported [3, 25, 69, 71]. Reports on biomass are scarce. In paddy fields the biomass of BGA can reach values of several tons per hectare [58]. The little data available suggest that BGA seem to develop more profusely in submerged soils than in other cultivated soils. An exceptionally high total algal biomass of about 40 t ha^{-1} was recorded by Reynaud and Roger [47] in a sandy soil in Senegal that was spontaneously watered by a permanent spring; such a soil exhibited a water regime related more to that of submerged soils than to that of upland soils.

3. Ecology

3.1. Physical factors

3.1.1. Light. Algae, as phototrophic microorganisms, are restricted to the photic zone and usually located in the upper 0.5 cm horizon. Yet algae also exist in deeper horizons, in a dormant condition as spores or filament fragments [10]. Light availability for soil algae depends upon the season and latitude, the cloud cover, the plant canopy, the vertical location of the algae in the photic zone and the turbidity of the water. Light intensity reaching the soil may vary from too low to excessive levels (10 to 110,000 lux).

In cultivated soils the screening effect of a growing crop canopy appears to cause a rapid decrease of light reaching the algae. Thus the canopy of transplanted rice decreased light by 50% when plants were 15 days old, 85% after one month and 95% after two months [29]. In Senegal, diatoms and unicellular green algae developed first and BGA developed later when the plant cover was dense enough to protect them from excessive light intensities, higher than 80 klux at 13:00 h (Fig. 1); the N_2-fixing algal biomass and the density of the plant cover were positively correlated [52].

In the laboratory, after one month of incubation of a submerged unplanted soil under a range of screens, BGA were dominant in the most heavily shaded one, and green algae and diatoms were dominant in the soil exposed to full sunlight [45]. A

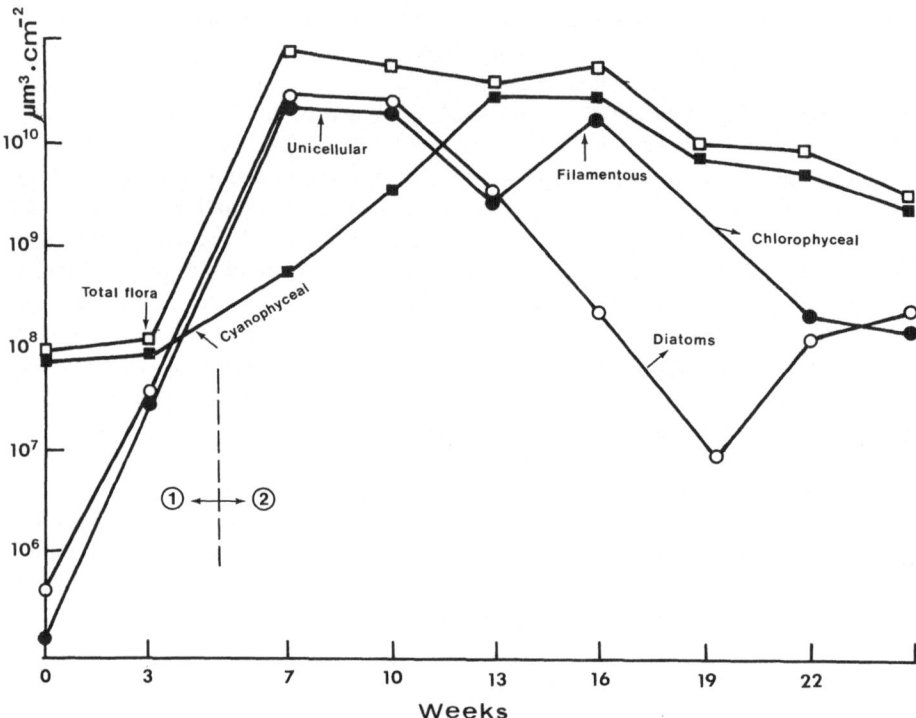

Fig. 1. Variation in biomass of different components of the algal flora during a vegetative cycle of rice in Senegal [52].

beneficial effect of the plant canopy shading the algae was also reported by Singh [63] in sugarcane fields, maize fields and grasslands in India.

As BGA are generally sensitive to high light intensities, they develop various protective mechanisms against it, namely:

— vertical migrations in the water of submerged soil;
— preferential growth in more shaded zones like embankments, under or inside decaying plant material [27] or a few millimeters, below the soil surface [17];
— migration into shaded zones (photophobotaxis) and aggregation providing a self-shading effect (photokinesis [46]);
— stratification of the strains in algal mats where N_2-fixing strains grow under a layer of eukaryotic algae more resistant to high light intensities [47].

However, some strains of BGA seem more resistant to high light intensities. *Cylindrospermum* sp. developed large biomasses in a harvested paddy field in Mali where light intensity was higher than 100 klux at 13:00 [72]. *Oscillatoria princeps* was also reported to grow profusely in full sunlight [46].

On the other hand, light deficiency may also be a limiting factor. In Japan, available light under the canopy was below the compensation point of the phytoplankton during the second part of the cycle [22]. In the Philippines, during the wet season when light was moderate acetylene reducing activity (ARA) was higher in bare soil than in planted soil [85].

3.1.2. Temperature. The optimal temperature for BGA growth is about 30–35 °C which is higher than that for the growth of eukaryotic algae. In submerged soils daily variations in the temperature are moderated by the buffering effect of flood-water; temperature is rarely a limiting factor for BGA in paddy fields, because the range of temperatures permitting the growth of BGA is larger than that required by rice; however, it influences both algal biomass composition and productivity. Low temperatures decrease productivity and favour eukaryotic algae. High temperatures favour both the phytoplankton productivity and BGA [56].

Daily changes in the temperature are more drastic in terrestrial habitats than in aquatic environments [42]. An inhibitory effect of high temperature was observed by Jones [24] in the Kikuyu grasslands in Africa where algal N_2 fixation was higher on overcast days than on hot sunny days. Stewart [65] indicated a correlation between the algal ARA response to temperature and the temperature of the habitats from which the algae were collected. For many tropical species, ARA is optimum between 30–35 °C, but a *Nostoc* sp. isolated from the algal crust on a sandy soil in Senegal still exhibited significant ARA at 60 °C (Reynaud, unpublished). High temperatures occurring in the surface of tropical upland soils may have a selective action on the algal flora, favouring BGA which are more tolerant to high temperatures than eukaryotic algae. For example, the dry spores of *Nostoc* sp. can tolerate 2 minutes of 100 °C, the wet spores 20 minutes at 60–70 °C, and the vegetative filaments 10 minutes at 40 °C [10].

3.1.3. Desiccation and remoistening. Algal growth is hindered by intermittent desiccation periods which occur during the dry season and even during drought periods that occur in the rainy season. BGA have a high capacity to withstand desiccation. *Nostoc muscorum* and *Nodularia harveyana* were isolated from a soil that had been dry for 79 years [10]. Resistance to desiccation has been attributed to various characteristics [41], namely, with respect to fatal plasmolysis, the lack of cell vacuoles, the ability of some genera to quickly take on an encysted form, the presence in some genera of a mucilaginous sheath that absorbs water quickly and retains it. This latter characteristic could explain the dominance of mucilaginous colonies of *Nostoc* spp. and *Cylindrospermum* spp. in the paddy field during the last part of the cultivation cycle when the soil dries [38, 45, 72]. The dominance of BGA comprised only about 30% of the algal flora [32], whereas in Senegal, where the dry season lasts about 8 months, spores of heterocystous BGA constituted more long periods of desiccation than to the ability to remain active at low humidities [8]. The floristic composition of desiccated soils can be related to the dryness of the biotope. In a paddy in Italy, where the dry period is relatively short, N_2-fixing BGA comprised only about 30% of the algal flora [32], whereas in Senegal, where the dry season lasts about 8 months, spores of heterocystous BGA constituted more than 95% of the algal flora at the end of the dry period. In Uttar Pradesh (India), a large number of Chlorophyceae occurred in low-lying fields, whereas BGA were found in larger numbers in paddies at higher elevations [38]. In arid soils, BGA have been reported as dominant species [31] and sometimes as the only species present [5, 10].

3.2. Biotic factors

Organisms that limit BGA growth are pathogens, antagonistic organisms and grazers. Of these, only grazers have been documented. The development of zooplankton populations, especially cladocerans, copepods, ostracods, mosquito larvae, etc. prevented the establishment of algal blooms within one or two weeks [73]. Snails form another group of algal grazers in submerged paddy fields: the biomass of snails can be as high as 1.6 t ha^{-1} in certain rice fields in the Philippines which explains the low population of algae [58].

3.3. Soil properties

Among the soil properties, pH is the most important factor determining the algal flora composition. In culture media the optimal pH for BGA growth seems to range from 7.5 to 10.0 and the lower limit is about 6.5 to 7.0 [20]. Under natural conditions BGA grow preferentially in environments that are neutral to alkaline; which explains that in paddies correlations occur between:
— water pH and BGA number [37];
— soil pH and number of spores of N_2-fixing BGA in the soil during the dry season [18];
— soil pH and BGA growth [36];
— soil pH and the N_2-fixing algal biomass (however, this relationship was conspicuous only in samples homogenous for stage of rice development, fertilization and plant cover density [53]).
    The beneficial influence of high pH on BGA growth is further demonstrated by the fact that the addition of lime increases BGA growth and N_2 fixation [58]. However, the presence of certain strains of BGA in soils with pH values between 5 and 6 have been reported. Durrel [15] demonstrated the presence of *Nostoc muscorum* and *Anabaena torulosa* in soils with pH ranging from 5 to 7. *Aulosira fertilissima* and *Calothrix brevissima* have been reported to be ubiquitous in Kerala rice fields with pH from 3.5 to 6.5 [1]. The development of a dense algal bloom on an acidic soil (pH 5.5) was observed after the surface application of straw [33]. Stewart [66] also reported that some tropical BGA exhibited ARA even at pH 4. The poor growth of N_2-fixing BGA, frequently observed in acidic soils, is probably due to the inability of BGA to compete with Chlorophyceae, which are favoured by acidic conditions.

3.4. Algal successions

The factors mentioned above are responsible for the algal successions that have been reported. An example of algal succession is provided by a study conducted in paddies in Senegal [52, 53]. Soils are acidic, and have an average pH value of 5.0 at the beginning of rice cultivation and 6.2 after 2 months of submersion. The rainy season is short (15 July—15 November) and rice fields are dry the rest of the year. High light intensities (70—80 klux) occur throughout the year. During the early part of the cultivation cycle (planting and tillering), the algal biomass increased and consisted mainly of diatoms and unicellular green algae (Fig. 1). From tillering to panicle initiation, the algal biomass reached its highest values, and filamentous green algae and non-N_2-fixing BGA were dominant. After panicle initiation the total biomass decreased; if the plant cover was sufficiently dense, heterocystous BGA developed, but if it was thin, filamentous green algae and homocystous BGA remained dominant. The observed variations in the algal flora were attributed to a decrease in light intensity and the N level resulting from the rice growth, and to an increase in pH value favouring BGA growth.

4. N_2 fixation

BGA are the only N_2-fixing microorganisms that exhibit a higher plant type of photosynthesis. Their N_2-fixing ability was first related to the presence of specialized non O_2-evolving cells called heterocysts in which the nitrogenase, a highly O_2-sensitive enzyme, is protected from O_2. It is now clearly demonstrated that the ability to fix N_2 is not confined to heterocystous BGA, but is also found in a number of BGA, which were not known to fix N_2 a few years ago, and which can fix N_2 under microaerobic or anaerobic conditions [67]; over 125 strains are presently known to fix N_2.

ARA is a reliable method for quantitative studies provided that one takes into account the following facts:
— the distribution of ARA is log-normal [54];
— N_2 fixation by BGA occurs not only in aerobic but also in anaerobic conditions [67];
— there are diurnal variations in ARA and also variations in ARA throughout the cultivation cycle. In submerged soils the variations exhibit typical patterns with one or two maxima. The occurrence of two maxima results from the inhibitory effect of excessive light during the middle of the day [45]. Similar variations were observed in upland soils but the inhibitory decrease of ARA during the middle of the day was attributed to an increase in temperature of the surface soil [24, 44]. Moreover, one should note that the duration of the incubation should be short [13] and that the conversion factor $C_2H_4:N_2$ reduced should be measured in each situation [40].

In submerged soils, light intensity appears to be the main factor governing variations in photosynthetic N_2 fixation, whereas in upland soils water availability becomes the main factor. Studying photosynthetic N_2 fixation in a sandy soil colonized by a thick algal mat, Reynaud and Roger [47] observed that the variations of both total algal biomass and ARA were governed not only by light but also by the distribution of precipitations.

The average of 38 quantitative evaluations of N_2 fixation in paddies recorded in the literature was 27 kg N ha^{-1} per crop [58]. The highest recorded value was 50–80 kg N ha^{-1} per crop [72].

In the savanna lands of southwest Nigeria, N_2-fixing Cyanobacteria such as species of *Scytonema, Tolypothrix* and *Nostoc* develop during the rainy season, particularly in previously burnt areas and may contribute significantly to the input of total nitrogen. Similarly in the open 'campina' areas among parts of the rain forests of tropical Brazil, N_2-fixing growth of blackish Stigonematales may be found [66]. In some conditions N_2-fixing BGA may significantly contribute to the N status of cultivated upland soils. Singh [63] reported a profuse growth of N_2-fixing BGA in sugarcane fields in India from beginning of the rainy season. Among the strains *Cylindrospermum licheniforme* was assumed to add about 90 kg N ha^{-1} in about 75 days. The same species was reported to very successfully colonize the clean soil surface of the weed-free maize fields. In grasslands of the Banaras Hindu University a thick and sometimes continuous BGA mat developed during the rainy season. The thicker the grass cover, the better the growth of BGA was. In protected and enclosed grasslands the total N content of the soil increased 88% in 5 months, an increase attributed mainly to the growth of *C. licheniforme*. From the above results it appears that BGA may contribute to maintain the N level of natural and cultivated ecosystems, but generally the extent of this contribution is poorly documented.

5. Relations between BGA and higher plants

5.1. Availability of fixed N_2 to higher plants

N_2 fixed by BGA is released either through exudation or through microbial decomposition after the death of the cells, but the process involved in the transfer of fixed N_2 to the plants is largely a mystery [30]. It has been shown in the laboratory, that BGA liberate large portions of their assimilated nitrogenous substances [17, 70]. However, the large amounts of nitrogenous substances recorded may result from methodological artifacts such as the osmotic shock occurring when resuspending the cells or any other physical damage of the algal material. No information on the exudation of fixed N_2 by BGA under field conditions is available, but it is clear that only part of it is available to rice, other parts being reincorporated by the microflora or volatilized. Nutrients released through microbial decomposition after the death of the algae appear to be the principal manner in which N is made available to the crop. Field experiments with *Tolypothrix tenuis* for 4 consecutive years indicated that only 1/3 of the field algae was decomposed and absorbed by rice plants in the first year: the rest remained as residual soil N, which could have been responsible for continued yield increases in the succeeding years [80]. The transfer of N from algae to higher plants has been investigated using ^{15}N-tracer techniques [34, 64]. Wilson (personal communication) recently recovered from a rice crop 39% and 51% of the N from ^{15}N-labeled *Aulosira* spp. spread on or incorporated into the soil, thus showing that BGA N is readily available to rice. A similar experiment is under way at the International Rice Research Institute (Philippines). Results of the analysis of the first crop show a recovery of 13% and 38% of the N from ^{15}N-labeled *Nostoc* sp. spread on or incorporated into the soil. This indicates that the availability of algal N for the plant varies either with the strain or with the physiological state of the algal material; i.e. N from an algal material rich in akinetes, not so easily decomposable, will be less available than that from vegetative cells which are more susceptible to decomposition.

5.2. Growth-promoting effect

Besides increasing N fertility, BGA have been assumed to benefit higher plants by producing growth-promoting substances. This hypothesis is based on the additive effects of BGA inoculations in the presence of nitrogenous fertilizers. Most of these results have been obtained with rice but similar results were observed also with vegetables such as radishes and tomatoes [51].

More direct evidence of hormonal effects has come primarily from treatments of rice seedlings with algal cultures or their extracts. Presoaking rice seeds with BGA cultures or extracts enhances germination, promotes the growth of roots and shoots, and increases the weight and protein content of the grain [19, 23]. It has also been established that algal growth-promoting substances are beneficial to other crops besides rice [11, 26, 39] and that the production of such substances is not confined to BGA. Whether these substances are hormones [19], vitamins [73], aminoacids [7] or any other components is still unknown.

5.3. Harmful effects

Blooms caused by filamentous algae can be harmful to rice, mainly due to a mechanical effect on the young plants [6]. However detrimental, effects [49,43,53,59] of BGA are incidental [12, 58]. Even when BGA produced a bloom at the beginning of the growing cycle, their effect on grain yield was rarely negative [9].

5.4. Epiphytism

Epiphytic BGA have been observed on wetland rice [57] deepwater rice [28] and on weeds growing in rice fields [27]. In wetland rice fields, epiphytic BGA on rice and weeds make a limited contribution to the N input but are important in providing an inoculum for the regeneration of the algal blooms that are periodically affected by adverse conditions. In deepwater rice, which offers a much greater biomass for colonization, the addition of N by epiphytic BGA is of agronomical significance [28]. BGA were found to grow preferentially on submerged decaying tissues. An endophytic growth inside the leaf sheath was also observed in deepwater rice. The results obtained did not confirm the existence or absence of biotic relationships between the algae and their hosts, but indicated that a mechanical effect in relation to the roughness of the support was involved in algal epiphytism and endophytism.

6. Role of BGA in soil colonization

Despite a worldwide distribution of BGA in terrestrial habitats their ecology has been less studied than that of heterotrophic microorganisms such as soil fungi and bacteria. This has sometimes given rise to the fallacious impression that they are unimportant soil microorganisms [17]. Terrestrial algae in fact play a major role in soil genesis and soil conservation. They constitute the initial successional stage on substrata which are poor in plant nutrients such as recent volcanic deposits, sand, and barren infertile soils denuded of macrovegetation. The interwoven algal growth consolidates the surface, leading to the formation of a soil crust which improves infiltration, may limit sheet erosion and affords a substratum upon which seeds of higher plants germinate. Algae produce a surface humus after death and dissolve certain soil minerals maintaining a reserve supply of elements in a semi-available form for higher plants [61]. Experiments on algal crusts from the Botanical Garden at the University of India indicated that these crusts (i) do not slow the rate of water infiltration [7], (ii) have a very efficient protective effect on erosion by buffeting rain, (iii) increased the moisture content of the soil underneath the algal stratum 10–15% [63]. The resistance of algal crusts to erosion is apparently the result of binding the soil surface particles into a non-erosible layer, which is also effective in breaking the force of falling water. The favourable effect of BGA on aggregation of the soil was demonstrated by Roychoudhury et al., [60] who observed a 50–70% increase in the water-stable aggregates after algal inoculation. This was attributed to the action of polysaccharides released by the algae and the pressure of filamentous BGA growing in the soil. In sugarcane fields in India, a profuse growth of Porphyrosiphon notarisii was reported to check the erosion of the soil and to help the crop stand erect on the intact ridges during the monsoon period [63]. The same author in a study of grasslands subjected to close and heavy grazing by sheep and cattle concluded that in such deteriorated grassland, where soils are light and sandy, a thick growth of BGA during the monsoon is perhaps the only check for erosion. *Porphyrosiphon notarisii* and *Microcoleus chthonoplastes* [8] were reported to be the most efficient colonizing strains growing as a mat that provides a suitable substratum for the germination of grass seeds. Booth [7] conducted a detailed investigation on the importance of BGA in badly eroded soils in south-central US. He concluded that (i) soil losses from plots with an algal stratum were greatly reduced as compared with the losses from bare areas, (ii) BGA constituted and initial stage in plant succession, and (iii) BGA colonization lasted for several years until higher perennial plants were able to form abundant ground cover. Singh [63] described the characteristic and common algal cover caused by *Aphanothece pallida*. Although the fresh thalli were extremely mulicaginous and fragile, they formed a compact dark grey stratum firmly adherent to the soil when dried; 0.3 g of algae were sufficient to bind 34.6 g of soil.

The floristic composition of algal crusts developing on cultivated sandy soils in Senegal was recently studied by Reynaud. The L.P.P. (*Lyngbia, Phormidium, Plectonema*) group [50] was dominant in most of the soils studied. A similar observation was made by Marathe and Anantany [31] in Indian arid soils. The dominance of the L.P.P. group in arid soils can be related to its ability to develop in drastic conditions a gel-like protoplasm and a thick mucilaginous sheath able to readily adsorb water and retain it. Among N_2-fixing genera, *Nostoc* was the most frequently observed, confirming the observation that it is one of the most consistently present N_2-fixing forms in arid soils [61]. Examples of the strains isolated from algal crusts on sandy Senegalese soils are given in Figs. 2 and 3.

7. Agronomical use of BGA

Most information on the effects of agronomical practices on BGA is related to paddies. Land preparation and management seem to have only incidental effects [58] and were reported not to interfere with the establishment and activity of inoculated BGA [2]. Pesticides depending on their nature and their concentration, could have inhibitory, selective or stimulatory effects on BGA. Experiments mainly with flask cultures suggest that BGA are generally more resistant to pesticides than other algae and are capable of tolerating pesticide levels recommended for field application [77]. Insecticides are generally less toxic to BGA than other pesticides [42] and have a secondary beneficial effect by controlling the grazers.

Organic manure may favour or depress BGA growth depending on its nature and mode of application. Incorporation of plant residues has frequently been reported to temporarily depress the algal populations and to have a negative effect on inoculated BGA (refer to Roger and Kulasooriya [58]). On the contrary the surface application of straw very significantly enhanced BGA growth and ARA [33].

The nature and the quantity of inorganic fertilizers as well as application techniques have a considerable influence on the algal flora. The surface application of NPK generally results in a profuse green algae growth. To prevent such a growth the incorporation of fertilizers into the soil is recommended. The deep placement

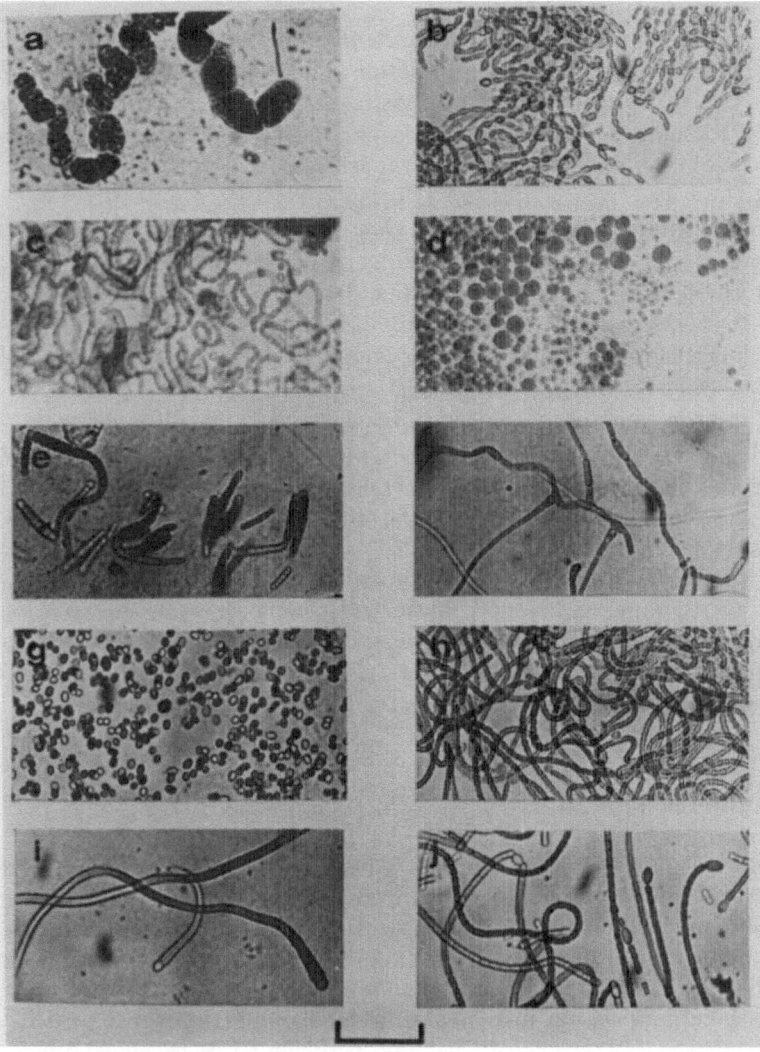

Fig. 2. BGA isolated from algal crusts in Senegal, in a medium without N. (a) *Nostoc* sp.; (b, c, h) *Anabaena* spp.; (d) *Dermocarpa*; (i, e) *Calothrix* sp.; (f) *Hapalosiphon* sp.; (g) *Gloeocapsa* sp.; (j) *Anabaenopsis* sp. (the scale represents 50 μm).

of urea supergranules prevents the dense growth of green algae that occurs when urea is surface broadcast; moreover it does not inhibit the growth and N_2 fixation in BGA [59]. Since the growth of N_2-fixing BGA in paddy fields is generally limited by P deficiency, P application alone or together with lime should be recommended (refer to Roger and Kulasooriya [58]).

In N-deficient conditions, N_2-fixing BGA are not hindered by competition from the other algae and they can develop profusely if the other environmental factors are not limiting. When nitrogenous fertilizers are applied their ARA is inhibited or

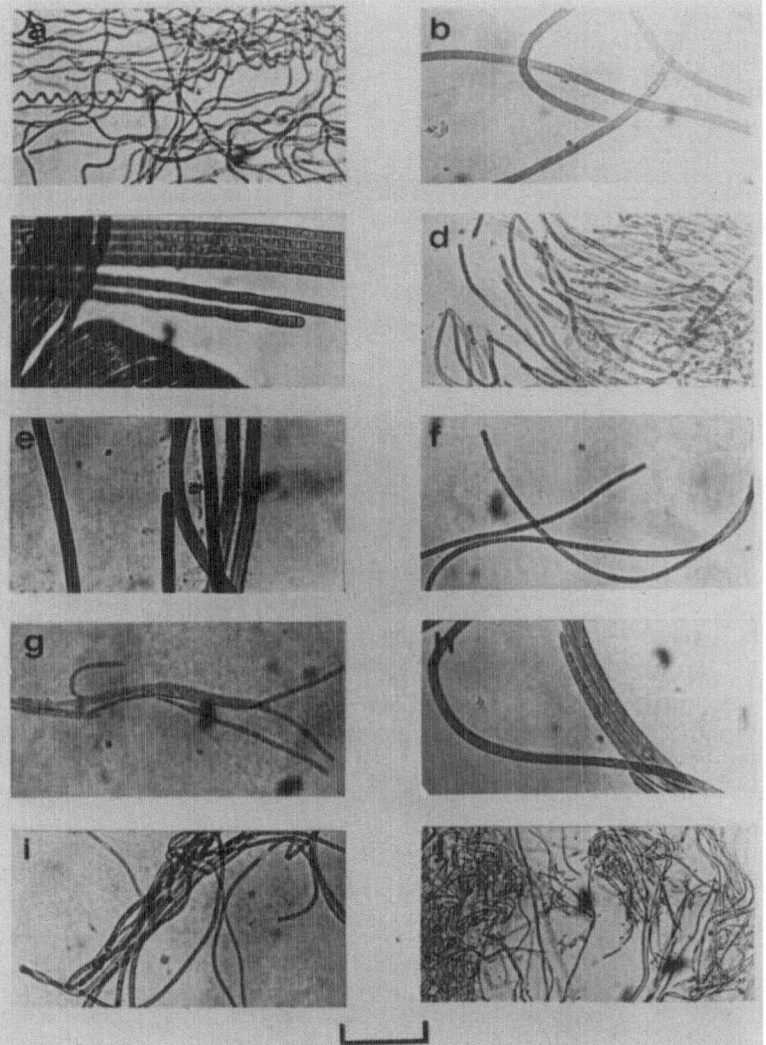

Fig. 3. BGA isolated from algal crusts in Senegal, on a medium with N. (a) *Pseudanabaena* sp.; (h, d) *Oscillatoria* spp.; (b, c, e, f, g, i, j) different species belonging to the LPP (*Lyngbia, Phormidium, Plectonema*) group (the scale represents 50 μm).

at least affected. Moreover, N fertilizers appear to enhance the growth of non-N_2-fixing algae thus increasing the competitive pressure against N_2-fixing forms [59]. However, this effect seems to be transient as suggested by pot experiments [49] and preliminary field data. Simultaneously the effect of N fertilizers on ARA is only partial. Venkataraman [79] reported that ARA was not depressed in a soil—rice—BGA system with less than 40 ppm ammonium-N in stagnant paddy water. Within mixed algal masses combined N may diffuse at a slower rate than the readily available dissolved N_2, and a local depletion in combined N may favour N_2-fixing algae. Thus in the fields, the lack of competition with N_2-fixing algae in the presence of mineral nitrogen may not be as clear-cut as was first thought.

Molybdenum, because of its function in nitrogenase, is required by N_2-fixing BGA. Subrahmanyan et al. [68], suggested the addition of sodium molybdate (0.25 kg ha^{-1}) to the soil to improve N_2-fixing algal growth. This addition has been shown to be beneficial in some situations.

Other nutrients (Fe, Mg, K, etc.) are required for optimal growth of BGA but their ecological implications as limiting factors or as factors affecting the composition of the algal community in paddies have not been documented.

7.1. Reclamation of Usar lands [62]

Usar soils are saline (solonchak) and alkaline (solonetz) unproductive soils, found extensively throughout India. They are characterized by impermeability, extreme hardness and the occasional presence of undesirable salts on the surface, all of which adversely affect the plant growth. The pH value is usually high throughout the profile. The subsoil water table is generally found between 3.0 and 4.5 meters below the surface. These soils are usually very deep. The method of reclamation arose from the observation [62] that in the alkaline Usar soils of northern India, while other plants fail to grow, blue-green algae form a thick stratum on the soil surface during the rainy season (July—September) and the retreating monsoon (December—January). These soils, characterized by an alkalization process, can be reclaimed by replacing sodium with calcium through the addition of chemical correctives such as gypsum, a rather expensive method. Another possibility is to help the reaction of calcium carbonate with the sodium clay by waterlogging the soil and by adding organic matter and N. The abundant growth of N_2-fixing BGA on waterlogged soils fulfills these requirements.

The following reclamation process was tested. During May and June, before the rains started, the land was divided into plots of less than 0.4 ha. The plots were enclosed by an earth embankment. After the first showers, N_2-fixing BGA formed a thick and compact stratum. Later, when the soil was waterlogged, N_2-fixing forms characteristic of rice fields developed and continued their active growth as long as the soil was waterlogged. In field trials, after a year of reclamation, a transplanted paddy crop produced a yield of 1,600—2,200 kg grain ha^{-1} and allowed the successful growth of sugarcane after 3 years. Laboratory experiments over a 3-year period indicated tremendous changes in the soil characteristics after algal growth compared with the control covered with a black cloth: pH decreased from 9.2 to 7.5. Large increase in organic matter content (69%), total N (46%), water-holding capacity (35%), and exchangeable calcium (31%), were observed. The most significant changes concerned the different forms of P. CO_2-soluble P was 0.45 in the control and 8.15 ppm in the treated plot with BGA. The role of BGA in converting the sodium clay into a type of calcium clay was not explained but a possible effect of oxalic acid excreted by the algae was suggested.

Subsequent improvements of the reclamation technique were proposed: doing the process several times in a year where irrigation water is available, contour farming (field operations such as plowing, etc. are done at right angles to the slope of the land) to retard water flow and increase water penetration; and introducing *Nostoc commune* where it does not occur naturally.

7.2. Algalization

BGA were among the first N_2-fixing agents recognized to be active in flooded rice soils. Since De [14] attributed the natural fertility of the tropical paddy fields to these organisms, many trials have been conducted to increase rice yield by inoculating the soil with BGA. This practice, also called algalization, a terminology introduced by Venkataraman [74], has been reported to have a beneficial effect on grain yield in different agroclimatic conditions. However, some reports also indicate failure of algalization. The conclusions of the review of Roger and Kulasooriya [58] on algalization are summarized thereafter.

7.2.1. Methodology. Most experiments on algalization were on a 'black box' basis, where only the last indirect effect (grain yield) of algalization was observed and the intermediate effects were not studied. There is little information on the qualitative and quantitative variations of the N_2-fixing algal flora and the N balance in inoculated paddy soils. Pot and field experiments have been conducted, usually on a single crop. The relative increase in grain yield over the control was an average 28% in pot experiments and 15% in field experiments. The better growth of BGA in pot experiments is probably attributable to the reduction of climatic disturbance and to the mechanical effect of the pot walls, where BGA frequently seem to grow preferentially and profusely. Pot experiments may therefore only be suitable for qualitative studies, since they overestimate the effects of BGA inoculation. Most of the field experiments on the other hand, were conducted in only one growing season and may underestimate the effects of algalization since the advantages of a slow N release from dead BGA may not be apparent in the first algalized crop.

7.2.2. Effects of algalization on rice. Algalization may affect plant size, its nitrogen content, and the number of tillers, ears, spikelets, and filled grains per panicle. The most frequently used criterion for assessing the effects of algalization has been better grain yield. Results of field experiments conducted mainly in India report an average yield increase of about 14% over the control, corresponding to about 450 kg grain ha^{-1} per crop where algal inoculation was effective. A higher grain yield increase was observed when algalization was in combination with lime, P and sometimes molybdenum application. Unfortunately it is not possible to separate the direct effect of PK fertilizers on rice from its indirect effect upon the growth of indigenous or introduced algae. The effects of algalization used with N fertilizers are controversial. Since biological N_2 fixation is known to be inhibited by inorganic N the beneficial effect of algalization in the presence of N fertilizers was most frequently interpreted as resulting from growth-promoting substances produced by algae or also by a temporarily immobilization of added N followed by a slow release through subsequent algal decomposition permitting a more efficient utilization of N by the crop (see Section 5.2.).

7.2.3. Effects of algalization on soil properties and microflora. Grain-yield measurements suggest that algalization produces both a cumulative and residual effect. This was attributed to a build up of both the organic N content and the number of BGA propagules in the soil, facilitating the reestablishment of the BGA biomass. Several reports indicate an increase in organic matter and organic N; algalization was also reported to increase: aggregation status of the soil [61], water-holding capacity [63], available P, total microflora, *Azotobacter, Clostridia,* and nitrofiers [21].

7.2.4. Limiting factors for algalization. Among the limiting factors responsible for the failure of algalization only pH and available P content of the soil have been studied. Low pH is an important limiting factor for algalization as demonstrated by unsuccessful trials in acidic soils of India, Sri Lanka and Japan. In Sri Lanka and Japan algalization was effective only when the soil was supplemented with calcium carbonate, the effect of inoculation being related to the amount of lime added. Since in some soils, algalization is inefficient in spite of the addition of lime and phosphate [36], available P content should probably not be the only factor limiting the effect of algalization; texture, organic matter content, CEC of saturated extracts, total N are probably not important limiting factors [36, 68]. Among the biotic factors that can possibly limit BGA growth inoculum, grazing by the zooplankton has been already mentioned. Other possible mechanisms involved such as antagonism, competition, etc. have been cited, but their role is not clear. Low temperatures, heavy rains, and cloudy weather have also been reported to limit the establishment of the inoculum.

7.2.5. Algalization technology
   7.2.5.1. Inoculum production and conservation. The methodology of BGA production has been reviewed by Watanabe and Yamamoto [83] and Venkataraman [76, 77]. Methods of inoculum production in artificially controlled conditions have been developed mainly in Japan. Different types of tank cultures and outdoor closed circulating systems have been described [82]. Algae are grown either in liquid culture, mixed with an inert material (support) and dried, or grown directly on the support and dried. Various supports such as sand, pumice stone, volcanic earth, and blocks of synthetic sponge have been tested to conserve and facilitate transportation of the algae. Such inocula reportedly maintained their capacity for growth unimpaired for at least two years.
   Producing the inoculum in artificially controlled conditions is well defined but

relatively expensive. On the contrary, the open air soil culture, used in India, is more simple, less expensive and easily adoptable by the farmers. It is based on the use of a starter culture that is a multistrain inoculum of *Aulosira, Tolypothrix, Scytonema, Nostoc, Anabaena* and *Plectonema*, provided by the 'All India Co-ordinated Project on Algae' [2]. This inoculum is multiplied by the farmer in shallow trays or tanks with 5–15 cm water, about 4 kg soil m^{-2}, 100 g triple superphosphate m^{-2} and insecticide. If necessary, lime is added to correct the soil pH to about 7.0–7.5. In 1 to 3 weeks, a thick mat develops on the soil surface and sometimes floats. Watering is stopped and water in the trays is allowed to dry up in the sun. Algal flakes are then scraped off and stored in bags for use in the fields. With such a method, the ultimate proportion of individual strains in the algal flakes is unpredictable. It is assumed that, because the inoculum is produced in soil and climatic conditions similar to those in the field, dominant strains will be the most adapted to the local conditions. The recorded rates of production of algal flakes in the open air soil culture range from 0.4 to 1.0 kg m^{-2} in 15 days, indicating that a 2m^2 tray can produce in 2–3 months enough algal material to inoculate 1 ha of rice field.

7.2.5.2. Inoculation of the rice field. The methods of field application have been reviewed by Venkataraman [77]. For transplanted rice the algal inoculum is generally applied 1 week after the transplanting. When rice is sown, seeds can be coated by mixing the algal suspension and 2–3 kg calcium carbonate per 10–20 kg seed and air-dried in the shade.

Recommendations for field application of dried algal inoculum (algal flakes) given by the 'All India Coordinated Project on Algae' [2] indicate that 8–10 kg of dry algal flakes applied 1 week after transplanting is sufficient to inoculate 1 ha; a large amount will accelerate multiplication and establishment in the field. Algalization can be used with high levels of commercial nitrogen fertilizer, but reduction of the N dose by one third is recommended. To benefit from the cumulative effect of algalization the algae should be applied for at least three consecutive seasons. Recommended pest-control measures and other management practices do not interfere with the establishment and activity of these algae in the fields.

7.2.5.3. Pay-off of algal technology. Trials in Indian station (in 1978), showed that adding 10 kg of algal culture ha^{-1} which costs about 4 US $, increases paddy yield; the average yield increase is worth 60 to 90 US $ [43]. The pay-off of algal technology based on the results of a large number of field experiments was discussed by the 'All India Coordinated Project on Algae' [2] and it was inferred that if algal technology was introduced into even 50% of the Indian rice area, it would supposedly result in a saving of about 3.8×10^5 t of N. In a large number of trials where the recommended level of N fertilizer (100 kg ha^{-1}) was complemented with algal application (10 kg ha^{-1}) an average increase of about 300 kg rice grains ha^{-1} was reported.

164

References

1. Aiyer, R.S. 1965 Comparative algological studies in rice fields in Kerala state. Agric. Res. J. Kerala 3, 100—104.
2. All India Coordinated Project on Algae. 1979 Algal biofertilizers for rice. India Agricultural Research Institute, New Delhi, 61 pp.
3. Araragi, M., Phetchawee, S. and Tantitanapat, P. 1979 Microflora related to the nitrogen cycle in the tropical upland farm soils. Soil Sci. Plant Nutr. 25, 235—244.
4. Arora, S.K. 1969 The role of algae on the availability of phosphorus in paddy fields. Riso 18, 135—138.
5. Barbey, C. and Coute, A. 1976 Croûtes à Cyanophycées sur les dunes du Sahel mauritanien. Bull. IFAN 38A, 732—736.
6. Bisiach, M. 1972 Laboratory algicidal screening for the control of algae in rice. Riso 21, 43—58.
7. Booth, W.E. 1941 Algae as pioneers in plant succession and their importance in erosion control. Ecology 22, 38—46.
8. Brock, T.D. 1975 Effect of water potential on a *Microcoleus* (Cyanophyeae) from a desert crust. J. Phycol. 11, 316—320.
9. Chapman, R.L., Bayer, D.E. and Lang, N.J. 1972 Observations on the dominant algae in experimental California rice fields. J. Phycol. Suppt. 8, 17.
10. Chapman, V.J. and Chapman, D.J. 1973 Soil algae and Symbiosis. In: The Algae, Ch. 17, pp. 381—387. Macmillan, London.
11. Dadhich, K.S., Varma, A.K. and Venkataraman, G.S. 1969 Effect of *Calothrix* inoculation on vegetable crops. Plant and Soil 31, 377.
12. Das, S.S. 1976 Algal weeds and their chemical control — a review. Indian J. Plant Prot. 4, 201—208.
13. David, K.A.V. and Fay, P. 1977 Effects of long-term treatment with acetylene on nitrogen fixing micro-organisms. Appl. Environ. Microbiol. 34, 640—646.
14. De, P.K. 1939 The role of blue-green algae in nitrogen fixation in rice fields. Proc. R. Soc. Lond. 127 B, 121—139.
15. Durrel, L.W. 1964 Algae in tropical soils. Trans. Amer. Micros. Soc. 83, 79—85.
16. Dutta, N. and Venkataraman, G.S. 1958 An exploratory study of the algae of some cultivated and uncultivated soils. Indian J. Agron. 3, 109—115.
17. Fogg, G.E., Stewart, W.D.P., Fay, P. and Walsby, A.E. 1973 The blue-green algae. Academic Press, London and New York, 459 pp.
18. Garcia, J-L., Raimbault, M., Jacq, V., Rinaudo, G. and Roger, P. 1973 Microbial activities in paddy fields in Senegal; influence of physico-chemical properties of soils and of the rhizosphere effect (in French, English summary) Rev. Ecol. Biol. Sol 11, 169—185.
19. Gupta, A.B. and Shukla, A.C. 1967 Studies on the nature of algal growth promoting substances and their influence on growth, yield, and protein content of rice plants. Labdev. J. Sci. Technol. Kanpur 5, 162—163.
20. Holm-Hansen, O. 1968 Ecology, physiology and biochemistry of blue-green algae. Ann. Rev. Microbiol. 22, 47—70.
21. Ibrahim, A.N., Kamel, M. and El-Sherbeny, M. 1971 Effect of inoculation with alga *Tolypothrix tenuis* on the yield of rice and soil nitrogen balance. Agrokem. Talagtan 20, 389—400.
22. Ichimura, S. 1954 Ecological studies on the plankton in paddy fields I. Seasonal fluctuations in the standing crop and productivity of plankton. Jpn. J. Bot. 14, 269—279.

23. Jacq, V. and Roger, P.A. 1977 Decrease of losses due to sulphate reducing processes in the spermosphere of rice by presoaking seeds in a culture of blue-green algae (in French, English summary). Cahiers ORSTOM. sér. Biol. 12, 101–108.

24. Jones, K. 1977 The effects of temperature on acetylene reduction by mats of blue-green algae in sub-tropical grassland. New Phytol. 78, 433–436.

25. Jurgensen, M.S. and Davey, C.B. 1968 Nitrogen-fixing blue-green algae in acid forest and nursery soils. Can. J. Microbiol. 14, 1179–1183.

26. Kaushik, B.D. and Venkataraman, G.S. 1979 Effect of algal inoculation on the yield and vitamin C content of two varieties of tomato. Plant and Soil 52, 135–137.

27. Kulasooriya, S.A., Roger, P.A., Barraquio, W.L. and Watanabe, I. 1980 Epiphytic nitrogen fixation on weeds in a rice field ecosystem. In: Proceedings of Nitrogen Cycling South-East Asian Wet Monsoonal Ecosystems. Wetselaar, R., Simpson, J.R. and Rosswall, T. (eds.), Australian Academy of Science, Canberra (in press).

28. Kulasooriya, S.A., Roger, P.A., Barraquio, W.L. and Watanabe, I. 1980 Biological nitrogen fixation by epiphytic microorganisms in rice fields. IRRI. IPRS No. 47. Feb. 1980, 10 pp.

29. Kurasawa, H. 1956 The weekly succession in the standing crop of plankton and zoobenthos in the paddy field, Part 1 and 2. Bull. Res. Sci. Japan 41–42, 86–98 and 45, 73–84.

30. Lowendorf, H.S. 1980 Biological nitrogen fixation in flooded rice. Agronomy paper 1305. Dpt. of Agronomy. Cornell Univ. Ithaca, N.Y.

31. Marathe, K.U. and Anantani, Y.S. 1972 Observations on the algae of some Indian arid soils. The botanique 3, 13–20.

32. Materasi, R. and Balloni, W. 1965 Some observations on the presence of autotrophic nitrogen-fixing microorganisms in paddy soils, (in French, English summary), Ann. Inst. Pasteur 109, 218–223.

33. Matsuguchi, T. and Ick-Dong Yoo. 1979 Stimulation of phototrophic N_2 fixation in paddy fields through rice straw applications. In: Proceedings of Nitrogen Cycling in South-East Asian Wet Monsoonal Ecosystems. Wetselaar, R., Simpson, J.R. and Rosswall, T. (eds.), Australian Academy of Science, Canberra (in press).

34. Mayland, H.F. and McIntosh, T.H. 1966 Availability of biologically fixed nitrogen-15 to higher plants. Nature, London 209, 420–421.

35. Mitra, A.K. 1951 The algal flora of certain Indian soils. Indian J. Agric. Sci. 21, 357.

36. Okuda, A. and Yamaguchi, M. 1952 Algae and atmospheric nitrogen fixation in paddy soils. II: Relation between the growth of blue-green algae and physical or chemical properties of soil and effect of soil treatments and inoculation on the nitrogen fixation. Mem. Res. Inst. Food Sci. 4, 1–11.

37. Okuda, A. and Yamaguchi, M. 1956 Nitrogen-fixing microorganisms in paddy soils. II: Distribution of blue-green algae in paddy soils and the relationship between the growth of them and soil properties. Soil and Plant Food 2, 4–7.

38. Pandey, D.C. 1965 A study of the algae from paddy soils of Ballia and Ghazipur districts of Uttar Pradesh, India. I. Cultural and ecological considerations. Nova Hedwigia 9, 299–334.

39. Pankratova, E.M. and Vakrushev, 1969 Utilization by higher plants of nitrogen fixed from the atmosphere by blue-green algae (in Russian, English summary). Mikrobiologiya 38, 1080–1084.

40. Peterson, R.B. and Burris, R.H. 1976 Conversion of acetylene reduction rates

to nitrogen fixation rates in natural populations of blue-green algae. Anal. Biochem. 73, 404—410.

41. Prescott, G.W. 1968 Ecology Soil. In: The Algae, pp. 320—323, Steers, W.C. and Glass, H.B. (eds.), Houghton Mifflin, Boston.

42. Raghu, K. and MacRae, I.C. 1967 The effect of the gamma isomer of benzene hexachloride upon the microflora of submerged rice soils. II. Effect upon algae. Can. J. Microbiol. 13, 173—180.

43. Rao, T.R. 1978 Blue-green algae boost rice yields. Intensive Agric. 16, 19—20.

44. Renaut, J., Sasson, A., Pearson, H.W. and Stewart, W.D.P. 1975 Nitrogen-fixing algae in Morocco. In: Nitrogen fixation by free-living microorganisms, pp. 229—246. Stewart, W.D.P. (ed.), Cambridge University Press, Cambridge.

45. Reynaud, P.A. and Roger, P.A. 1978 N_2-fixing algal biomass in Senegal rice fields. Ecol. Bull. Stockholm 26, 148—157.

46. Reynaud, P.A. and Roger, P.A. 1978 Photophobotaxis and photokinesis among *Oscillatoria* sp. 77 S23 (in French, English summary). Cah. ORSTOM, sér. Biol., 13, 157—164.

47. Reynaud, P.A. and Roger, P.A. 1981 Seasonal variations of algal flora and of N_2-fixing activity in a waterlogged sandy soil. Rev. Ecol. Biol. Sol. 18, 9—27.

48. Reynaud, P.A. 1980 Cyanobacteria and *Azolla* nitrogen fixation: Agronomical potentialities in tropical Africa. In: Proceedings of Recycling of Organic Matter in West Africa. FAO/SIDA, LOME, Togo (in French, English summary).

49. Reynaud, P.A. 1980 Fate of Cyanobacteria inoculums during the two first months of rice growth on a waterlogged soil (in French, English summary). Cah. ORSTOM. sér. Biol. 43, 53—60.

50. Rippka, R., Deruelles, J., Waterbury, J.B., Herdman, M. and Stanier, R.Y. 1979 Generic assignments, Strain histories and Properties of pure Culture of Cyanobacteria. J. Gen. Microbiol. 111, 1—61.

51. Rodgers, G.A., Bergman, B., Henriksson, E. and Udris, M. 1979 Utilization of blue-green algae as biofertilizers. Plant and Soil 52, 99—107.

52. Roger, P. and Reynaud, P. 1976 Dynamics of the algal populations during a culture cycle in a Sahel rice field (in French, English summary). Rev. Ecol. Biol. Sol, 13, 545—560.

53. Roger, P. and Reynaud, P. 1977 Algal biomass in rice fields of Senegal: relative importance of Cyanophyceae that fix nitrogen (in French, English summary). Rev. Ecol. Biol. Sol 14, 519—530.

54. Roger, P.A., Reynaud, P.A., Rinaudo, G.E., Ducerf, P.E. and Traore, T.M. 1977 Log-normal distribution of acetylene-reducing activity *in situ* (in French, English summary). Cah. ORSTOM, sér. Biol. 12, 133—140.

55. Roger, P.A. and Reynaud, P.A. 1978 Enumeration of the algae in submerged soil: law of the distribution of organisms and the density of sampling (in French, English summary). Rev. Ecol. Biol. Sol 15, 219—234.

56. Roger, P.A. and Reynaud, P.A. 1979 Ecology of blue-green algae in paddy fields. In: Nitrogen and rice, pp. 289—309. International Rice Research Institute, Los Banos.

57. Roger, P.A., Kulasooriya, S.A., Barraquio, W.L. and Watanabe, I. 1979 Epiphytic nitrogen fixation on lowland rice plants. In: Proceedings of Nitrogen Cycling in South-East Asian Wet Monsoonal Ecosystems. Wetselaar, R., Simpson, R.J. and Rosswall, T. (eds.), Australian Academy of Science, Canberra (in press).

58. Roger, P.A. and Kulasooriya, S.A. 1980 Blue-green algae and rice. The International Rice Research Institute, Los Baños. Philippines, 112 pp.
59. Roger, P.A., Kulasooriya, S.A., Tirol, A.C. and Craswell, E.T. 1980 Deep placement: a method of nitrogen fertilizer application compatible with algal nitrogen fixation in wetland rice soils. Plant and Soil 57, 137–142.
60. Roychoudhury, P., Kaushik, B.D., Krishnamurthy, G.S.R. and Venkataraman, G.S. 1979 Effect of blue-green algae and *Azolla* application on the aggregation status of the soil. Current Science 48, 454–455.
61. Shield, L.M. and Durrell, L.W. 1964 Algae in relation to soil fertility. Bot. Rev. 30, 93–128.
62. Singh, R.N. 1950 Reclamation of Usar lands in India through blue-green algae. Nature, London. 165, 325–326.
63. Singh, R.N. 1961 Role of blue-green algae in nitrogen economy of Indian agriculture. Indian Council of Agricultural Research, New Delhi, 175 pp.
64. Stewart, W.D.P. 1967 Transfer of biologically fixed nitrogen in a sand-dune-slack region. Nature, London 214, 603–604.
65. Stewart, W.D.P. 1978 Nitrogen fixing Cyanobacteria and their association with eukaryotic plants. Endeavour 2, 170–179.
66. Stewart, W.D.P., Sampaio, M.J., Isichei, A.O. and Sylvester-Bradley, R. 1978 Nitrogen fixation by soil algae of temperate and tropical soils. In: Limitation and potentials for biological nitrogen fixation in the tropics, pp. 41–63. Döbereiner, J. *et al.* (eds.), Plenum Press, New York and London.
67. Stewart, W.D.P., Rowell, P., Ladha, J.K. and Sampaio, M.J.A. 1979 Blue-green algae (Cyanobacteria) – some aspects related to their role as sources of fixed nitrogen in paddy soils. In: Nitrogen and Rice, pp. 263–285. International Rice Research Institute, Los Baños, Philippines.
68. Subrahmanyan, R., Manna, G.B. and Patnaik, S. 1965 Preliminary observations on the interaction of different rice soil types to inoculation of blue-green algae in relation to rice culture. Proc. Indian Acad. Sci. B 62, 171–175.
69. Suzuki, T. and Kawai, K. 1971 Soil of the Sambor. Cambodia. Bull. Natl. Agric. Sci. B 22, 211–304.
70. Taha, E.E.M. and Al Refai, A.E. 1962 Physiological and biochemical studies on the nitrogen-fixing blue-green algae: 1. On the nature of cellular and extracellular substances formed by *Nostoc commune*. Arch. Mikrobiol. 41, 307–312.
71. Tchan, Y.T. and Beadle, N.C. 1955 Nitrogen economy in semi arid plant communities II. The non-symbiotic nitrogen-fixing organisms. Proc. Linn. Soc. N.S.W. 80, 97–104.
72. Traore, R.M., Roger, P.A., Reynaud, P.A. and Sasson, A. 1978 N_2-fixation by blue-green algae in a paddy field in Mali (in French, English summary). Cah. ORSTOM sér. Biol. 13, 181–185.
73. Venkataraman, G.S. 1961 The role of blue-green algae in agriculture. Sci. Cult. 27, 9–13.
74. Venkataraman, G.S. 1966 Algalization. Phytos. 5, 164–174.
75. Venkataraman, G.S. and Neelakanthan, S. 1967 Effect of the cellular constituents of the nitrogen fixing blue-green alga *Cylindrospermum musicola* on the root growth of the rice seedlings. J. Gen. Appl. Microbiol. 13, 53–61.
76. Venkataraman, G.S. 1969 The cultivation of algae. Indian Council of Agricultural Research, New Delhi, 319 pp.
77. Venkataraman, G.S. 1972 Algal biofertilizers and rice cultivation. Today and Tomorrow's Printers, Faridabad (Haryana), 75 pp.
78. Venkataraman, G.S. 1975 The role of blue-green algae in tropical rice culti-

vation. In: Nitrogen fixation by free-living microorganisms, pp. 207–218. Stewart, W.D.P. (ed.), Cambridge Univ. Press.

79. Venkataraman, G.S. 1979 Algal inoculation of rice fields. In: Nitrogen and rice, pp. 311–321. International Rice Research Institute, Los Baños.

80. Watanabe, A. 1956 On the effect of the atmospheric nitrogen-fixing blue-green algae on the yield of rice (in Japanese). Bot. Mag. 69 (820/821), 530–535.

81. Watanabe, A. 1959 Distribution of nitrogen-fixing blue-green algae in various areas of south and east Asia. J. Gen. Appl. Microbiol. 5, 21–29.

82. Watanabe, A. 1959 On the mass culturing of a nitrogen fixing blue-green algae *Tolypothrix tenuis*. J. Gen. Appl. Microbiol. 5, 85–91.

83. Watanabe, A. and Yamamoto, Y. 1970 Mass culturing preservation and transportation of the nitrogen fixing blue-green algae. In: Proc. 2nd Symposium on Nitrogen Fixation and Nitrogen Cycle Sendai, Japan, pp. 22–28.

84. Watanabe, A. and Yamamoto, Y. 1971 Algal nitrogen fixation in the tropics. Plant and Soil (special volume), 403–413.

85. Watanabe, I., Lee, K.K., Alimagno, B.V., Sato, M., Del Rosario, D.C. and De Guzman, M.R. 1977 Biological N_2-fixation in paddy field studied by *in situ* acetylene-reduction assays. IRRI Res. Pap. Ser. 3, 1–16.

6. *Azolla–Anabaena* symbiosis – its physiology and use in tropical agriculture

I. WATANABE

1. Introduction

Azolla is a water fern widely distributed in aquatic habitats like ponds, canals, and paddies in temperate and tropical regions. This plant has been of interest to botanists and Asian agronomists because of its symbiotic association with a N_2-fixing blue-green alga and rapid growth in nitrogen-deficient habitats.

Recently, the interest in this plant–alga association has been renewed by the demand for less fossil energy-dependent agricultural technology.

Reviews on updating information were made by Moore [20], Watanabe [42], and Lumpkin and Plucknett [19]. A bibliographic list was published by the International Rice Research Institute [15].

2. Biology and physiology of Azolla–alga relation

Azolla belongs to the Azollaceae, a heterosporous free-floating fern, and is close to the family Salviniaceae. There are six extant species of *Azolla* (Table 1) and 25 fossil species are recorded [14]. These are divided into two subgenera: *Euazolla*, a New World azolla, and *Rhizosperma*. Species differentiation is based on the morphology of the sexual organ.

The number of septa in the glochidia was used as a taxonomic tool to differentiate *Euazolla*. This criterion was questioned by taxonomists because of variations within a given species [10]. In the subgenus *Rhizosperma*, the glochidia are replaced by a root-like structure emerging from the massulae in the microsporangium. In *A. nilotica*, neither the glochidia nor the root-like structure is present on the massulae (Fig. 1).

Because the sporocarps are usually absent in naturally grown azolla, it is difficult to identify species.

Four species of *Euazolla* originated from the New World, but currently these are widely spread in the temperate regions of Europe and Asia. *A. pinnata* is widely distributed in the subtropical and tropical regions of Asia and has been used for agricultural purposes in Asia. *A. nilotica* is a giant water fern, about 10 cm long, and is distributed in central Africa.

The azolla plant has a branched floating stem that bears alternately arranged overlapping leaves and true roots. Each leaf has two lobes – the ventral or lower

Y.R. Dommergues and H.G. Diem (eds.), Microbiology of Tropical Soils
and Plant Productivity. ISBN 978-94-009-7531-6.

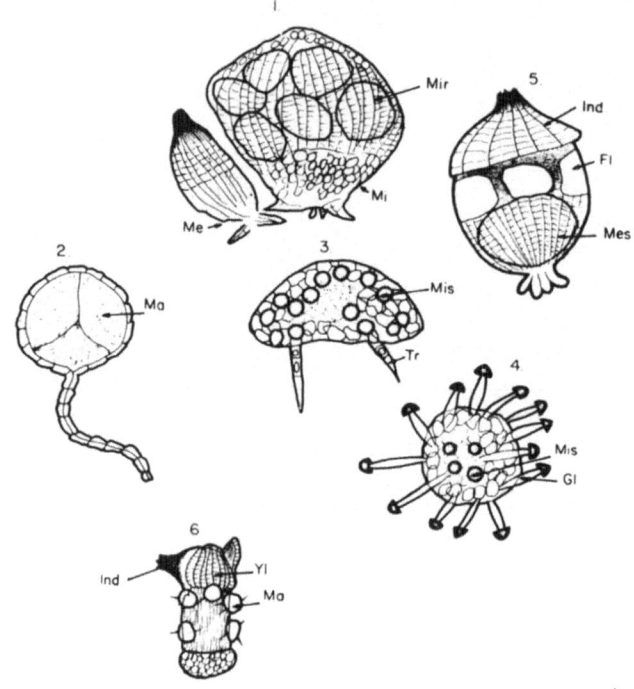

Fig. 1. Sexual organs of *Azolla*.

1. Megasporocarp and Microsporocarp — Me: megasporocarp; Mi: microsporocarp; Mir: microsporangium
2. Microsporangium — Ma: massulae
3. Massulae of *A. pinnata* — Tr: trichome; Mis: microspore
4. Massulae of *A. filiculoides* — Gl: glochidium
5. Megasporocarp — Ind: Indisum; Fl: float; Mes: megaspore
6. Germinating megasporocarp — Yl: young leaf; Ma: massulae

Table 1. Species of *Azolla*

Subgenus	No. of floats on megasporocarps	Species	Major distribution before dispersal by men
Euazolla	3	*A. filiculoides*, Lamarck	Southern South America Western North America
		A, caroliniana, Willd.	Eastern North America Central America
		A. mexicana, Presl.	Northern South America Western North America
		A. microphylla, Kaulfuss	Tropical and subtropical America
Rhizosperma	9	*A. pinnata*, R. Brown	Tropical-subtropical Asia and coastal Africa
		A. nilotica, De Laisne	Upper Nile and Sudan Central Africa

lobe and the dorsal or upper lobe. The dorsal lobes are chlorophyllous and aerial, the ventral lobes are partly submerged, thin, and achlorophyllous. The frond is about 1 to 3 cm long. In optimum condition, the lateral branch of stems of *A. filiculoides* and *A. nilotica* and sometimes, of *A. caroliniana*, partly becomes aerial and new shoots grow upward, thus giving a higher biomass than the flatly growing ones. The roots occur at branch nodes on the ventral surface of the stem. They are about 2 to 10 cm long, depending on species, have hairs and a sheathing root cap that falls off with age. In shallow water, the roots adhere to the soil surface and absorb nutrients from the soil.

In natural conditions, azolla multiply by vegetative reproduction. Under certain circumstances, the formation of sexual organs is observed. A new generation is formed from the fertilized embryo. Although a sporophytic life cycle is described (Fig. 2), little is known about the conditions for spore formation and its ecological significance. High temperature (early summer) in temperate regions and low temperature in tropical and subtropical regions (*A. pinnata*) have been reported to induce sporocarps. In southern China, some local strains of *A. pinnata* form spores abundantly in June and July, and to a lesser extent in September and October. In northern Vietnam, spores are formed in March–April. Formation of sexual organs seems to be associated with high density of azolla population.

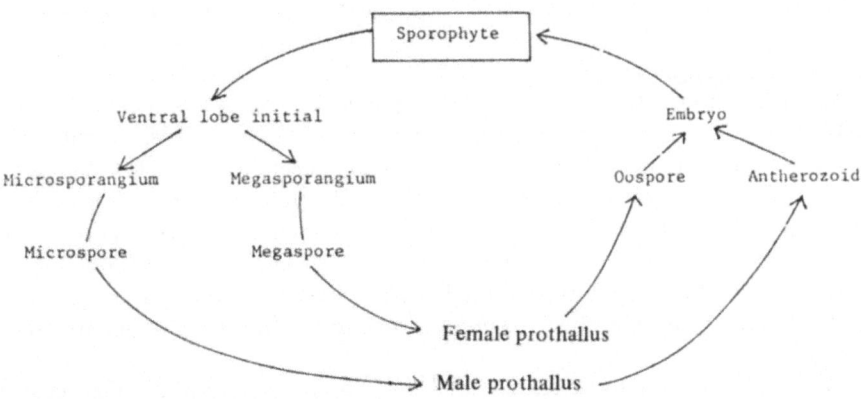

Fig. 2. Heterosporous life cycle of *Azolla*.

When sporocarps are formed, the vegetative growth rate is retarded [2, 3]. Sporocarps are borne by shoot stalks on the first ventral lobe initial of a lateral branch and occur in pairs (Fig. 1).

Microsporocarps, the male organs, are larger than the mega(macro)-sporocarps and are brown yellow or brownish red. They contain many microsporangia and within the periplasmodium of a microsporangium, 32 or 64 microspores develop and aggregate into massulae. From the massulae, glochidia develop. Microspores germinate and release antherozoids which fertilize the oospore in the megaspore.

Megasporocarps, the female organ, are smaller and produce only one megaspore. In mature megasporocarps, the megaspore is covered by 3 or 9 floats together with a columella. One oospore is formed from a megaspore. The germination of embryo and the subsequent growth of young seedling is slow. It takes 1 or 2 months from germination for an azolla to grow as big as a vegetatively growing fern with branches. The germinating megasporocarps are attached by many massulae. The fertilized megaspores can withstand desiccation and can survive more than 1 year in dry condition [46]. Therefore, the fertilized megasporocarps can be used for storage of azolla germplasm collection. Because of the slow growth of a new azolla seedling, the megasporocarps are unlikely to be used as seeding material for agricultural use. Light is necessary for the germination of sporocarps.

The symbiont alga is found in the cavity formed in the proximal portion of the dorsal lobes. The cavity has a mouth which opens toward the ventral side of the dorsal lobes. This mouth probably acts as the site of gas exchange between the atmosphere and the symbiont.

The symbiont alga was named *Anabaena azollae* Strasburger, but Fjerdingstad [7] recently claimed that the alga is actually an ecoform of *Anabaena variabilis*. The isolation of *Anabaena azollae* from azolla has been frequently reported. But none has succeeded in the re-inoculation of the isolated symbiont to alga-free azolla and nothing is known about the free-living state of this symbiont. *Anabaena azollae* was claimed to be associated with the microsporocarps and megasporocarps. It is possible that *Anabaena* in young azolla seedlings originated from the algal cells in megasporocarps.

During the differentiation of dorsal lobe primordia, the cavity occupied by the symbiont is created by epidermal cell growth. Several algal cells sheltered in the shoot apex are entrapped by the enclosing epidermal cells and colonized the cavity [17]. The hair-like cell of plant origin is seen in the youngest cavity. Algal cells in the youngest lobes do not have heterocysts. The frequency of heterocysts increases to about 30% in the 15th leaf as the lobes are traced back from the shoot apex to the basal parts. After the 20th leaf, the heterocysts begin to be senescent [12, 13]. The N_2-fixing ability of each leaf is proportional to the frequency of heterocysts in each leaf.

The frequency of heterocysts is higher in symbiotic alga in azolla than in free-living heterocystous blue-green algae. Hair-like cells are seen in all cavities, which are believed to be the site material exchange between the host plant and the alga [26]. The interaction between both partners is still poorly understood. The growth of azolla and alga may be synchronized. The alga-free azolla has also the cavity and hair-like structure.

The N_2-fixing ability of the symbiont is demonstrated in acetylene reduction assay and $^{15}N_2$. The endophyte algae which are mechanically isolated from the fern have the ability to fix $^{15}N_2$ or to reduce acetylene, although the rate is lower than in association with the fern [23, 24, 25, 27].

The isolated alga excretes about half of the fixed N_2 as ammonium. Because most of the ammonia-assimilating enzyme glutamine synthetase are found in the fern parts of the association, the algae in the association provide fixed N_2 to the host plant mainly as ammonia and the plant converts it to amino acids [25]. N_2 fixation is associated with photosynthesis. Evidences show that the blue-green algae in the association catch solar energy and use this for nitrogenase reaction [28]. It is not known if the energy and reductant necessary for nitrogenase reaction are partly provided by the fern.

The N_2-fixing system in the fern—alga association is not strongly inhibited by ammonium, nitrate, and urea. Azolla growing in 2.5 mM ammonium can maintain its growth rate similarly in its absence and N_2 fixation is inhibited by approximately 30%. N_2-fixing activity is quickly recovered upon the transfer to N-free medium (Ito and Peters, unpublished). The N_2-fixing system in the endophyte may be, by some unknown mechanisms, protected from the inhibitory action of combined nitrogen. This is a unique nature of the *Azolla—Anabaena* symbiosis unlike the legume—rhizobium symbiosis which is more sensitive to combined nitrogen.

3. Environmental factors affecting growth and N_2-fixing activities

The growth rate, maximum biomass, and N_2-fixing activity in optimum conditions provide the estimate of the potential of *Azolla—Anabaena* symbiosis for agricultural use.

In optimum light and temperature conditions in the laboratory, Peters *et al.* [29] obtained about 2.0 days doubling time or less for *A. filiculoides, A. caroliniana, A. mexicana,* and *A. pinnata.* Doubling time of 2 days corresponds to 0.34 g^{-1} day^{-1}. Growth in the liquid medium by Talley and Rains [39] with *Azolla filiculoides* and Watanabe *et al.* [42] with *Azolla pinnata* showed that the maximum growth rate was about 2.5 days doubling time.

N_2-fixing rate is estimated by the relative growth rate and nitrogen content of the fern. Assuming 0.277 daily relative growth rate (2.5 days doubling time) and 4% N content in dry matter, daily N_2-fixing rate is calculated as 11.0 mg N g^{-1} dwt. Assuming 4:1 electron ratio of N_2 fixation to acetylene reduction and 12-hr light period a day, this N_2-fixing rate corresponds to 130 μmol C_2H_4 g^{-1} dwt h^{-1}. The reported value of ARA (acetylene reduction activity) fluctuated from 20 to 200 [4].

The growth curve of azolla approximately follows a logistic curve until the biomass reaches the maximum. The growth is characterized by the initial growth rate and the maximum biomass. The growth rate is retarded as the plant density increases [2].

The maximum biomass or nitrogen accumulation reported by researchers is summarized in Table 2. Because *Azolla filiculoides* grows upward from the water surface and forms a thick mat, the maximum biomass is higher than that of other azolla species. No data on the maximum biomass of *A. nilotica,* which forms 10—15 cm long stem, are available. The maximum daily N_2-fixing activity of *A. filiculoides* per unit area was about 2.8 kg N ha^{-1}, whereas the maximum daily N_2-fixing rate of *A. pinnata* grown in open paddy fields in the Philippines was 3.1 kg N ha^{-1} (Watanabe, unpublished). Average daily N_2-fixing rates, measured from

Table 2. Maximum biomass and average N_2-fixing rate

Species	Condition	Maximum biomass		Days	Average N_2-fixing rate ($kg\ ha^{-1}\ day^{-1}$)	References
		Dry matter ($kg\ ha^{-1}$)	N content ($kg\ ha^{-1}$)			
A. filiculoides	Fallow paddy, USA	1700	52	35	1.5	Talley and Rains [38]
	Shallow pond, USA	1820	105	?		Talley et al. [37]
	Paddy soil in pots	5200	128	50	2.6	Tuzimura et al. [41]
	Fallow paddy, USA	2300	93	46	2.0	Talley and Rains [39]
A. mexicana	Ponds, USA	830	39	39	1.0	Talley et al. [37]
	Paddy field, USA	1100	38	?		Talley and Rains [38]
A. pinnata	Fallow paddy, Philippines	900–1200	48	30–25	1.9–1.6	Watanabe (unpublished)
	Phytotron					
	26°(d)/18° C(n)	2170	96	37	2.6	Watanabe (unpublished)
	33°(d)/25° C(n)	1500	33	22	1.5	Watanabe (unpublished)
	37°(d)/29° C(n)	1120	30	23	1.3	Watanabe (unpublished)
	(var africana)	640	26	15	1.8	Roger Reynaud [31]
	greenhouse					
A. filiculoides	26° (d)/18° C(n)	3200	126	51	2.5	Watanabe (unpublished)
A. caroliniana	26° (d)/18° C(n)	3190	146	41	3.6	Watanabe (unpublished)

d = day; n = night

inoculation to harvest, are presented in Table 2. The values fluctuate from 1.0 to 2.6 kg N ha^{-1}. Watanabe *et al.* [44] reported that 26 crops of azolla yielded 450 kg N ha^{-1} for 330 days in an open paddy field. Singh [34] reported an annual production of 333 ton fresh weight ha by weekly harvest and estimated annual nitrogen production at 840 kg N ha^{-1}. Shen *et al.* [33] reported 93–152 kg N ha^{-1} for 45 days.

From these figures, the high potential of azolla as a N$_2$-fixing crop is easily realized. The fixing rate is almost comparable to the figure of forage legumes [22].

Among environmental factors affecting the growth and N$_2$-fixing activity of azolla, temperature, light, humidity, and mineral elements are described.

3.1. Temperature

The optimum temperature of *A. pinnata, A. mexicana,* and *A. caroliniana,* when grown in constant temperature under 15 klux artificial light [29] is about 30 °C. *A. filiculoides* requires 25 °C. The response of nitrogenase activity to temperatures ranging from 10° to 42 °C also shows that *A. filiculoides* likes lower temperature than *A. pinnata* [4]. Although *A. pinnata* is widely distributed in the tropics, it grows better in cooler seasons. In northern Vietnam, the growth of *A. pinnata* is best in January when the average air temperature is 17 °C. In Varanasi, India, *A. pinnata* grew from July to December but was absent from the ponds [11] in hot summer (April to June). In southern China, azolla grows most abundantly from February to May. In the Philippines, the growth is poorest in April and May when monthly average temperature exceeds 32 °C [44]. Watanabe *et al.* [43] reported that in controlled temperature with 8 °C differences between day and night, the growth of *A. pinnata* in culture solution was about the same at 26 °C (day)/18 °C (night), 29 °C/21 °C, and 32 °C/24 °C, but was reduced by about 50% at 35°/27 °C (average 31 °C). The maximum biomass is more adversely affected by higher temperature than the growth rate at low plant density [42]. *A pinnata* dies progressively at temperatures higher than 40 °C and lower than 5 °C. Most of the experiments to examine the response of the fern to temperature were conducted at constant day temperature or without the shift of temperature from day to night. *A. filiculoides* could not grow at 40 °C (day)/30 °C (night) temperature. However, if the ferns are grown at lower temperatures and then subjected to a step-wise increase in temperature simulating dawn to midday of the diurnal cycle, the nitrogenase activity increases with temperature up to 40 °C and remains high at 45 °C. Similarly, *A. filiculoides* continues to fix N$_2$ in the field during hot (40–45 °C) afternoons [39]. Temperature response varies with light intensity [1]. The lower the temperature, the lower the optimum light intensity for the growth of nitrogenase [2, 39]. The temperature response is also dependent on the source of nitrogen [2]. In the tropics, the poor growth of *A. pinnata* in hot summers (average monthly temperature exceeding 30 °C) is a problem to be overcome for the agricultural use of the fern–alga association.

3.2. Light

Short periods of exposure experiments on various light intensities showed that light saturation to nitrogenase is about $250 \mu E \, m^{-2} \, sec^{-1}$ (20 klux) by Talley and Rains [39], 5 klux by Peters [23] and 10 to 5 klux by Watanabe [42]. For long-term experiments, the fern requires higher light intensity than in the short-term exposure experiments, because the growing fronds overlap each other. Ashton [2] observed that the growth rate of *A. filiculoides* increased with increasing light intensity to a maximum in 50% sunlight (49 klux), but further increase of light intensity retarded the growth rate. Talley and Rains [39] however, did not observe retardation of growth of *A. filiculoides* under artificial light when light intensity was increased from $500 \mu E \, m^{-2} \, sec^{-1}$ to 1000 (ca. 80 klux) when the temperature during illumination was higher than $25 \, ^{\circ}C$. The apparent discrepancy may be due to the plant density, temperature, and light source. Although shading reduces not only light intensity but also temperature of water and air during sunny midday, experiences tell that shading is beneficial for the growth of *A. pinnata* during hot summer. *A. pinnata*, *A. mexicana*, and *A. caroliniana* have been observed to turn red in strong sunlight and remain green in shading.

3.3. Humidity

Optimum relative humidity is reported to be 85–90% [47]. At a relative humidity lower than 60%, azolla becomes dry and fragile and more susceptible to adverse condition.

3.4. Mineral requirements

The mineral composition of azolla is summarized in Table 3. Although reported figures vary greatly due to excess uptake, it would be reasonable to assume that macronutrient contents in their sufficiently supplied levels are as follows (in percent to dry weight): N: 4–5, P: 0.5, K: 1.0–2.0, Ca: 0.5, Mg: 0.5, Fe: 0.1.

In batch cultures of azolla in nutrient solution the levels of nutrients to induce mineral deficiency symptoms are reported [41]. Yatazawa *et al.* [45] carefully examined the threshold concentration of macroelements in nutrient solution by using the inoculum that was precultured in the nutrient-deficient solution. Threshold concentrations for *A. pinnata* growth were 0.03, 0.04, 0.4 and 0.5 mmol liter^{-1} for P, K, Mg and Ca, respectively.

Phosphorus-deficient azolla turns reddish brown, fronds become fragile, and roots are elongated and easily detached. *Anabaena* cells become pale green and deformation of vegetative cells and heterocysts occur. The browning of dorsal lobes starts from the newer leaves in *A. mexicana*, but from the older leaves at the basal parts of stems in other species. Reddening of dorsal lobes is more intense in

Table 3. Mineral composition of *Azolla*

Species	Conditions	Percent in dry matter						References
		N	P	K	Ca	Mg	Fe	
A. filiculoides	Water culture	–	0.79	6.5	0.25	0.30	–	Tuzimura et al. [41]
A. filiculoides	Soil culture	–	0.95	1.6	1.0	0.36	–	Tuzimura et al. [41]
A. filiculoides	Naturally grown	4.5	0.5	2.0	0.97	–	0.1	Buckingham et al. [5]
A. pinnata	Naturally grown	4.5	0.5	24.5	0.4	0.5	0.06	Singh [34]
			–0.9		1.0	–0.6	0.26	
A. filiculoides	Naturally grown	2	0.1	2.0	0.6	0.4	0.3	Lumpkin and Plucknett [19]
		–3.4	–0.4	2.6	–0.8	–0.6	0.5	
A. pinnata	Water culture	5.0	1.0	3.2	0.2	0.7	0.08	Watanabe (unpublished)
	(deficient)	–	(0.08)	(0.4)	(0.05)	–	(0.016)	

calcium-deficient azolla, than in phosphorus-deficient plants. The fronds turn fragmented and algal cells are lost from the cavity in the extreme Ca-deficient plant.

In potassium-deficient plants, yellowish browning also occurs. In iron-deficient azolla, chlorophyll content decreases and the plant turns yellowish [43].

Effect of microelements are also reported by Yatazawa et al. [45] and Johnson et al. [16].

Azolla grows in aquatic habitats and absorbs nutrients mainly from the water. In shallow water the plant root attaches to the soil and plant absorbs nutrients from the soil. Because phosphorus content in soil solution or in paddy water is generally too low to meet the requirement by azolla, the addition of phosphorus is necessary for better growth of azolla [44].

To determine the minimum level of phosphorus in water medium, continuous flow culture was used. At 2 μmol P 1^{-1}, *Azolla pinnata* from Bangkok grew normally, but at 1 μmol P 1^{-1}, the azolla suffered phosphorus-deficiency symptoms [35]. The growth of various species and strains of azolla and the difference in supporting growth at 1 μmol 1^{-1} were observed and compared (Subudhi and Watanabe, unpublished).

4. Agricultural use of Azolla

A book on Chinese agricultural techniques — the essence of feeding people, written in 540 A.D. — describes the cultivation and use of azolla in rice fields. At the beginning of the 17th century (end of Ming dynasty), many local records reported the use of azolla as manure [18].

Azolla used in China originated in Fujiang and Guangdong and spread to rice fields in other provinces, south of the Yantze River. Since the foundation of the People's Republic of China, azolla techniques as green manure for rice and as animal feed have been greatly encouraged and have spread north of the Yantze River. In double cropping of rice in central and southern China, azolla is grown only before the early rice (March—May).

In northern Vietnam, A. *pinnata* has been used as a green manure crop for centuries. It is believed that a peasant in La Van village of Thai Binh province discovered and domesticated the azolla. Before the revolution, some families in the village knew the techniques of rearing azolla starter from April to November. The villagers began selling starter stocks of azolla to regional propagators at high prices [21]. Nowadays, the maintenance and multiplication of starters, and the further propagation from the starters, are systematically conducted and distributed from regional azolla multiplication centers to farmers' cooperatives. Azolla is grown in January and February in paddy fields before spring rice and then incorporated [40].

5. Green manure for rice

Because of its high N_2-fixing ability and nitrogen content, azolla has high potential as green manure crop for wetland rice. It is grown either before or after transplanting rice.

One crop of *A. pinnata* contains about 20–40 kg N ha^{-1}. A Vietnamese study showed that incorporation of 1 ton fresh azolla increased the average rough rice yield by 28 kg in 1958–1967 [6]. If 20 t fresh weight of azolla ha^{-1} is produced before transplanting (this figure is reasonable) about 0.5 t ha^{-1} rice yield increase will be obtained.

In 1975, a conference on azolla for southern China summarized findings from 1,500 experiments in 7 provinces (Chiangxu, Guandong, Fujien, etc). Azolla as manure increased rice yield by 600–750 kg ha^{-1}. Ninety percent of 422 field experiments in Chekiang Province reported an average increase in rice grain yield of 700 kg ha^{-1} or 18.6% [18].

Lately, interest in azolla as green manure for rice was resumed in other south Asian countries. The International Rice Research Institute organized INSFFER (International Network of Soil Fertility and Fertilization Evaluation for Rice) collaborative network activity to test the effect of azolla as green manure. Scientists from 5 countries joined the network and field experiments were conducted at 12 sites in 1979. The results are summarized in Table 4. Positive responses of azolla incorporation either before or after transplanting rice over no nitrogen control were obtained in 10 sites. Growing of azolla before or after the rice produced a rice yield increase equivalent to that obtained from 30 kg N ha^{-1} as urea or ammonium sulphate. Growing of azolla and its incorporation before and after rice increased

Table 4. Effects of azolla and nitrogen fertilizer on rice yield, International Network of Soil Fertility and Fertilizer Evaluation for Rice (INSFFER) trials[a]

Treatment	Average grain yield (t ha^{-1})	Index
1 No nitrogen	2.6	100
2 30 kg N ha^{-1} chemical fertilizer	3.2	122
3 60 kg N ha^{-1}	3.7	141
4 Azolla grown before transplanting, incorporated	3.2	122
5 Azolla grown after transplanting, incorporated	3.1	118
6 Azolla grown after transplanting	3.1	119
7 30 kg N ha^{-1} + Azolla before transplanting, incorporated	3.7	143
8 30 kg N ha^{-1} + Azolla after transplanting, incorporated	3.5	134
9 Azolla grown before and after transplanting, incorporated	3.6	139

[a] Conducted in 8 sites in Thailand, 2 in India, and 1 each in China and Nepal

rice yield equivalent to that obtained from 60 kg N ha^{-1} as chemical N fertilizer. Whether azolla was incorporated or not after it covers fully the paddy surface in the rice canopy did not affect rice yield.

An average N_2-fixing activity of $1-2$ kg N ha^{-1} day^{-1}, which is shown by the *Azolla–Anabaena* complex, is sufficient to meet the nitrogen requirement of rice if azolla is grown for the period of one rice cropping.

By widening the distance between rice rows, azolla was grown continuously in the rice canopy [18]. This technique was examined at the International Rice Research Institute. Wide rows (53 cm) were alternated with narrow rows (13 cm). Distance between hills was 6.6 cm. Azolla was grown six or four times and incorporated into the soil after water was drained. A total of $100-70$ kg N ha^{-1} were contained in the incorporated azolla. Grain yield was almost equivalent to that obtained from $70-100$ kg N ha^{-1} chemical nitrogen fertilizer (Watanabe, unpublished).

The decomposition of azolla is rapid and nitrogen efficiency of azolla is almost comparable or slightly inferior to that of urea or ammonium sulphate [34, 43]. Principally, azolla is grown in the fields, where rice is grown after the azolla harvest or together with azolla as described above. Alternately, azolla is grown continuously year round in paddy fields or in the adjacent ponds. Excess of azolla, after it has been used for rice, can be composted for dryland or vegetable crops.

6. Management practices

The inoculum of azolla must be healthy and fresh. It is continuously multiplied in the inoculum preparation plots or ponds. The inoculum density is an important factor in the efficient production of azolla. Singh [34] recommends 2 tons fresh weight ha^{-1} as the inoculum size. In Vietnam, 5 tons ha^{-1} or more is recommended. When inoculum density is low, azolla is overgrown by algae and weeds.

In Vietnam, the half-saturation method is recommended. The saturated density of *A. pinnata* is about $10-20$ t fresh weight ha^{-1}. First, the available inoculum is spread in an area to keep the density at 0.5 kg fresh weight m^{-2}. After one week, the surface is fully covered. Then, half of the azolla is transferred to the open area which has about the size as the area where the azolla was taken (see Fig. 3). After one week or so, both areas are fully covered by azolla. Again, half of the azolla is taken from both areas and transferred to the field about the size of the area where the azolla was taken. By repeating this procedure, the area covered by azolla is exponentially expanded. Azolla is notably responsive to phosphorus fertilizer and requires a continuous supply of water-soluble phosphorus for rapid propagation. Split application of superphosphate is more efficient in promoting azolla growth than basal application and 1 kg P_2O_5 ha^{-1} every 4 days is recommended [44]. The Vietnamese recommended $5-10$ kg superphosphate ($1-2$ kg P_2O_5 ha^{-1}) every 5 days. Singh [34] recommended $4-6$ kg P_2O_5 ha^{-1} every week. Superphosphate

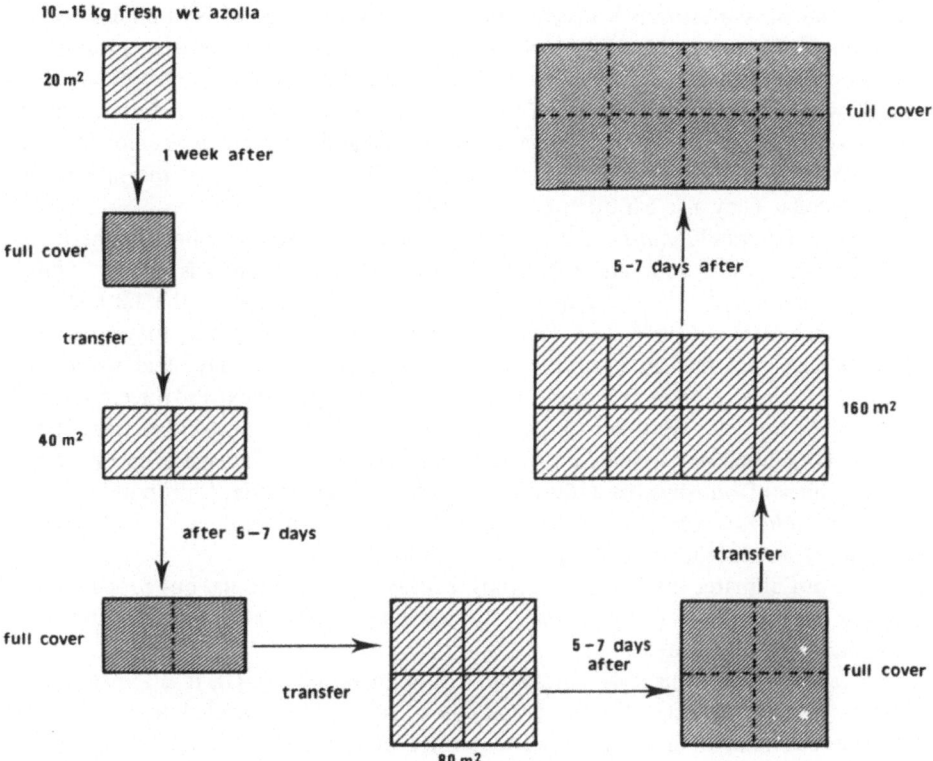

Fig. 3. Schematic of the half-saturation method for *A. pinnata.*

must be applied on the surface of azolla. In China, 3% (w/w) suspension of super-phosphate with $0.2\,l\,m^{-2}$ is recommended.

In a suitable condition each kg of P_2O_5 results in 2 kg additional nitrogen in the azolla biomass [44].

The application of potassium is recommended in light mineral soils. Talley *et al.* [37] obtained significant azolla growth response from a single application of 0.8 kg Fe ha^{-1} as ferric chelate. When soil is flooded before azolla inoculations, iron from the reduced layer becomes available to azolla growth on floodwater.

Water control is critical. Water depth should be kept at about 3–5 cm. When water depth is less than 3 cm, the roots attach to the soil and mineral deficiency is recovered. In hot summer, keeping the water shallow to allow azolla roots to attach to the soil is effective in maintaining healthy azolla. When azolla is incorporated, water is drained and azolla is easily turned under by rake.

Insect damage is serious, particularly in hot summer (higher than 30 °C), because generation time of insects decreases at higher temperature.

Tran and Dao [40] identified the main insect pests of azolla in Vietnam as larvae of *Pyralis, Nymphula,* and *Chironomus* species. In China these genera are also the major and most destructive insect pests [18]. Two species of Nymphula, brown Nymphula (*N. tarbata*) and black Nymphula (*N. swindol*), are important. Adults of brown Nymphula have yellow brown wings and those of black Nymphula have black wings. Their eggs are laid on the back side of ventral lobes and are hatched after 4 or 6 days at 30°C. The larvae attack azolla and form burrows embedded by the silk and fragmented azolla leaves and roots. They come out of the sack while eating azolla. They also eat other aquatic plants.

The gray pyralid-*Pyralis* sp. lays whitish oval eggs on the periphery of the space between the dorsal and ventral lobes. The newly hatched larva is grayish white, but becomes grayish blue or green in later stage. The duration of the 1st and 2nd instar stage at about 30°C is 3 days. At the end of the 2nd instar, they construct burrows and eat the shoot of the fern. Pupae live for 4–5 days and so do the adults. Total development takes 20 days (Lin Shi-he, Guangxi Province, personal communication). In southern China, this pyralis is the most destructive one. There are 8–10 generations a year, the generations overlap. The damage is easily visible by the lines of burrows. In addition to *Pyralis* and *Nymphula*, Chironomid (*Polypedilium, Microspector, Tendipes,* smout beetle (*Bagous* sp.), and snail (*Radix swinhoer*) are recorded as pests in southern China.

In a hot climate, the damage by fungus attack is also serious, but probably the fungus attack follows the damage by insects or desiccation. *Sclerotium* and *Rhizoctonia* are recorded [32].

Because insect damage is serious in a hot climate (> 30°C), it is a major constraint in growing azolla in the tropics. A close watch is needed to predict the sudden outbreak of insect pests. The surface of azolla mats should be carefully examined for the presence or traces of insect burrows. The collecting lamp is also useful for the prediction of lepidopterous insects.

The application of 3–4 kg a.i. carbofuran or ferithion per ha is effective, particularly when the application is made with phosphorus fertilizer. Dipping the inoculum in 1000–3000 times diluted insecticide solution is also effective.

In China, bacterial insecticides have also been tried to control azolla pests [18].

7. Other uses of azolla

Azolla can act as a weed suppressor, because the azolla mat covering the surface depresses the growth of weeds.

Rains and Talley [30] reported that early development of *A. filiculoides* eliminated *Cyperus difformus* and *Polygonum* species from the paddy, but not *Echinochloa crusgalli* which is taller. Total biomass was reduced by the azolla cover, although *Echinochloa* predominated. In the International Rice Research Institute's fields where azolla has been continuously grown, sporadic growth of *Monochoria vaginalis* (non-submerged weed) is noticed.

In Hawaii, taro growers used *A. filiculoides* as weed suppressor [8]. But there are opinions that azolla is harmful to rice. Fujiwara *et al*. [9] mentioned that a Japanese farmer who used azolla as green manure observed that a thick mass of azolla covered the young shoot of rice when water was flooded, resulting in the death of rice seedlings. The azolla mat at the early stage of rice, when temperature is still cool, lowers water temperature, resulting in the slight depression of tillering. In direct-seeded rice, quick growth of azolla may be harmful for young rice seedlings.

Azolla has also been used as food to fish and domestic animals in Asia and Africa, because it is rich in protein [36]. It is fed to cattle, pigs, ducks, and fish. Buckingham *et al*. [5] analyzed the nutritional value of *A. filiculoides* and concluded that protein in azolla had a low nutritive value for growing rat. Addition of lysine, methionine, and histidine was effective in improving quality. The high (39%) neutral fibre content of azolla was a major limiting factor in the efficient use of azolla as protein source for monogastric animals. *In vitro* digestability data showed it would be a useful source for ruminants. However, the azolla samples were taken from the creek in an urban area and nothing was described about the age of the fern. Fibre content of their samples was much higher than that reported by Fujiwara *et al*. [9]. Change of nutritive value according to the age of azolla needs to be examined. Singh [34] also suggested the possibility of azolla as human food.

References

1. Ahmad, Ghias-ud-din. 1941 Effect of light intensity and temperatures on the growth of *Azolla filiculoides*. J. Indian Bot. 20, 213–226.
2. Ashton, P.J. 1974 Effect of some environmental factors on the growth of *Azolla filiculoides* Lam. In: Orange River Progress Report, pp. 124–138. Inst. for Environmental Sci., Univ. O.F.S. Bloemfontein, South Africa.
3. Ashton, P.J. 1977 Factors affecting the growth and development of *Azolla filiculoides* Lam. In: Proc. Second Natl. Weed Conf. South Africa, Cape Town, pp. 249–268.
4. Becking, J.H. 1979 Environmental requirements of *Azolla* for use in tropical rice production. In: Nitrogen and Rice, pp. 345–373. Int. Rice Res. Inst.
5. Buckingham, K., Ela, S.W., Morris, J.G. and Goldman, C.R. 1978 Nutritive value of nitrogen-fixing aquatic fern *Azolla filiculoides*. J. Agric. Food. Chem. 26, 1230–1234.
6. Dao, T.T. and Do, A. 1970 A few ideas on equating different kinds of nitrogenous fertilizer with pure nitrogen. Agric. Sci. Technol. Vietnam 93: 168–174 (in Vietnamese).
7. Fjerdingstad, E. 1976 *Anabaena variabilis*, status azollae. Arch. Hydrobiol. Suppl. 49. Algol Studies 17, 377–381.
8. Fosberg, F.R. 1942 Use of Hawaiian ferns. Am. Fern. J. 32, 15–23.
9. Fujiwara, A., Tsuboi, I. and Yoshida, F. 1947 Atmospheric N_2-fixing activity of *Azolla*. Nogaku 1, 361–363 (in Japanese).
10. Godfrey, R.K., Reinert, G.W. and Houk, R.D. 1961 Observations of microscopic material of *Azolla caroliniana*. Am. Fern. J. 51, 89–92.
11. Gopal, B. 1967 Contribution of *Azolla pinnata* R. Br. to the productivity of temporary ponds at Varanasi. Trop. Ecol. 8, 126–129.

184

12. Hill, D.J. 1975 The pattern of development of *Anabaena* in the *Azolla*—*Anabaena* symbiosis. Planta 122, 179—184.
13. Hill, D.J. 1977 The role of Anabaena in the *Azolla*—*Anabaena* symbiosis. New Phytol. 78, 611—616.
14. Hills, L.V. and Gopal, B. 1967 *Azolla primaeva* and its phylogenetic significance. Can. J. Bot. 45, 1179—1191.
15. International Rice Research Institute. 1979 International Bibliography on *Azolla*, 66 pp.
16. Johnson, G.V., Mayeux, P.A. and Evans, H.J. 1966 A cobalt requirement for symbiotic growth of *Azolla filiculoides* in the absence of combined nitrogen. Plant Physiol. 41, 852—855.
17. Konar, R.N. and Kapoor, R.K. 1972 Anatomical studies on *Azolla pinnata*. Phytomorphology 22, 211—233.
18. Liu, C.C. 1979 Use of Azolla in rice production in China. In: Nitrogen and Rice, pp. 375—394. Int. Rice Res. Inst.
19. Lumpkin, T.A. and Plucknett, D.A. 1980 *Azolla*: botany, physiology and use as a green manure. Economic Bot. 34, 111—153.
20. Moore, A.W. 1969 *Azolla*: biology and agronomic significance. Bot. Rev. 35, 17—34.
21. Nguyen Cong Tieu. 1930 The *Azolla* plant cultivated for use as a green fertilizer. Bull. Econ. Indochine 33, 335—350 (in French).
22. Nutman, P.S. 1976 IBP field experiments on nitrogen fixation by nodulated legumes. In: Symbiotic nitrogen fixation, pp. 211—237. Nutman, P.S. (ed.), Cambridge Univ. Press.
23. Peters, G.A. 1975 The *Azolla*—*Anabaena azollae* relationship III. Studies on metabolic capabilities and a further characterization of the symbiont. Arch. Microbiol. 103, 113—122.
24. Peters, G.A. and Mayne, B.C. 1974 The *Azolla*—*Anabaena azollae* relationship II. Localization of nitrogen fixing activity as assayed by acetylene reduction. Plant Physiol. 53, 820—824.
25. Peters, G.A., Toia, Jr., R.E. and Lough, S.M. 1977 The *Azolla*—*Anabaena azollae* relationship V. $^{15}N_2$ fixation, acetylene reduction and H_2 production. Plant Physiol. 59, 1021—1025.
26. Peters, G.A., Toia, Jr., R.E., Raveed, D. and Levine, N.J. 1978 The *Azolla*—*Anabaena azollae* relationship VI. Morphological aspects of the association. New Phytol. 80, 583—593.
27. Peters, G.A., Mayne, B.C., Ray, T.B. and Toia, Jr., R.E. 1979 Physiology and biochemistry of the *Azolla*—*Anabaena* symbiosis. In: Nitrogen and Rice. pp. 325—343. Int. Rice Res. Inst.
28. Peters, G.A., Ray, T.B., Mayne, B.C. and Toia, Jr., R.E. 1980 *Azolla*—*Anabaena* association: morphological and physiological studies. In: Nitrogen fixation, Vol. 2, pp. 293—309. Newton, W.E. and Orme-Johnson, H.H. (eds.), Univ. Park Press, Baltimore.
29. Peters, G.A., Toia, Jr., R.E., Evans, W.R., Christ, D.K., Mayne, B.C. and Poole, R.E. 1980 Characterization and comparisons of five N_2-fixing *Azolla*—*Anabaena* associations. I. Optimization on growth conditions for biomass increase and N content in a controlled environment. Plant Cell Env. 3, 261—269.
30. Rains, D.W. and Talley, S.N. 1979 Use of *Azolla* in north America. In: Nitrogen and Rice, pp. 419—433. Int. Rice Res. Inst.
31. Roger, P.A. and Reynaud, P.A. 1979 Premières données dur l'écologie d'*Azolla africana* en zone Sahelienne. Oecol. Plant. 14(1), 75—84.

32. Shahjahan, A.K.M., Miah, S.A., Nahar, M.A. and Majid, M.A. 1980 Fungi attack azolla in Bangladesh. Int. Rice Res. Newsletter (IRRI): 5(1), 17—18.
33. Shen, Z.H., Leu, S.E., Chen, K.Z. and Gi, S.A. 1963 A preliminary study of the nitrogen fixation of *Azolla pinnata*. Turang Tongbao (Pedology Bulletin) Peking 4, 45—48 (in Chinese).
34. Singh, P.K. 1979 Use of *Azolla* in rice production in India. In: Nitrogen and Rice, pp. 407—418. Int. Rice Res. Inst.
35. Subudhi, B.P.R. and Watanabe, I. 1979 Minimum level of phosphate in water for growth of *Azolla* determined by continuous flow culture. Curr. Sci. 48, 1065—1066.
36. Subudhi, B.P.R. and Singh, P.K. 1978 Nutritive value of the water fern *Azolla pinnata* for chicks. Poult. Sci. 57, 378—380.
37. Talley, S.N., Talley, B.J. and Rains, D.W. 1977 Nitrogen fixation by *Azolla* in rice fields. In: Genetic Engineering in Nitrogen Fixation, pp. 259—281. Hollander, A. (ed.), Plenum Pub. Co.
38. Talley, S.N. and Rains, D.W. 1980 *Azolla* as a nitrogen source for temperate rice. In: Nitrogen Fixation, Vol. 2, pp. 311—320. Newton, W.E. and Orme-Johnson, W.H. (eds.), Univ. Park Press.
39. Talley, S.N. and Rains, D.W. 1980 *Azolla filiculoides* Lam as a fallow season green manure for rice in a temperate climate. Agron. J. 72, 11—18.
40. Tran, Q.T. and Dao, T.T. 1973 Azolla a green compost. Vietnamese studies, Agric, Problem 38, 119—127.
41. Tuzimura, K., Ikeda, F. and Tukamoto, K. 1957 *Azolla imbricata* as a green manure for rice. Nippon Dojo Hiryo Gaku Zasshi (J. Sci. Soil Manure) 28, 275—278 (in Japanese).
42. Watanabe, I. 1978 Azolla and its use in lowland rice culture. Tsuchi to Biseibutsu (Soil and Microbes), Tokyo 20, 1—10.
43. Watanabe, I., Espinas, C.R., Berja, N.S. and Alimagno, B.V. 1977 Utilization of the *Azolla—Anabaena* complex as nitrogen fertilizer for rice. Int. Rice Res. Paper Ser. No. 11.
44. Watanabe, I., Berja, N.S., Rosario, D.C. del 1980 Growth of *Azolla* in paddy field as affected by phosphorus fertilizer. Soil Sci. Plant Nutr. 26, 301—307.
45. Yatazawa, M., Tomomatsu, N., Hosoda, N., Nunome, K. 1980 Nitrogen fixation in *Azolla—Anabaena* symbiosis as affected by mineral nutrient status. Soil Sci. Plant Nutr. 26, 415—426.
46. Yu, L.H. 1979 Preliminary observations on sexual reproduction of *Azolla*. Zhejiang Nongye Kexue, No. 4, 19—22 (in Chinese).
47. Zhejiang Academy Agric. Sci. 1975 Cultivation, propagation and utilization of *Azolla*. Agriculture Publ. Beijing, 127 pp. (in Chinese).

7. Denitrification in rice soils

J.L. GARCIA and J.M. TIEDJE

1. Introduction

Denitrification is defined as the reduction of nitrate or nitrite to gaseous nitrogen (usually N_2O and N_2) and is usually catalyzed by bacteria. Though other organisms are known to produce N_2O [119], denitrifiers are thought to be those in which the reduction of nitrogenous oxides is coupled to electron transport phosphorylation thereby providing energy to the cell. Most of the nitrate used by denitrifiers is reduced to gas which is not the case for other N_2O producing organisms [111, 118]. The pathway of denitrification is thought to be $NO_3^- \rightarrow NO_2^- \rightarrow [Enz{-}NO] \rightarrow N_2O \rightarrow N_2$. Whether free NO is an intermediate is still in dispute, but evidence is now stronger for at least an enzyme bound NO intermediate [32, 50]. Denitrifiers often excrete the intermediates NO_2^- and N_2O, but significant production of NO is rarely reported. Perhaps owing to the renewed interest in denitrification a number of recent reviews have been prepared on the microbiological, biochemical, physiological, and ecological aspects of denitrification [25, 26, 30, 34, 35, 37, 43, 63, 67, 79, 80, 81, 87]. This text is focused only on those aspects of denitrification of relevance to rice culture.[*]

Denitrification is of interest in paddy rice culture because nitrogen-fertilizer losses from this cropping system are well known. Abichandani and Patnaik [1] estimated these losses to be 20 to 40% in India, while losses of 30 to 50% were reported by Mitsui [74] in Japan. From a recent review of a number of ^{15}N balance studies by Craswell and Vlek [20], it appears that N losses in these ranges are typical. In the past it was reasoned that denitrification was the most significant loss mechanism, but ammonia volatilization and leaching losses have recently been shown to be more significant in some situations. Even in the non-fertilized rice paddy, which still is the major practice in the world, denitrification losses apparently occur. Wetselaar [132], in his summary and evaluation of N budget of traditional Asian wet season culture, estimated denitrification losses of 3 to 34 kg N ha^{-1} yr^{-1} with an average of 18.5 kg N lost ha^{-1} yr^{-1}[**]. When all nitrogen inputs were maximum and outputs were minimum (except denitrification), the maximum estimated denitrification loss became 129 kg N ha^{-1} yr^{-1}. This latter value is unlikely because high inputs would be from nitrogen fixation which would

---

[*] The literature review for this chapter was completed in February, 1981.
[**] In his estimates NH_3 volatilization, leaching, predatory harvest, overflows, plant harvest, and wind erosion were separately evaluated, and thus not lumped with the denitrification loss.

Y.R. Dommergues and H.G. Diem (eds.), Microbiology of Tropical Soils and Plant Productivity. ISBN 978-94-009-7531-6.
© *1982 Martinus Nijhoff/Dr W. Junk Publishers, The Hague/Boston/London.*

not likely lead to high denitrification because of the slow transfer of biologically fixed nitrogen to nitrate. In these same estimates he calculated a net gain of 50 kg N ha^{-1} crop^{-1} in these steady-state agro-systems. His average denitrification loss is equivalent to 37% of the nitrogen accretion. Thus even in the N-stressed, non-fertilized system, denitrification losses appear to be significant and should translate into a loss in productivity.

2. Environmental factors affecting denitrification

The major environmental factors that control denitrification are *oxygen*, adequate *energy sources* (i.e., available organic matter), and a supply of *nitrate* [30]. Oxygen is the preferred electron acceptor for denitrifiers; its presence both inhibits the activity and prevents synthesis of the denitrifying enzymes [79, 86]. Increasing available carbon [49, 97] and nitrate both favor denitrification. However, high organic matter can also cause immobilization [19] thereby reducing denitrification.

Rice soils generally have two of these major environmental factors controlling denitrification — limited oxygen due to flooding, and available organic matter released from the decomposing straw. Thus the major limiting factor is the supply of nitrate. Nitrate fertilizers are no longer used for this obvious reason. Ammonium produced by mineralization or from fertilizers must be oxidized by nitrification before nitrate is available to denitrification. Nitrification is an obligately oxygen requiring process so it can only occur in aerobic environments. In rice culture, these are: (1) the aerobic layer at the surface of the soil, (2) the aerobic zone surrounding the oxygen excreting rice root, and (3) the aerobic period caused by intermittent flooding. The first two environments and the reactions which describe the rice paddy nitrogen cycle are diagrammed in Fig. 1.

For a better understanding of denitrification and other nitrogen losses from paddy soils, it is essential to understand this unique habitat. Figure 1 is meant to aid this understanding. Paddy soils are characterized by two distinct layers [2, 74, 85, 122, 137]: an oxidized (aerobic) surface layer generally a few millimeters to 1 to 2 cm deep. The depth depends on the amount of organic matter, since this results in oxygen consumption by respiring microorganisms, and by the structure of the soil which influences the rate of diffusion of oxygen into the mud. Below the aerobic layer is a reduced (anaerobic) layer. If ammonium fertilizer is applied to the surface layer, it is first nitrified (k_3) and the nitrate or nitrite produced then can diffuse down (k_4) where it is reduced by denitrification (k_7) to N_2O and N_2. A similar aerobic–anaerobic interface exists around the rice roots (Fig. 1, inset) which can also result in denitrification as is discussed later. The alternate wetting and drying conditions create an ideal environment for denitrification. The nitrate formed during the dry period is rapidly lost through denitrification when the soil is reflooded and a stimulation of decomposition occurs [83, 94, 102].

Factors other than oxygen, energy, and nitrate which influence denitrification rates are temperature, pH, salinity, and perhaps iron concentration. Denitrifying

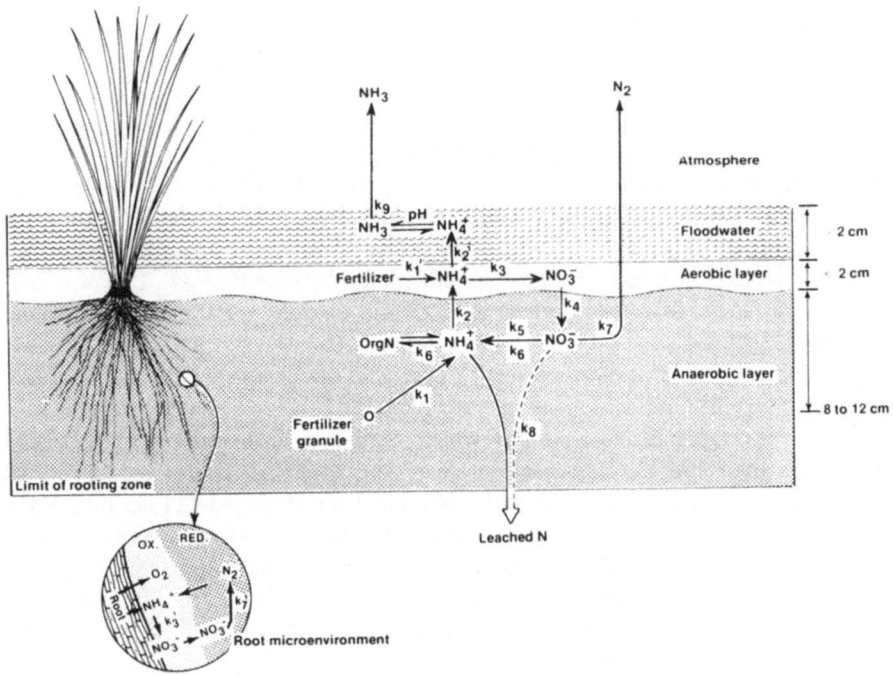

Fig. 1. Diagram of paddy rice environment which emphasizes the three major losses of nitrogen (N_2, NH_3 and leaching), the two sites of denitrification — the rhizosphere and just below the aerobic layer — and the preceeding mechanisms which control these losses. The rates depicted are: k_1, release of NH_4^+ from fertilizer; k_2, diffusion of NH_4^+ to aerobic layer; k_3, nitrification; k_4, diffusion of NO_3^- to anaerobic layer; k_5, dissimilatory reduction of NO_3^- to NH_4^+; k_6, immobilization reactions. The loss mechanisms are: k_7, denitrification; k_8, percolation rate of N; k_9, volatilization rate of NH_3. k_n, refers to the same reaction but in a different microenvironment.

activity is strongly correlated with temperature in soil. It is negligible at 3 to 10 °C but the rate of nitrate reduction rises with temperature to reach a maximum at 65 °C and ceases at 85 °C [78]. In rice soil Garcia [41] showed the existence of two maxima during N_2O reduction which were 37 and 65 °C; the higher maxima is no doubt due to the existence of thermophilic *Bacillus* species [46].

The rate of denitrification is low in acidic conditions and rises with pH to reach an optimum activity in the neutral range [78]. In the acid sulfate rice soils, which are prevalent in certain areas of the world, denitrification has been shown to be slow with 70% of 100 ppm NO_3^-–N still remaining after one week of anaerobic incubation [10]. Salinity appears to depress denitrification [49]. In these studies a low correlation was found between the denitrifying activity and the initial number of denitrifying bacteria in Senegal rice soils. In preliminary studies by Sexstone and Tiedje (unpublished data), ferrous iron in concentrations as low as 10^{-4} *M* inhibited the overall denitrification sequence by *Pseudomonas fluorescens*, a common denitrifier in nature.

Environmental factors have also been shown to have a large influence on the N gases produced. Increasing concentrations of NO_3^- [9, 33, 78], NO_2^- [33], O_2 [8. 33, 78], S^{2-} [115, 117] increase the proportion of N_2O produced. Decreasing pH [33, 39, 70, 78, 133] and temperature [65, 78] also increase the N_2O production. However a decrease in N_2O is caused by more available carbon [78] and a longer anaerobic period [31]. In fact, most soils become sinks for N_2O if they are anaerobic for more than 1 to 2 days [31]. This is particularly significant for rice since it suggests that rice paddies should not evolve much N_2O. This was confirmed in a recent field study where a maximum of only 1.4% of the NO_3^--N was emitted as N_2O [24].

NO is rarely found in nature except under acid conditions where it is produced from chemical decomposition of nitrite [77, 125]. It is known, however, that sulfide causes some increase in NO which originates from biological reduction [115]. Furthermore, at the aerobic—anaerobic interface in marine sediments (which are high in sulfide), substantial NO has been detected [113]. In rice soil where there is sulfate intrusion or ammonium sulfate fertilizers are heavily used, one might expect some NO to be produced. Garcia [45] has reported some NO to be biologically produced by rice soils but the role of sulfide in its production was not investigated. Because of the chemical reactivity of NO, however, its life in the paddy soil should not be long.

3. Denitrifying bacteria

A number of methods for enumerating denitrifying bacteria have been reported and are summarized elsewhere [119]. Most of these are based on the most probable number (MPN) technique [36, 40, 124, 129]. Recent reevaluation of the MPN technique in Tiedje's laboratory has shown that some false positive tubes can occur due to nitrate and nitrite removal by ammonium producing organisms and that false negative tubes also occur due to minute bubbles in Durham tubes and/or failure to totally remove the nitrate and nitrite. To avoid these problems a modified method has been recommended in the Revised Methods of Soil Analysis [119]. This method recommends using a medium of 5.0 mM NO_3^- (instead of 9.9) in nutrient broth to reduce the number of false negatives. To avoid false positives, a confirmatory procedure is suggested in which acetylene is added and tubes with N_2O concentrations over 20% are judged as positive for denitrifiers. The latter procedure requires more time and equipment and is probably not necessary when comparative results within the same experimental design are intended.

Comprehensive lists of genera of denitrifying bacteria are compiled in recent reviews [30, 63, 80]. According to Gamble *et al.* [38], *Pseudomonas fluorescens* biotype II and the 'alcaligenes-like' group are the most commonly encountered numerically dominant denitrifiers. Virtually all recognized denitrifying bacteria are strict aerobes and are incapable of growing anaerobically by fermentative means. Recently, however, many denitrifying strains of the genus *Bacillus* have been isolated from rice soils in Senegal and are able to ferment carbohydrates [46]. All of these organisms tolerate high concentrations of nitrite during growth and some can use NO as the electron acceptor for growth [46, 88]. They were isolated in a complex medium with $5 \, gl^{-1}$ yeast extract and $10 \, gl^{-1}$ trypticase. A thermophilic denitrifying *Bacillus* was also isolated [46]. The taxonomic study of these *Bacillus* isolates demonstrates that they are a new and unique group of this genus [88, 89]. Perhaps there are other fermentative denitrifiers. Indirect evidence suggests that *Propionibacterium pentosaceum* produces N gas from nitrite [123]. Furthermore, Kaspar *et al.* have noted denitrifying activity in deep lake sediments, anaerobic sludge with long retention times and in a continuous culture of a methanogen growing on acetate [64]. In these habitats devoid of O_2 and NO_3^- survival could only be by fermentative growth.

4. Methods of analysis of denitrification in soils

Early studies of denitrification depended on manometric assays [71] since this was the only means to measure the gaseous products. But soil samples had to be dried before introduction into respirometer vessels which produced biological modifications, and the method was not sensitive enough to measure natural denitrification rates.

Mass spectrometry has also been employed to trace the fate of ^{15}N fertilizer [20]. The classical approach has been to establish a ^{15}N balance for the soil and to assume the losses are due to denitrification [53]. A more recent approach by Rolston *et al.* [99] was to measure denitrified gas flux directly from a field soil. Following addition of $^{15}NO_3^-$ to microfield plots $^{15}N_2$ and N_2O flux were measured to establish a field denitrification rate. This is the only reported direct measurement of denitrification in the field; unfortunately it is rather insensitive as rates below 1 kg N ha^{-1} day^{-1} cannot be measured. Thus it is doubtful that this method is directly applicable to denitrification studies on rice. The disadvantage of ^{15}N for many researchers working on rice is that the equipment and isotope are expensive and the technology rather complicated. Nonetheless, ^{15}N balance studies remain the best method for field assessment of nitrogen losses.

Gas chromatography was quickly recognized to be a key method for studies on denitrification since it is simple, rather inexpensive and precise, and repetitive analyses can be made. The key development was by Hollis [55] in 1966 when he introduced the porous polyaromatic beads for chromatographic separation of nitrogen, oxygen, nitric oxide, nitrous oxide, and carbon dioxide. Helium is usually used as the carrier gas and thermoconductivity wires or microthermisters for detectors [67]. Barbaree and Payne [7] devised methods for using columns packed with these beads in their assays for products of denitrification. Subsequently, gas chromatographs equipped with helium ionization [27, 108] ^{63}Ni electron capture

[100, 107], and ultrasonic detectors [8a] have been shown to measure N_2O concentrations of 1 ppm (v/v) or less. These methods were made even more useful with the discovery in 1973 of Fedorova et al. [29] who showed that acetylene blocked the reduction of N_2O to N_2. Balderston et al. [6] and Yoshinari and Knowles [138] used this method to inhibit denitrification at the N_2O stage and showed that it could be used as an assay for rate of denitrification [66, 112, 139].

During a study of the sequential products of denitrification in vitro in Senegal rice soils, Garcia [39] noted that the rate of formation and reduction of N_2O was directly related to the rate of nitrate reduction and to the soil organic carbon content. This lead the author to develop a new method to estimate the rate of denitrification in soil based on the rate of reduction of a supplied quantity of N_2O [91]. This gaseous substrate can be introduced into a water-saturated soil without previous drying of soil. A battery-operated portable chamber was devised for use in the field for incubations of soil samples at 37 °C in 250 ml serum flasks immediately after collecting the sample. At regular intervals, gas samples are withdrawn into evacuated gas vials. These samples are returned to the laboratory for analysis of the remaining N_2O by gas chromatography. Calculations are based on the N_2O: Kr ratio [44].

The constant rate of N_2O reduction observed during the first few hours represents the initial denitrifying activity present in soil at the time of sampling. After 4 to 6 hours the rate increases sharply to a new constant rate and is indicative of the final denitrification capacity [41, 47]. A similar result was obtained by Smith and Tiedje [110] using the acetylene inhibition method for the assay of denitrification. The existence of the two distinct phases of denitrification was confirmed in the two laboratories by the use of chloramphenicol. The activity measured by these methods after long incubation represents only a small fraction of the potential activity in the soil because the electron acceptor and sometimes the electron donor becomes rate limiting [49]. The limitation of the N_2O-reduction method is that nitrate must not be present as a competing electron acceptor [9]. But in waterlogged rice soils the level of nitrate is generally very low. This method also suffers from the classical limitations of requiring complete anoxia and high levels of electron acceptor, but it has the advantage over the nitrate disappearance method in that it reflects only dissimilatory nitrogen transformations.

The acetylene inhibition method at least initially appeared to offer more promise for field measurements of denitrification in rice paddy soils. A gasing apparatus such as the one developed by Ryden et al. and used to measure denitrification rates in irrigated California soils looked promising for field studies [100]. The most serious limitation to the acetylene approach is the finding that acetylene also inhibits nitrification [56, 130] which prevents further nitrate formation and thus accurate measurements of denitrification. A lack of complete inhibition of N_2O reduction by acetylene has also been noted on a few occasions [136, 117]. This seems to be a function of the intensity of microbial activity and can often be overcome by adding more acetylene to be in proportion to microbial activity [64]. As a further caution, it has been shown that high sulfide can relieve the acetylene inhibition [117], and that acetylene can be decomposed in anaerobic soils at fairly high rates (up to $100\,\mu\text{mole}$ decomposed by 20 g soil in 4 days) [131]. Anaerobic oxidation of acetylene has recently been confirmed in sediments and was shown to be metabolized to acetate and then converted to methane [23]. These findings make it clear that adequate inhibition of N_2O reduction by acetylene must be carefully demonstrated for each habitat.

Field methods using acetylene which are more appropriate for the rice paddy appear to be those developed for aquatic studies. Chan and Knowles [15] devised a chamber which could be inserted into sediment and enclosed the overlying water. The enclosed water was circulated by pump. Acetylene (to yield 5 to 10% final concentration) and ethane (1%, internal standard) were injected into the pumped waterline to distribute the hydrocarbons. Water samples were withdrawn periodically to measure the change in N gases and ions. When nitrate was added N_2O accumulation could be measured, but in the absence of this amendment N_2O accumulation did not occur except to a limited extent in one case where there was a natural measurable nitrate concentration. The acetylene inhibition may not have been complete since only 14 and 6% of the added nitrate was recovered as N_2O-N. At first thought this type of approach would seem applicable to measurement of denitrification in rice paddies. However, as Chan and Knowles found, N_2O accumulation was undetectable or very low. The likely problem is that in habitats with extremely low nitrate pools, any denitrification requires prior nitrification which would be inhibited by the acetylene treatment.

Another approach which reduces this problem by being a more short-term *in situ* measurement is the direct injection technique used by Sørensen *et al.* [114] in marine sediments. They injected $100\,\mu\text{l}$ of acetylene-saturated water at 1 cm intervals in small sediment cores. The cores were incubated and then extracted and the accumulated N_2O measured by gas chromatography. They were able to measure denitrification rates, but this process was a significant fraction of the total electron flow only in the winter when nitrate concentrations were higher in the interstitial water. This technique should also be applicable to rice soils, but it is likely to work only if denitrification is not intimately linked to the nitrate supply from nitrification.

The last method is the use of the radioactive isotope of nitrogen, ^{13}N, which has a half-life of 10 min. Gersberg *et al.* [52] and Tiedje *et al.* [120, 121] have used ^{13}N in short-term denitrification studies. The ^{13}N is usually generated by irradiation of water with 12 to 15 meV proton beams from a cyclotron. A gas chromatograph-proportional counter was developed to separate and measure ^{13}N-N_2, ^{13}N-N_2O and ^{13}NO, and a radio-high pressure liquid chromatograph was developed to separate and measure $^{13}NO_3^-$, $^{13}NO_2^-$ and $^{13}NH_4^+$ [120]. ^{13}N has the advantage in denitrification studies of allowing direct measurement of N_2, is very sensitive, and allows isotope exchange experiments [32]. One need only add ^{13}N at femtogram quantities (68 fg = 0.1 mCi) for an experiment; this should not alter the indigenous nitrate concentration even in anaerobic habitats with extremely low nitrate concentrations. Gersberg used this technique with slurries of rice soils [52] and found denitrification rates of 1 to $6 \mu g$ N liter^{-1} hour^{-1} which was about 10 times higher than the rates he reported for oligotrophic lake sediments [51]. The need for highly sophisticated equipment and for production of the isotope at the site of its use due to its short half-life, limits the general application of this method. Despite these limitations, ^{13}N holds some promise for denitrification studies in habitats such as rice where the nitrate pool is low and its turnover rapid. A recent summary of all ^{13}N denitrification work and methods has been prepared by Tiedje et al. [121].

5. Rice rhizosphere effect on denitrification

In non-submerged soil it is well known that denitrification activity in the rhizosphere is stimulated by plant roots. Woldendorp [134, 135] obtained evidence that rates of organic carbon excretion and consumption of oxygen by roots and microorganisms could be sufficient to create microenvironments in which denitrification would occur. Smith and Tiedje [109] later confirmed that denitrification enzyme concentration increased with distance to roots. Interestingly, however, they showed that plants could decrease denitrification compared to unplanted soil if nitrate was limiting. If nitrate was high, however, plants stimulated denitrification as had been shown before. This suggests that plants can compete well with denitrifiers for limiting nitrate.

In the case of rice, however, it is not known how this different environment of anaerobic soil and an oxygen-excreting root (Fig. 1, inset) affects denitrification. Another marked difference in the rice rhizosphere is the very high concentration of ferrous iron which precipitates as FeOOH forming iron coatings representing up to 14% of the root dry weight [16, 17]. Since Fe^{3+} and Fe^{2+} both show some inhibitory effect on denitrification and N_2O reduction in lab studies [68, 105], it is conceivable these ions could also have an inhibitory effect on denitrification in the rice rhizosphere. The question of denitrification in the rhizosphere was approached by examining the potential denitrifying activity and the number of denitrifiers in paddy soils from Senegal. In these studies Garcia [40] showed that rice has a stimulatory effect on these parameters. This stimulation is restricted to the thin layer of soil adhering to the roots and is especially evident in soils low in organic carbon. Using the N_2O reduction method a positive rhizosphere effect was demonstrated for 3-week old rice seedlings which had been grown in 15 g waterlogged soil in glass tubes [42]. The study was made during the period of active photosynthesis and presumably oxygen exudation by comparing induced activities of N_2O reductase in planted and nonplanted soils. This effect may be attributed to the development of anaerobic zones, the presence of root exudates and the large number of denitrifiers in the rhizosphere. These results were confirmed by use of a simple device developed to study metabolic gases in the rice rhizosphere [92].

The positive effect of the rice rhizosphere on denitrification was also confirmed with experiments made in rice fields in Senegal [44] and in pot cultures [44, 48]. The rhizosphere effect on potential denitrifying activity was maximum in the first stages of growth and diminished progressively with age of plant.

In the rice rhizosphere the nature of fertilizers or the manner of application have a lesser effect than for nonplanted soil [48] as was previously shown by De and Digar [28] and by Broadbent and Tusneem [11]. They demonstrated that rice culture diminished the losses of nitrogen compared to nonplanted soil. This could result from NH_4^+ absorption by rice plants before it was oxidized and lost through denitrification.

6. Description of the components of the nitrogen cycle in paddy rice (Fig. 1).

6.1. Ammonium volatilization (k_9)

This loss is considered in this chapter on denitrification because some of the nitrogen loss reported in the past has likely been due to ammonia volatilization. This loss has only recently been quantified. Ammonium volatilization has been shown to be first order (k_9) with respect to NH_3 (aq) (ammonia concentration in the flood water) [127]. Basically NH_3 will diffuse into an atmosphere as long as NH_3 (aq) is present in flood water. The rate of volatilization increases with wind as well as with any factor which increases the ammonia in solution such as fertilizer input $(k_1,$ and $k_2)$, temperature and pH [10]. pH acts according to the equation [127]

$$NH_4^+ \rightleftharpoons NH_3 (aq) + H^+ \rightarrow NH_3 (air)$$

The floodwater is largely buffered by the bicarbonate system and usually does not closely reflect the soil pH [126]. The greatest effect on the pH is due to algal photosynthesis and respiration [73], $nCO_2 + nH_2O \rightleftharpoons (CH_2O) n + nO_2$. During active photosynthesis, CO_2 (aq) is depleted driving the pH values up to 8.5 to 10. At night respiration produces an excess of CO_2 forcing the pH down to 6.5 to 7.5. It is not uncommon for pH values to change two to three pH units per daily cycle. The consequence of this is that during high pH periods the NH_3 (aq) concentration greatly increases thereby causing a proportional increase in loss by ammonium volatilization [73, 126]. Losses as high as 20 to 50% of fertilizer N have been reported due to volatilization. Factors such as water chemistry, form and location of fertilizer, and plant cover which affects wind directly above the water surface and light reaching the algae are the major parameters affecting this loss [10, 126, 127].

In some nitrogen-fertilizer experiments in the past algal growth has not been controlled and thus it is difficult to determine how important ammonia losses were to the poor N recovery. One pot experiment where ammonia was trapped, a significant amount of the N loss was accounted for by the trapped ammonia [126]. Since denitrification losses still cannot be measured directly, it is necessary that future N-budget experiments accurately control or assess ammonia volatilization if N losses are to be equated to denitrification losses.

6.2. Leaching losses (k_8)

As for ammonia volatilization, leaching losses are sometimes not adequately assessed and can contribute to the N loss in budget studies. In two recent studies these losses were evaluated. Vlek *et al.* [128] reported that leaching losses in lighter texture soils for deep placed ^{15}N-labeled fertilizer were substantial if percolation rates were high. The losses were particularly high for the supergranules (10% loss for 4.4 mm water/day to 88% loss for 18.3 mm water/day). The leach form was ammonium. These authors recommended not to use deep placement of supergranules if percolation rates exceeded 5 mm/day particularly if the cation-exchange capacity of the soil is low. In another study, Rao and Prasad [90] found less than 20% of the N leached from several fertilizers placed in experimental pots and subjected to the same leaching rate as found in the field for the Delhi region. About half of the N leached from ammonium fertilizers in this study was in the nitrate form. By use of nitrification inhibitors they achieved a modest reduction in the amount of N leached.

It is not clear how important leaching losses might have been in past field and pot nitrogen-budget studies. Since these losses are more easily recognized and tend to be smaller under the typical study conditions than for ammonia volatilization, they probably have not been major contributors to the lack of nitrogen recovery.

6.3. Immobilization (k_6) and dissimilatory (k_5) reduction of nitrate to ammonium

These reactions are of interest because they compete for nitrate with the two nitrate loss mechanisms and thereby conserve nitrogen. Assimilatory and dissimilatory nitrate reduction both produce ammonium but the latter occurs at a faster rate since it is linked to electron transport while the former is slower since it parallels growth. The assimilatory process is inhibited by ammonium and the dissimilatory one by oxygen [18]. In flooded rice soils the ammonium concentration often exceeds 50 ppm which is usually considered sufficient to shut off the assimilatory process [18]. Furthermore, many organisms capable of the dissimilatory process are common in soil [14, 111, 118]. Therefore, it is Tiedje's assumption that the dissimilatory mechanism is the dominant one responsible for any nitrate reduced to ammonium in the rice habitat.

The reduction of $^{15}NO_3$ to $^{15}NH_4^+$ + organic-N (org-N) in rice soils has been studied by several authors and the percentages of ^{15}N found in the reduced form range from a few percent to 39% in the absence of a carbon amendment [12, 19, 69]. Additions of available carbon do substantially enhance the reduction but rice straw, which is more slowly degraded, resulted in only 6% reduction to ammonium [12]. However, non-amended rice soil preincubated anaerobically 20 days, which should be similar to the natural rice habitat resulted in 24% reduced N of which 20.5% was free ammonium [12]. Higher amounts of conversion to reduce nitrogen have been achieved by increasing the density of clostridia (83%) [14], a prevalent soil organism capable of this process, by preincubating with glucose and adding extra glucose (70%) [12], or by maintaining the redox at −200 mv (42%) [13]. Typically during the reduction, free ammonium accumulates and then drops to lower levels as the ammonium is converted to organic nitrogen. However in the studies above where high yields were obtained, much of the reduced nitrogen remained as free ammonium throughout the experiment.

All of the above studies have been made with anaerobic soil slurries to which nitrate was added; this does not reflect the way in which nitrate would naturally enter the anaerobic zone (Fig. 1) since the gradient from high to low redox was avoided. In a very recent study Buresh and Patrick [13] added nitrate periodically to aerobic water overlying estuarine sediment. In this way nitrate was allowed to diffuse into the anaerobic zone much like it should in nature. They found 15% of the $^{15}NO_3^-$ recovered as NH_4^+ whereas 28% was recovered as NH_4^+ if the NO_3^- was directly added to the sediment. Since stronger reducing conditions favor the reduction to ammonium, these results suggest that as nitrate enters the transition zone to the anaerobic layer, the ammonium-producing reaction is less favored compared to denitrification. The generally low recoveries of nitrate in the ammonium form in rice soils and the probably inhibitory effect of the transition zone causes us to conclude that this fate of nitrate is generally minor. It does remain an intriguing process, however, since one could perhaps manage this competition to better favor the conserving reaction over denitrification.

6.4. Other transfers of nitrogen (k_1, k_2, k_3, k_4)

k_1 refers to the rate of release of N from the fertilizer form applied. Most commercial forms have a rapid release but there is special interest for rice culture in use of coatings or materials with low dissolution rates to slow this reaction. These fertilizers will be discussed in the last section. Rates k_2, k_3, k_4 can be discussed as a group since they are integrated precurser steps for denitrification. Patrick, Reddy and Phillips first emphasized the importance of considering each of the sequential steps as a unit to evaluate controls on denitrification losses. Very recently the same authors have been able to obtain experimental values for each of these constants for several rice soils in the Louisiana and Mississippi rice-growing regions [96]. Their rates were: NH_4^+ diffusion (k_2) 0.059 to 0.216 cm^2 day^{-1}, NH_4^+ oxidation (k_3) 1.2 to 3.5 μg g^{-1} day^{-1}, NO_3^- diffusion (k_4) 0.96 to 1.91 cm^2 day^{-1}, and NO_3^- reduction (first order, $k_{5,6,7}$) 0.315 to 0.520 day^{-1}. A quantitative study focused on mechanisms such as this one allows one to evaluate which steps are rate limiting so that management strategies can be soundly developed. In this case the rate-limiting steps were ammonium diffusion into the aerobic layer and nitrification. Nitrate diffusion and reduction were both more than 10 times faster than the above.

7. Fertilization and management of paddy rice

In a recent historical review on denitrification Mitsui [75] noted that there has been an enormous number of field studies to minimize this loss since the time when losses attributed to denitrification in paddy soils were first recognized by Shioiri. In earlier periods the recommendation was to apply nitrogen to the dry field and plow it under or mix it with the furrow slice. The irrigation water was introduced soon after; loss of urea by leaching was less than the loss due to nitrification (which occurred before irrigation) and subsequent denitrification. Subsequently it was recognized that deep placement of ammonium fertilizers in flooded soils further improved nitrogen efficiency [1, 3, 54, 60, 72, 74, 82, 95, 116]. The reason for this should be apparent from Fig. 1; with deep placement losses are restricted to leaching or by diffusion of ammonium to the surface (k_2). Though the rate constants were not known then, intuition suggested that this rate should be slow in puddled soils, as has now been proved. This strategy was feasible because rice plants have long been recognized to assimilate ammonium forms of nitrogen. A second approach to controlling losses has been to reduce the rate of release of ammonium from the applied material (k_1 or k_1) [103]. This has been attempted by use of coatings and low solubility materials. A third approach has been to incorporate nitrification inhibitors with the fertilizer in order to reduce rate k_3. The fertilizer materials that have been evaluated most recently are summarized in Table 1;

Table 1. Most common fertilizers recently evaluated for paddy rice culture

Readily available
    AS (ammonium sulfate)
    GU (granular urea)

Large granule for deep placement
    USG (urea supergranule, 1 to 3g)
    MBU (mudball urea)

Slow release
    SCU (sulfur-coated urea)[a]
    SPCU (silicate polymer-coated urea)
    SC-USG (sulfur-coated urea supergranule)
    IBDU (isobutylene diurea)[b]

Nitrification inhibitors incorporated with fertilizer
    NSU (N-Serve blended urea)[c]
    NCU (Neem cake-coated urea)[d]

[a] Other less successful or inexpensive coatings which have been investigated are plastics, waxes, resins
[b] Other materials with low solubility products are ureaform and oxamide
[c] Other commercial nitrification inhibitors exist [54], principally thiazoles, triazines and thioureas, but none seem as promising as N-Serve
[d] Neem cake is the most effective natural nitrification inhibitor; it is obtained from the kernels of the neem tree (*Azadirachta indica*) which is a native, widely distributed tree in India and other semi-arid tropical regions [90]. Neem is coated on urea by use of coaltar

they are grouped according to the strategies which are employed to reduce the rate constants and thus the losses. There can also be combinations of these strategies to further reduce losses, e.g. deep placement of SC-USG [21]. The results from testing of some of these strategies are further described below.

The initial difficulty with deep placement was how to accomplish it in a practical manner. An early approach was to pack urea in a peat or mudball which then could be inserted by hand to the desired depth. Experiments with ^{15}N-labeled peat ball fertilizer showed that deep placement of nitrogen at the panicle formation stage resulted in 86% of the nitrogen being taken up by the rice plant; surface broadcasting gave only 50% nitrogen uptake [75]. A second approach was mechanized deep placement of liquid ammonia or liquid fertilizers as was used in Japan. Also, fertilizer applicators attached to engine-driven cultivators, and deep-point applicators of paste fertilizer attached to engine-driven transplanters have been evaluated [75]. In 1974 IRRI [60] recommended enclosing concentrated urea in small mudballs and placing them about 10 cm deep in the soil beside newly transplanted seedlings. This process produced a yield of 8 t ha^{-1} of rice with 60 kg of nitrogen applied per hectare compared with 6.6 t ha^{-1} using 100 kg N ha^{-1} when applied as a topdressing. However, so many mudballs are needed (62,500 ha^{-1}) that the technique is probably feasible only in areas where land holdings are small and labor is plentiful [61].

IRRI [60] has also developed inexpensive, manually-operated machines to apply both fertilizer and insecticides. The chemicals are applied in a band below the soil surface between the new seedlings. To eliminate the process of making mudballs the International Fertilizer Development Center (IFDC), at Muscle Shoals, Alabama, USA, developed supergranules and briquets of varying quantities of urea. When tested at IRRI briquets equalled mudballs in efficiency of fertilizer recovery by the plant [62]. Another N-application method is mixing of urea with moist soil and incubating the mixture for 24 hours before application so losses can be limited by converting the released nitrogen to NH_4^+ [91]. At IRRI, foliar application of urea in split doses was also tried but the results were not favorable [91].

A slow release nitrogen fertilizer developed by the Tennessee Valley Authority, sulfur-coated urea (SCU), has been tested in rice culture and is a promising fertilizer for the tropical regions. The advantages of SCU have been more obvious in soils with intermittent flooding and drying [91]. Anjaneya [4] obtained equivalent yields with 100 kg SCU-N ha^{-1} and 150 kg urea-N ha^{-1}. Similar results were obtained at the IRRI Center [62] with SCU at 84 kg N ha^{-1} compared to split applications of urea at 120 kg N ha^{-1} as part of the study of the International Network on Fertilizer Efficiency for Rice (INFER). From 1975 to 1977, 84 trials were conducted in 11 countries to obtain an insight into the relationship between N source, N management, and N efficiency under a variety of environmental conditions. In a recent summary of INSFFER (International Network on Soil Fertility and Fertilizer Efficiency in Rice), and other comparative studies, Craswell and Vlek [22] report that use of supergranules resulted in a significant increase in rice yield over that with broadcast prilled urea in 42% of the studies and was slightly better or equal in most of the rest.

With measurement of rates of N_2O reduction used for estimating the denitrifying capacity of potted rice-growing soils enriched with urea applied on the surface or at depth, or with SCU applied one time on the surface before transplanting rice seedlings, Garcia [48] showed that the best grain yield was obtained with deep placement of urea; with SCU the yield was equivalent to that obtained with non-coated urea spread in greater quantities and at two stages of growth. Similarly, other slow release nitrogen fertilizers like oxamide and IBDU were also found to increase the yield of rice by preventing gaseous losses [62, 93].

Losses of nitrogen by denitrification are also reduced if the nitrification of ammonium is prevented. A number of chemicals are known to inhibit nitrification and the two most popular are N-Serve [2-chloro-6-(trichloromethyl) pyridine] and AN (2-amino-4-chloro-6-methyl pyrimidine). Increase yields of rice by the use of N-Serve and AN have been reported [84, 93, 137]. Other studies have shown slight to moderate reduction in N loss by use of N-Serve [90, 102, 106] or neem-coated fertilizer [106]. The effectiveness of inhibitors in the field is not well established, however. A 22% increase in rice grain yield due to treatment of urea with an acetone extract of neem kernels was obtained in India [91]. The cake, seed, bark, and leaf of Karanj (*Pongamia glabra*) also possess nitrification-inhibiting properties [19]. Some chemicals have been reported to specifically inhibit denitrification, such as pesticides [5. 76], the azides and some aromatic ketonitriles [58, 59] but these products are expensive and not particularly promising for agricultural use.

Very recent studies on the most promising N-fertilizer materials have provided quantitative data on yield and nitrogen recoveries and losses. Some of these results are summarized in Table 2. The trends are clear and consistent. Yields and nitrogen

Table 2. Summary of recent studies done on the effectiveness of different fertilizer materials

Fertilizer (Ref)	Grain yield			Fertilizer-N recovered in plant			NH_3 volatilized [126]	N leached [90]	N lost [90] [c]
	[21]	[90]	[106] [a]	[21] [b]	[90]	[106] [b]			
	g pot^{-1}		g ha^{-1}	% of applied			%	%	%
Basal urea (BU)	38	19	44	28	25	27	48	11.5	43
Split urea	47	–	–	48	–	–	–	–	–
Basal $(NH_4)_2SO_4$	47	21	–	54	31	–	17	13.7	31
Basal-SCU	40	24	46	56	40	32	<3	7.1	15
USG (deep)	45	30	–	69	60	–	3	12.6	18
SC-USG (deep)	52	–	–	84	–	–	–	–	–
BU + N-Serve	–	26	49	–	41	38	–	9.1	20
BU + Neem	–	23	49	–	38	40	–	9.8	25

[a] Field experiment

[b] Recovery is of ^{15}N label

[c] Since label was not used losses are sum of denitrification, NH_3 volatilization, and any net immobilization; leaching was accounted for in previous column

recovery by plants go up and losses go down for materials employing slow release, deep placement, and/or inhibitors. The best material seems to be the combination of deep placement and slow release, the sulfur-coated urea supergranule (SC-USG). In other studies Savant and De Datta [103, 104] have shown that the release of ammonium from sulfur-coated materials was much slower than from USG or prilled urea. The apparent diffusion coefficient of ammonia from USG was 2.2×10^{-2} cm^2 day^{-1} versus 1.6×10^{-3} cm^2 day^{-1} for SCU [104]. As a result soil ammonium concentrations increased for four weeks for the sulfur-coated materials where other materials showed an ammonium maximum immediately and then declined. From this study it appears that the reduction of k_1 results in yield improvement (Table 2). The second benefit of the SC-USG fertilization method is that it did not suppress growth of blue-green algae nor their acetylene reducing activity [98]. Fertilizers that release ammonium into the floodwater favor green algae which do not fix N_2.

Though denitrification activity of paddy soils has been widely demonstrated by gas chromatographic methods, quantitation of actual N losses by this mechanism has not been reported. In Table 2, the last column on N losses indirectly suggest denitrification is important since N-Serve reduced losses from urea-N by 23%; this value may be taken as an estimate of denitrification loss since in this study leaching and volatilization have been separately accounted for.

8. Conclusion

To minimize N losses slow-release nitrogen fertilizers are promising for the tropical regions. The objective of these fertilizers is to make the amount of N released coincide with the nitrogen requirement of growing plants, especially the tillering and heading stages, and thereby reducing N losses. There are two ways to achieve this objective: (1) use of chemicals with slow rates of dissolution such as IBDU. Field experiments in Japan showed that IBDU produced 20% more rice than ammonium sulfate applied in two split doses [91], (2) coating of conventional fertilizers to reduce their dissolution rates; of these products SCU has been the most widely tested for rice. The response to SCU was generally superior to the response to urea and to split applications of N [22].

Before new fertilizers can be recommended, the economics must show that the new technique is worth the extra cost. Average data for INFER trials showed that the extra yield resulting from use of SCU more than covered the 60% greater cost of this material. Most recent results suggest that the combination of controlled release especially by sulfur-coating and deep placement should substantially improve the efficiency of fertilizer nitrogen. For readers interested in the more agronomic aspects of efficient fertilizer use in rice crops, they are referred to the recent and comprehensive reviews by Craswell and Vlek [20, 22].

202

References

1. Abichandani, C.T. and Patnaik, S. 1955 Mineralizing action of lime on soil nitrogen in waterlogged rice soils. Int. Rice Comm. Newslett. 13, 11–13.
2. Alberda, T. 1953 Growth and root development of lowland rice and its relation to oxygen supply. Plant and Soil 5, 1–28.
3. Alecksic, Z., Broeshart, M. and Middelboe, V. 1968 Shallow depth placement of $(NH_4)_2SO_4$ in submerged rice soils as related to gaseous losses of fertilizer nitrogen and fertilizer efficiency. Plant and Soil 29, 338–342.
4. Anjaneya, S.D. 1974 Relative efficiency of fertilizer nitrogen with and without nitrification inhibitor on IR8 rice. Andhra Agric. J. 19, 139–144.
5. Azad, M.I. and Khan, A.A. 1968 Reduction of nitrogen losses through denitrification from paddy soil by the application of pesticides. West Pak. J. Agr. Res. 6, 128–133.
6. Balderston, W.L., Sherr, B. and Payne, W.J. 1976 Blockage by acetylene of nitrous oxide reduction in *Pseudomonas perfectomarinus*. Appl. Environ. Microbiol. 31, 504–508.
7. Barbaree, J.M. and Payne, W.J. 1967 Products of denitrification by a marine bacterium as revealed by gas chromatography. Mar. Biol. 1, 136–139.
8. Betlach, M.R. and Tiedje, J.M. 1981 A kinetic explanation for accumulation of nitrite, nitric oxide, and nitrous oxide during bacterial denitrification. Appl. Environ. Microbiol. 42 (in press).
8a. Blackmer, A.M. and Bremner, J.M. 1977 Gas chromatographic analysis of soil atmospheres. Soil Sci. Soc. Amer. J. 41, 908–912.
9. Blackmer, A.M. and Bremner, J.M. 1978 Inhibitory effect of nitrate on reduction of N_2O to N_2 by soil microorganisms. Soil Biol. Biochem. 10, 187–191.
10. Bouwmeester, R.J.B. and Vlek, P.L.G. 1981 Rate control of ammonia volatilization from rice paddies. Atmos. Environ. 15, 131–141.
11. Broadbent, F.E. and Tusneem, M.E. 1971 Losses of nitrogen from some flooded soils in tracer experiments. Soil Sci. Soc. Amer. Proc. 35, 922–926.
12. Buresh, R.S. and Patrick, W.H. 1978 Nitrate reduction to ammonium in anaerobic soil. Soil Sci. Soc. Amer. J. 42, 913–918.
13. Buresh, R.J. and Patrick, W.J. 1981 Nitrate reduction to ammonium and organic nitrogen in an estuarine sediment. Soil Biol. Biochem. (in press).
14. Caskey, W.H. and Tiedje, J.M. 1979 Evidence for Clostridia as agents of dissimilatory reduction of nitrate to ammonium in soils. Soil Sci. Soc. Amer. J. 43, 931–936.
15. Chan, Y.K. and Knowles, R. 1979 Measurement of denitrification in two freshwater sediments by an *in situ* acetylene inhibition method. Appl. Environ. Microbiol. 37, 1067–1072.
16. Chen, C.C., Dixon, J.B. and Turner, F.T. 1980 Iron coatings on rice roots: morphology and models of development. Soil Sci. Soc. Amer. J. 44, 1113–1120.
17. Chen, C.C., Dixon, J.B. and Turner, F.T. 1980 Iron coatings on rice roots: mineralogy and quantity influencing factors. Soil Sci. Soc. Amer. J. 44, 635–639.
18. Cole, J.A. and Brown, C.M. 1980 Nitrite reduction to ammonia by fermentative bacteria: a short circuit in the biological nitrogen cycle. FEMS Microbiol. Letters 7, 65–72.
19. Craswell, E.T. 1978 Some factors influencing denitrification and nitrogen immobilization in a clay soil. Soil Biol. Biochem. 10, 241–245.

20. Craswell, E.T. and Vlek, P.L.G. 1980 Research to reduce losses of fertilizer nitrogen from wetland rice soils. Paper presented at The Fertilizer Assoc. of India (FAI) Annual Seminar – 'Fertilizers in India in the Eighties' – Dec. 4–6, 1980, New Delhi, India.
21. Craswell, E.T. and Vlek, P.L.G. 1979 Greenhouse evaluation of nitrogen fertilizers for rice. Soil Sci. Soc. Amer. J. 43, 1184.
22. Craswell, E.T. and Vlek, P.L.G. 1979 Fate of fertilizer nitrogen applied to wetland rice. In: Nitrogen and Rice, pp. 175–192. IRRI, Los Baños. Philippines.
23. Culbertson, C.W., Zehnder, A.J.B. and Oremland, R.S. 1981 Anaerobic oxidation of acetylene by estuarine sediments and enrichment cultures. Appl. Environ. Microbiol. 41, 396–403.
24. Denmead, D.T., Freney, J.R. and Simpson, J.R. 1979 Nitrous oxide emission during denitrification in a flooded field. Soil Sci. Soc. Amer. J. 43, 716–718.
25. Delwiche, C.C. 1981 Nitrification, denitrification and atmospheric nitrous oxide. Academic Press, Inc., New York.
26. Delwiche, C.C. and Bryan, B.A. 1976 Denitrification. Ann. Rev. Microbiol. 30, 241–262.
27. Delwiche, C.C. and Rolston, D.E. 1976 Measurement of small nitrous oxide concentrations by gas chromatography. Soil Sci. Soc. Am. J. 40, 324–327.
28. De, P.K. and Digar, S. 1955 Influence of the rice crop on the loss of nitrogen gas from waterlogged soils. J. Agric. Sci. 45, 280–282.
29. Fedorova, R.I., Milenkhina, E.I. and Il'Yukhina, N.I. 1973 Evaluation of the method of 'gas metabolism' for detecting extraterrestrial life. Identification of nitrogen-fixing microorganisms. Izv. Akad. Nauk SSSR, Ser. Biol. 6, 797–806.
30. Firestone, M.K. 1981 Denitrification. In: Nitrogen in Agricultural Soils, pp. 289–326. Stevenson, F.I., Bremner, J.M., Hauck, R.D. and Keeney, D.R. (eds.), Agron. Monogr. No. 22, Amer. Soc. Agron. Madison, WI.
31. Firestone, M.K., Firestone, R.B. and Tiedje, J.M. 1979 Nitric oxide as an intermediate in denitrification: evidence from nitrogen-13 isotope exchange. Biochem. Biophys. Res. Commun. 91, 10–16.
32. Firestone, M.K. and Tiedje, J.M. 1979 Temporal changes in nitrous oxide and dinitrogen from denitrification following onset of anaerobiosis. Appl. Environ. Microbiol. 38, 673–679.
33. Firestone, M.K., Firestone, R.B. and Tiedje, J.M. 1980 Nitrous oxide from soil denitrification: factors controlling its biological production. Science 208, 749–751.
34. Focht, D.D. 1978 Methods for analysis of denitrification in soils. In: Nitrogen in the Environment, Vol. 2, pp. 429–490. Nielson, D.R. and Mac-Donald, J.G. (eds.), Academic Press, New York.
35. Focht, D.D. 1979 Microbial kinetics on nitrogen losses in paddy soils. In: 'Nitrogen and Rice', pp. 119–134. I.R.R.I., Los Baños, Laguna, Philippines.
36. Focht, D.D. and Joseph, H. 1973 An improved method for the enumeration of denitrifying bacteria. Soil Sci. Soc. Amer. Proc. 37, 698–699.
37. Focht, D.D. and Verstraete, W. 1977 Biochemical ecology of nitrification and denitrification. Adv. Microb. Ecol., Vol. I, pp. 135–214. Alexander, M. (ed.), Plenum Press, New-York.
38. Gamble, T.N., Betlach, M.R. and Tiedje, J.M. 1977 Numerically dominant denitrifying bacteria from world soils. Appl. Environ. Microbiol. 33, 926–939.
39. Garcia, J.-L. 1973 Séquence des produits formés au cours de la dénitrification

204

dans les sols de rizières du Sénégal. Ann. Microbiol. (Inst. Pasteur) 124 B, 351–362.
40. Garcia, J.-L. 1973 Influence de la rhizosphère du riz sur l'activité dénitrifiante potentielle des sols de rizière du Sénégal. Oecol. Plant. 8, 315–323.
41. Garcia, J.-L. 1974 Réduction de l'oxyde nitreux dans les sols de rizières du Sénégal: mesure de l'activité dénitrifiante. Soil Biol. Biochem. 6, 79–84.
42. Garcia, J.-L. 1975 Effet rhizosphère du riz sur la dénitrification. Soil Biol. Biochem. 7, 139–141.
43. Garcia, J.-L. 1975 La dénitrification dans les sols. Bull. Inst. Pasteur. 73, 167–193.
44. Garcia, J.-L. 1975 Evaluation de la dénitrification dans les rizières par la méthode de réduction de N_2O. Soil Biol. Biochem. 7, 251–256.
45. Garcia, J.-L. 1976 Production d'oxyde nitrique dans les sols de rizière. Ann. Microbiol. (Inst. Pasteur) 127 A, 401–414.
46. Garcia, J.-L. 1977 Analyse de différents groupes composant la microflore dénitrifiante des sols de rizière du Sénégal. Ann Microbiol. (Inst. Pasteur) 128 A, 433–446.
47. Garcia, J.-L. 1977 Evaluation de la dénitrification par la mesure de l'activité oxyde nitreux-réductase. Etude complémentaire. Cah. ORSTOM, Sér. Biol. 12, 89–95.
48. Garcia, J.-L. 1977 La dénitrification en sol en rizière: influence de la nature et du mode d'épandage des engrais azotés. Cah. ORSTOM, Sér. Biol. 12, 83–87.
49. Garcia, J.-L., Raimbault, M., Jacq, V., Rinaudo, G. and Roger, P. 1974 Activités microbiennes dans les sols de rizière du Sénégal: relations avec les propriétés physicochimiques et influence de la rhizosphère. Rev. Ecol. Biol. Sol. 11, 169–185.
50. Garber, E.A.E. and Hollocher, T.C. 1981 ^{15}N-tracer studies on the role of NO in denitrification. J. Biol. Chem. 256, 5459–5465.
51. Gersberg, R.M. 1977 Denitrification studies in Castle Lake, California utilizing the radioisotope nitrogen-13. Ph.D. Thesis, Univ. of California, Davis, pp. 112. D.A. No. 78-03612.
52. Gersberg, R., Krohn, K., Peek, N. and Goldman, C.R. 1976 Denitrification studies with ^{13}N-labeled nitrate. Science 192, 1229–1231.
53. Hauck, R.D. 1979 Methods for studying N transformations in paddy soils: review and comments. – In: Nitrogen and Rice, pp. 73–94. I.R.R.I., Los Baños, Laguna, Philippines.
54. Hauck, R.D. 1980 Mode of action of nitrification inhibitors. In: Nitrification Inhibitors – Potentials and Limitations, pp. 19–32. Amer. Soc. Agron., Madison, WI.
55. Hollis, O.L. 1966 Separation of gaseous mixtures using porous polyaromatic beads. Anal. Chem. 38, 309–316.
56. Hynes, R.K. and Knowles, R. 1978 Inhibition by acetylene of ammonia oxidation in *Nitrosomonas europaea*. FEMS Microbiol. Lett. 4, 319–321.
57. I.A.E.A. 1970 Rice fertilization. Tech. Repts. Ser. 108, 1–6.
58. Ichikawa, T., Seto, S. and Goto, K. 1968 Denitrification resistant fertilizers. Japan, 71 29, 765, 3 p. (in Chem. Abstr. Biochem. Sec. 78, 83316 s).
59. Ichikawa, T., Seto, S. and Goto, K. 1968 Aromatic ketonitriles as inhibitors of nitrogen loss from soil. Japan, 72 04, 965, 3 p. (in Chem. Abstr. Biochem. Sec. 78, 70719 a).
60. I.R.R.I. 1974 Annual Report, 11–13.
61. I.R.R.I. 1975 Annual Report, 44–46.

62. I.R.R.I. 1977 Research Highlights for 1977, p. 13.
63. Jeter, R.M. and Ingraham, J.L. 1981 Denitrifying procaryotes. In: The Procaryotes. Starr, M.P. (ed.), Springer Verlag, New York. Vol. I, pp. 913–925.
64. Kaspar, H.F., Tiedje, J.M. and Firestone, R.B. 1981 Denitrification and dissimilatory nitrate reduction to ammonium by digested sludge. Can. J. Microbiol. 27, 878–885.
65. Keeney, D.R., Fillery, I.R. and Marx, G.P. 1979 Effect of temperature on the gaseous nitrogen producers of denitrification in a silt loam soil. Soil Sci. Soc. Amer. J. 43, 124–1128.
66. Klemedtsson, L., Svensson, B.H., Lindberg, T. and Rosswall, T. 1977 The use of acetylene inhibition of nitrous oxide reductase in quantifying denitrification in soil. Swed. J. Agr. Res. 7, 179–186.
67. Knowles, R. 1981 Denitrification. In: Soil Biochemistry, Volume 5, pp. 323–369. Paul, E.A. and Ladd, J.N. (eds.), Marcel Dekker, Inc., NY.
68. Komatsu, Y., Takagi, M. and Yamaguchi, M. 1978 Participation of iron in denitrification in waterlogged soil. Soil Biol. Biochem. 10, 21–26.
69. MacRae, I.C. Ancajas, R.R. and Salandanan, S. 1968 The fate of nitrite nitrogen in some tropical soils following submergence. Soil Sci. 105, 327–334.
70. Matsubara, T. and Mori, T. 1968 Studies on denitrification. IX. Nitrous oxide, its production and reduction to nitrogen. J. Biochem. 64, 863–871.
71. McGarity, J.W. 1961 Denitrification studies on some south Australian soils. Plant and Soil 14, 1–21.
72. Mikkelsen, D.S. and Finfrock, D.C. 1957 Availability of ammoniacal nitrogen to lowland rice as influenced by fertilizer placement. Agron. J. 49, 296–300.
73. Mikkelsen, D.S., De Datta, S.K. and Obcemea, W.N. 1978 Ammonia volatilization losses from flooded rice soils. Soil Sci. Soc. Amer. J. 42, 725–730.
74. Mitsui, S. 1954 Inorganic nutrition, fertilization and soil amelioration for lowland rice. Yokendo Press, Tokyo, 107 p.
75. Mitsui, S. 1977 Recognition of the importance of denitrification and its impact on various improved and mechanized applications of nitrogen to rice plant. In: Proc. Int. Seminar of Soil Environment and Fertility Management in Intensive Agriculture (SEFMIA), pp. 259–268. Soc. Sci. Soil Manure, Japan, Tokyo.
76. Mitsui, S., Watanabe, I., Honma, M. and Honda, S. 1964 The effect of pesticides on the denitrification in paddy soil. Soil Sci. Plant Nut. 10, 15–23.
77. Nelson, D.W. and Bremner, J.M. 1970 Gaseous products of nitrite decomposition in soils. Soil Biol. Biochem. 2, 203–215.
78. Nommik, H. 1956 Investigations on denitrification in soil. Acta Agr. Scand. 6, 195–228.
79. Payne, W.J. 1973 Reduction of nitrogenous oxides by microorganisms. Bacteriol Rev. 37, 409–452.
80. Payne, W.J. 1981 Denitrification. Wiley Interscience, NJ (in press).
81. Payne, W.J. and Balderston, W.L. 1978 Denitrification. Microbiology. 1978, pp. 339–342. Schlessinger, D. (ed.), A.S.M., Washington, DC. and subsequent papers therein.
82. Patrick, W.H., Jr. and Miears, R.J. 1960 Depth of placement and source of nitrogen fertilizer as factors in the production of rice. Ann. Prog. Rep. Rice Exp. Sta. Crowley, La. 51, 90–95.
83. Patrick, W.H. Jr. and Wyatt, R. 1964 Soil nitrogen loss as a result of alternate submergence and drying. Soil Sci. Soc. Amer. Proc. 28, 647–653.
84. Patrick, W.H. Jr., Peterson, F.J. and Turner, F.T. 1968 Nitrification inhibitors for lowland rice. Soil Sci. 105, 103–105.

85. Pearsall, W.H. and Mortimer, C.H. 1939 Oxidation-reduction potentials in waterlogged soils, natural waters and muds. J. Ecol. 27, 483–501.
86. Pichinoty, F. 1965 L'inhibition par l'oxygène de la dénitrification bactérienne. Ann. Inst. Pasteur. 109 (suppl.), 248–255.
87. Pichinoty, F. 1973 La réduction bactérienne des composés oxygénés minéraux de l'azote. Bull. Inst. Pasteur. 71, 317–395.
88. Pichinoty, F., Garcia, J.-L., Mandel, M., Job, C. and Durand, M. 1978 Isolement de bactéries utilisant en anaérobiose l'oxyde nitrique comme accepteur d'électrons respiratoire. C.R. Acad. Sc. (Paris) 286, Sér. D, 1403–1405.
89. Pichinoty, F., Mandel, M. and Garcia, J.-L. 1979 The properties of novel mesophilic denitrifying *Bacillus* culture found in tropical soil. J. Gen. Microbiol. 115, 419–430.
90. Prakasa Rao, E.V.S. and Prasad, R. 1980 Nitrogen leaching losses from conventional and new nitrogenous fertilizers in low-land rice culture. Plant and Soil 57, 383–392.
91. Prasad, R. and De Datta, S.K. 1979 Increasing fertilizer nitrogen efficiency in wetland rice. In: Nitrogen and Rice, pp. 465–484. I.R.R.I., Los Baños, Laguna, Philippines.
92. Raimbault, M., Rinaudo, G., Garcia, J.-L. and Boureau, M. 1977 A device to study metabolic gases in the rice rhizosphere. Soil Biol. Biochem. 9, 193–196.
93. Rajale, G.B. and Prasad, R. 1973 Nitrification inhibitors and slow-release nitrogen fertilzers for rice (*Oryza sativa* L.). J. Agric. Sci., Camb. 80, 479–487.
94. Reddy, K.R. and Patrick, W.H., Jr. 1975 Effect of alternate aerobic and anaerobic conditions on redox potential, organic matter decomposition and nitrogen loss in a flooded soil. Soil Biol. Biochem. 7, 87–94.
95. Reddy, K.R. and Patrick, W.H., Jr. 1977 Effect of placement and concentration of applied NH_4^+-N on nitrogen loss from flooded soil. Soil Sci. 123, 142–148.
96. Reddy, K.R., Patrick, W.H. and Phillips, R.E. 1980 Evaluation of selected processes controlling nitrogen losses in a flooded soil. Soil Sci. Soc. Amer. J. 44, 1241–1246.
97. Redman, F.H. and Patrick, W.H., Jr. 1965 Effect of submergence on several biological and chemical soil properties. Agric. Expt. Station, Louisiana State Univ., Bull. 592, 1–28.
98. Roger, P.A., Kulasooiya, S.A., Tirol, A.C. and Craswell, E.T. 1980 Deep placement: A method of nitrogen fertilizer application compatible with algal nitrogen fixation in wetland rice soils. Plant and Soil 57, 137–142.
99. Rolston, D.E., Hoffman, D.L. and Toy, D.W. 1978 Field measurement of denitrification: I. Flux of N_2 and N_2O. Soil Sci. Soc. Amer. J. 42, 863–869.
100. Ryden, J.C., Lund, L.J. and Focht, D.D. 1978 Direct in-field measurement of nitrous oxides flux from soils. Soil Sci. Soc. Amer. J. 42, 731–737.
101. Sahrawat, K.L. 1980 Is nitrate reduced to ammonium in waterlogged acid sulfate soils? Plant and Soil 57, 147–149.
102. Sahrawat, K.L. 1980 Soil and fertilizer nitrogen transformations under alternate flooding and drying moisture regimes. Plant and Soil 55, 225–233.
103. Savant, N.K. and De Datta, S.K. 1979 Nitrogen release patterns from deep placement sites of urea in a wetland rice soil. Soil Sci. Soc. Amer. J. 43, 131–134.
104. Savant, N.K. and De Datta, S.K. 1980 Movement and distribution of ammonium-N following deep placement of urea in a wetland rice soil. Soil Sci. Soc. Amer. J. 44, 559–565.

105. Sexstone, A.J. and Tiedje, J.M. 1980 Influence of iron on denitrification. Agron. Abstr. Amer. Soc. Agron. 80, 160.

106. Sharma, S.N. and Prasad, R. 1980 Effect of rates of nitrogen and relative efficiency of sulfur-coated urea and nitrapyrin-treated urea in dry matter production and nitrogen uptake by rice. Plant and Soil 55, 389–396.

107. Simmonds, P.G. 1978 Direct determination of ambient carbon dioxide and nitrous oxide with a high-temperature ^{63}Ni electron-capture detector. J. Chrom. 166, 593–598.

108. Smith, K.A. and Hall, K.C. 1973 Gas chromatographic measurement of dissolved oxygen and trace levels of nitrous oxide using a helium ionization detector. Letcombe Laboratory, Ann. Rep. 53–55.

109. Smith, M.S. and Tiedje, J.M. 1979 The effect of roots on soil denitrification. Soil Sci. Soc. Amer. J. 43, 951–955.

110. Smith, M.S. and Tiedje, J.M. 1979 Phases of denitrification following oxygen depletion in soil. Soil Biol. Biochem. 11, 261–267.

111. Smith, M.S. and Zimmerman, K. 1981 Nitrous oxide production by non-denitrifying soil nitrate reducers. Soil Sci. Soc. Amer. J. (in press).

112. Smith, M.S., Firestone, M.K. and Tiedje, J.M. 1978 The acetylene inhibition method for short-term measurement of soil denitrification and its evaluation using nitrogen-13. Soil Sci. Soc. Amer. J. 42, 611–615.

113. Sørensen, J. 1978 Occurrence of nitric and nitrous oxides in a coastal marine sediment. Appl. Environ. Microbiol. 36, 809–813.

114. Sørensen, J., Jorgensen, B.B. and Revsbech, N.P. 1979 A comparison of oxygen, nitrate, and sulfate respiration in coastal marine sediments. Microbiol. Ecol. 5, 105–115.

115. Sørensen, J., Tiedje, J.M. and Firestone, R.B. 1980 Inhibition by sulfide of nitric and nitrous oxide reduction by denitrifying Pseudomonas fluorescens. Appl. Environ. Microbiol. 39, 105–108.

116. Stefanson, R.C. 1975 Denitrification from nitrogen fertilizer placed at various depths in the soil–plant system. Soil Sci. 121, 353–363.

117. Tam. T.Y. and Knowles, R. 1979 Effects of sulfide and acetylene on nitrous oxide reduction by soil and by Pseudomonas aeruginosa. Can. J. Microbiol. 25, 1133–1138.

118. Tiedje, J.M. 1981 Use of nitrogen-13 and nitrogen-15 in studies in the dissimilatory fate of nitrate. In: Genetic Engineering of Symbiotic Nitrogen Fixation and Conservation of Fixed Nitrogen. Lyons, J.M., Valentine, R.C., Phillips, D.A., Rains, D.W. and Huffaker, R.C. (eds.), Plenum Press, NY (in press).

119. Tiedje, J.M. 1981 Denitrification. In: Methods of Soil Analysis, 2nd edition. Amer. Soc. Agron. Madison, WI (in press).

120. Tiedje, J.M., Firestone, R.B., Firestone, M.K., Betlach, M.R., Smith, M.S. and Caskey, W.H. 1979 Methods for the production and use of nitrogen-13 in studies of denitrification. Soil Sci. Amer. J. 43, 709–716.

121. Tiedje, J.M., Firestone, R.B., Firestone, M.K., Betlach, M.R., Kaspar, H.F. and Sørensen, J. 1981 Use of nitrogen-13 in studies of denitrification. In: Short-lived Radionuclides in Chemistry and Biology, Chapter 15. Root, J.W. and Krohn, KA. (eds.), Advances in Chemistry, No. 197. Amer. Chem. Soc., Washington, D.C. (in press).

122. Tusneem, M.E. and Patrick, W.H., Jr. 1971 Nitrogen transformations in waterlogged soil. Louisiana State Univ. Agric. Expt. Station Bull. 657, 1–75.

123. Van Gent-Ruijters, M.L.W., De Vries, W. and Stouthamer, A.H. 1975 Influence of nitrate on fermentation patterns, molar growth yields and

synthesis of cytochrome b in *Propionibacterium pentosaceum*. J. Gen. Microbiol. 88, 36—48.

124. Valera, C.L. and Alexander, M. 1961 Nutrition and physiology of denitrifying bacteria. Plant and Soil 15, 268—280.

125. Van Cleemput, O. and Patrick, W.H., Jr. 1974 Nitrate and nitrite reduction in flooded gamma-irradiated soil under controlled pH and redox potential conditions. Soil Biol. Biochem. 6, 85—88.

126. Vlek, P.L.G. and Craswell, E.T. 1979 Effect of nitrogen source and management on ammonia volatilization losses from flooded rice-soil systems. Soil Sci. Soc. Amer. J. 43, 352—358.

127. Vlek, P.L.G. and Stumpe, J.M. 1978 Effect of solution chemistry and environmental conditions on ammonia volatilization losses from aqueous systems. Soil Sci. Soc. Amer. J. 42, 416—421.

128. Vlek, P.L.G., Byrnes, B.H. and Craswell, E.T. 1980 Effect of urea placement on leaching losses of nitrogen from flooded rice soils. Plant and Soil 54, 441—449.

129. Volz, M.G. 1977 Assessing two diagnostic methods for enumeration of nitrate reducing and denitrifying bacteria in soil—plant root associations. Soil Sci. Soc. Amer. J. 41, 337—340.

130. Walter, H.M., Keeney, D.R. and Fillery, I.R. 1979 Inhibition of nitrification by acetylene. Soil Sci. Soc. Amer. J. 43, 195—196.

131. Watanabe, I. and Deguzman, M.R. 1980 Effect of nitrate on acetylene disappearance from anaerobic soil. 12, 193—194.

132. Wetselaar, R. 1981 Nitrogen inputs and outputs of an unfertilized paddy field. In: Terrestrial Nitrogen Cycles. Processes, Ecosystem Strategies and Management Impacts. Clark, F.E. and Rosswall, T. (eds.), Ecol. Bull (Stockholm) 33, 573—583.

133. Wijler, J. and Delwiche, C.C. 1954 Investigations on the denitrifying process in soil. Plant and Soil 5, 155—169.

134. Woldendorp, J.W. 1963 The influence of living plants on denitrification. Meded. Landbouwhogesch., Wageningen 63, 1—100.

135. Woldendorp, J.W. 1975 Nitrification and denitrification in the rhizosphere. Bull. Soc. Bot. Fr., Coll. Rhizosphere 122, 89—107.

136. Yeomans, J.C. and Beauchamp, E.G. 1978 Limited inhibition of nitrous oxide reduction in soil in the presence of acetylene. Soil Biol. Biochem. 10, 517—519.

137. Yoshida, T. and Padre, B.C., Jr. 1974 Nitrification and denitrification in submerged Maahas clay soil. Soil Sci. Plant Nutr. 20, 241—247.

138. Yoshinari, T. and Knowles, R. 1976 Acetylene inhibition of nitrous oxide reduction by denitrifying bacteria. Biochem. Biophys. Res. Comm. 69, 705—710.

139. Yoshinari, T., Hynes, R. and Knowles, R. 1977 Acetylene inhibition of nitrous oxide reduction and measurement of denitrification and nitrogen fixation in soil. Soil Biol. Biochem. 9, 177—184.

8. Endomycorrhizae in the tropics

V. GIANINAZZI-PEARSON and H. G. DIEM

1. Introduction

The term 'endomycorrhiza' is used to describe mutualistic symbiotic associations between certain fungi and plant roots, in which the fungal partner grows mainly inside the root cortex and penetrates the cells of the host root.

Endomycorrhizae include three groups: ericoid mycorrhizae, orchidaceous mycorrhizae and vesicular-arbuscular (VA) mycorrhizae. This review is restricted to VA mycorrhizae which are widely distributed in the world and potentially very important both economically and ecologically.

Although VA mycorrhizae were first observed before the turn of the 20th century [59, 103], research on the fungal symbionts began much later with the development of techniques for maintaining these fungi in pot cultures [136]. Considerable advances were made following extraction of their spores from soil [65] and after the discovery that plants inoculated with VA fungi grew better due to increased phosphate uptake [24, 66]. Increased absorption of potassium, sulphate, zinc and strontium-90 by mycorrhizal plants has also been shown experimentally [33, 68].

General information concerning VA mycorrhizae may be found in the recent excellent review by Hayman [90]. Some aspects of the biochemistry of VA mycorrhizae were recently discussed by Gianinazzi-Pearson and Gianinazzi [73] and the relevance of VA mycorrhizae to plant nutrition in agriculture has been reviewed by Tinker [214].

At the international workshop held in Kumasi (1978), different aspects concerning the role of VA mycorrhizae in tropical ecosystems and in tropical agriculture were discussed by Bowen [33], Black [31] and Mosse and Hayman [143]. Since a detailed review on the ecology of VA mycorrhizae in relation to tropical environments is lacking, the aim of this chapter is to summarize recent advances in our knowledge in this field. Some considerations are also given to the possible use of VA mycorrhizae in tropical agriculture with the hope that their utilization for improving plant production can become a practical reality for many tropical countries.

2. Morphology of VA Mycorrhizae

VA infection does not apparently change the external morphology of roots; the internal morphology can be readily observed after clearing and staining the

Y.R. Dommergues and H.G. Diem (eds.), Microbiology of Tropical Soils and Plant Productivity. ISBN 978-94-009-7531-6.
© 1982 Martinus Nijhoff/Dr W. Junk Publishers, The Hague/Boston/London.

infected roots [160]. Although the morphology of the infection can vary slightly depending on both the host plant and the VA fungus [2, 67], certain features are generally observed. Development of a VA infection begins with formation of an ill-defined appressorium (Fig. 2) on the root surface by external hyphae originating from spores or other infected roots in the soil. Like pathogenic fungi, VA fungal hyphae penetrate within or between the epidermal cells from the appressorium then spread inter- and intra-cellularly within the cortex along the root length, sometimes forming coils within the outer cortical cells. Apart from the mycelium, two structures are typically formed by VA fungi within roots:
— arbuscules (Fig. 3), which are formed intra-cellularly by repeated bifurcation of an infecting hypha and where nutrient exchanges most probably occur between the host and the symbiont. The arbuscules are relatively short-lived and they degenerate to form a granular mass of fungal wall material within the host cell;
— vesicles (Figs. 1 and 3), which are most often ovoid swellings with lipidic contents usually formed terminally on hyphae and which are thought to act as temporary storage organs.

Little is known about the factors controlling the formation and the degeneration of these two fungal structures.

Outside the mycorrhizal roots a loose network of external hyphae, continuous with the internal mycelium, invades and explores the soil. These external hyphae take up nutrients, especially phosphorus, from the soil and translocate them to the internal mycelium where they are released from the fungal structures into the root cells [73, 214]. VA fungi generally form resting spores in soil, either singly, or in hypogeous or epigeous sporocarps. However, sometimes true spores of the fungal symbiont also occur within the root cortex [150]. Figure 4 shows the dead cortex of Sudan grass roots completely occupied by numerous spores of *Glomus mosseae* after being inoculated with this fungus. Present-day taxonomy of VA fungi is based entirely on the morphology of these spores and sporocarps.

3. The ecology of VA mycorrhizae in the tropics

3.1. The occurrence of VA mycorrhizae

As reported in what are probably the first records of VA mycorrhizae in the tropics by Janse [103] in Java and by Johnston [109] in Trinidad, the VA association is by far the most common kind of mycorrhiza. VA mycorrhizae are in fact widely distributed geographically and throughout the plant kingdom [90], but not all plant families form them. In temperate regions members of the Chenopodiaceae, Cyperaceae and Cruciferaceae are generally non-mycorrhizal [96] and in the tropics

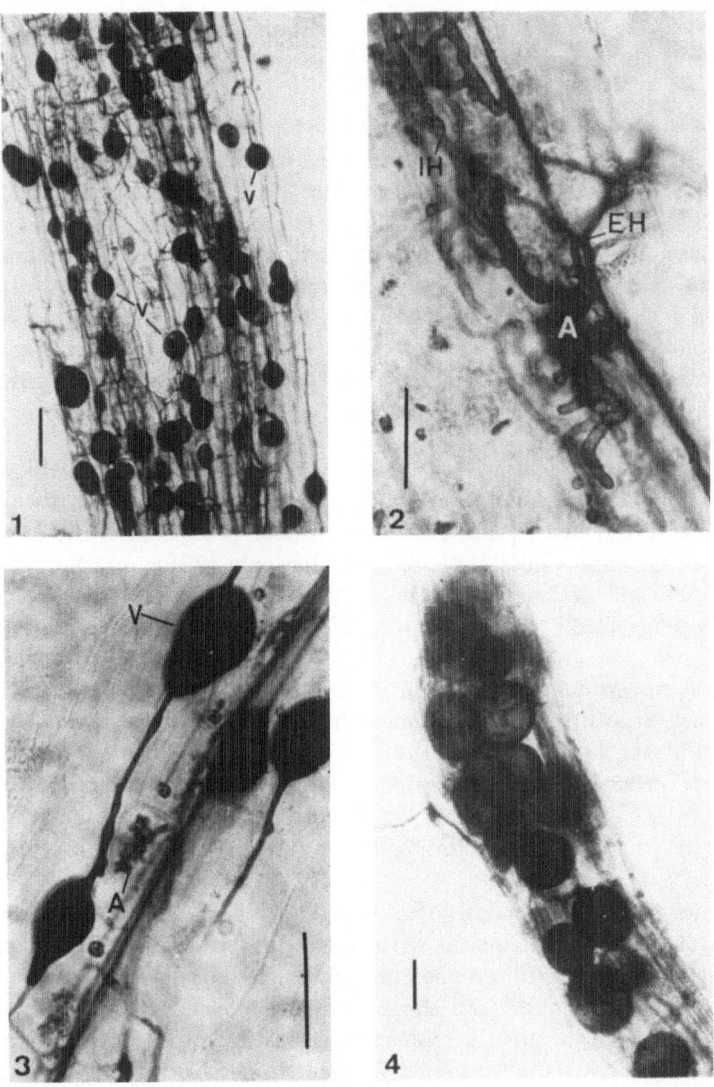

Plate 1. Morphology of VA mycorrhizae.
(Fig. 1) General view of *Vigna unguiculata* root infected with *Glomus mosseae* and showing a network of internal hyphae with vesicles (V); bar = 100 μm; (Fig. 2) Part of cortex of *V. unguiculata* root showing early stage of infection, appressorium (A), external (EH) and internal (IH) hyphae; bar = 50 μm; (Fig. 3) Part of cortex of *V. unguiculata* root showing vesicles (V) and arbuscules (A) within host cells; bar = 50 μm; (Fig. 4) Segment of Sudan grass root showing internal spores of *Glomus mosseae*; bar = 100 μm.

VA mycorrhizae are sometimes lacking in natural and semi-natural vegetation, for example in a number of plant families belonging to the xerophytic groups [111].

3.1.1. Forest tree species. Many species of forest trees in the tropics can be infected by VA fungi [177]. Under lowland rain forest condition in Nigeria [173, 177], in Sri Lanka [52], in Java [181], in Costa Rica [101, 102], in Cuba [95], in Philippines [217] and in Brazil [212], a wide range of forest tree species form VA mycorrhizae. Most attention has been focused on forest trees in the humid tropics and there remain many gaps in our knowledge concerning VA mycorrhizae associated with trees growing in the arid or semi-arid tropics. Studies carried out in Senegal on mycorrhizal associations of *Azadirachta indica, Casuarina equisetifolia* and *Acacia senegal* revealed variabilities in the VA infections of trees growing under dry conditions [53].

3.1.2. Perennial and annual crop plants. VA mycorrhizae occur on almost all perennial crops of economic importance in the tropics, such as avocado [75], cocoa [118, 148], citrus [123, 125, 213], rubber [222], cassava [97, 218, 227], papaya [172], finger millet [12], tea and pineapple [225], oil palm, litchi, sugarcane, coffee and coconut (see Janos [102] and Redhead [176]) and diverse tropical legumes [163] (see Section 4.3.2).

3.1.3. Host susceptibility to VA infection. It is generally admitted that there is a marked lack of host specificity amongst the different VA fungal isolates or species. However, Mosse [140], Jehne [104] and Cornet and Diem (unpublished data) found that certain VA fungi may be preferentially associated with a particular plant species.

Different causes have been proposed to explain the susceptibility of a plant to VA infection:

— physiological: the principal effect of VA mycorrhizae on plant growth is phosphate mediated and plant species or cultivars which are highly P-dependent tend to be strongly susceptible to VA infection although some plant species or cultivars which are less P-dependent can also be strongly infected. Recent observations by Janos [102] indicate that in some tropical forest tree growth responses to mycorrhizae are correlated to seed dry weight, and Hall [81] suggested that plant growth rates may influence P absorption and therefore responses to mycorrhizae.

— anatomical: according to Baylis [26] mycotrophy is largely a feature of woody and herbaceous plants lacking root hairs (magnolioid roots). In their survey on tropical forest trees, Janos [101] in Costa-Rica and St John [208] in Brazil have shown a significant relationship between magnolioid root characteristics and VA infection. Tropical herbage legumes such as *Centrosema* and *Stylosanthes* are sparse in root hairs and so potentially sensitive to VA infection [41]. *Leucaena leucocephala* [147] and many species of *Acacia* (Cornet and Diem, unpublished data) which have few root hairs are strongly mycorrhiza-dependent in P-deficient soils.

3.2. The VA fungi observed

Most attention has been devoted to the survey of VA fungi in temperate zones of both hemispheres and it is only during the past few years that attempts have been made to record VA mycorrhizal species found in the tropics. There is now evidence that species of four genera of the Endogonaceae (*Glomus, Gigaspora, Acaulospora, Sclerocystis*) form VA mycorrhizae in the tropics.

Principal records of different species of VA mycorrhizal fungi have been made in the humid tropics [176, 177]. Waidyanatha [219] working on *Hevea* plantations in Sri Lanka observed a wide range of spore types; most belonged to the genus *Glomus*, several to *Sclerocystis* species and some were species of *Acaulospora* and *Gigaspora*. In an extensive survey, Herrera and Ferrer [95] reported a similar distribution of endogonaceous genera in Cuban soils where the number of species of Endogonaceae per soil sample was higher than that recorded in other countries. Nadarajah [148] found that in Malaysian soils most spores also belonged to *Glomus* species and very few to other genera.

In the few studies on the taxonomic distribution of VA fungi in the dry tropics, the endogonaceous spores that have been isolated from arid or semi-arid soils belong to the three genera *Glomus, Gigaspora* and *Sclerocystis* [53, 56, 175]; typical spores of *Acaulospora* species have not been reported in these soils. Bulbous-based spores (*Gigaspora* spp.) are amongst the most common and spores closely resembling *Glomus mosseae* are also abundant. In conclusion, the current surveys of endogonaceous spores seem to indicate the world-wide distribution of several species and genera [177].

According to Diem *et al.* [53], it is impossible to indicate whether the taxonomic distribution of VA fungi varies widely in terms of environmental factors encountered in different tropical regions but Herrera and Ferrer [95] hypothesized that indigenous VA fungi may be strongly affected by local conditions. In general. *Glomus* species seem to be predominant both in the humid tropics and in the semi-arid areas, and certain *Gigaspora* species appear to be well adapted to dry and hot conditions [197, 198]. Herrera and Ferrer [95] have suggested that *Sclerocystis* may be considered as a tropical genus whilst *Acaulospora* species are apparently less common in semi-arid zones [53].

The taxonomical study of VA fungi in the tropics remains a vast area for research; new species of VA fungi have already been described in Nigeria [157], India [69], and probably several VA fungi found in the Ivory Coast [182], Sri Lanka [52] and Cuba [95] are new. Recently Nicolson and Schenck [151] have compiled an inventory of endogonaceous fungi including new species from the region of Florida which to some extent may be considered a subtropical zone.

4. Major factors affecting VA mycorrhizae

4.1. Climatic factors

Although VA mycorrhizae are formed in the soil and special attention should there-fore be given to soil factors, climatic factors may also be important since they can act on soil characteristics, control the physiology of the host plant and con-sequently influence the relationships between the plant and its endophyte.

4.1.1. Light. Microorganisms living closely in symbiosis with plant roots obtain their carbon energy source from the host plant and thus rely on both the photo-synthetic ability of the plant and the translocation of photosynthates to the root. For such systems, light is obviously a limiting factor [55]. The stimulatory effect of light on the development of VA mycorrhizae has been shown by Furlan and Fortin [62] and Hayman [87]. Shading not only reduces root infection and spore production [68] but also the plant response to VA mycorrhiza [132, 152]. This is probably due to a reduced spread of internal hyphae within root tissues, and con-sequently restricted growth of external hyphae in the soil. Redhead [174] postulated that day length may play an important role in VA mycorrhiza develop-ment and this seemed to be confirmed by Daft and El-Giahmi [44]. However, shading did not have a negative effect on the development of mycorrhizae of *Khaya grandifolia* [174] which suggests that the effect of light on VA mycorrhizae depends on the photosensitivity of the species of host plant. In fact, using plant species which exhibit different light requirements, Johnson [106] showed that infection level of *Coprosma leptospermum* and *Microlaena* sp. did not decrease with shading whilst in *Griselinia* sp. and *Parsonsia* sp., they were significantly lower in heavy shade.

Because of the beneficial effect of light, it could be expected that under the high solar radiation generally occurring in the tropics, VA infection levels would be high and growth responses marked. However, the favourable effect of light may be reduced in some cases, e.g. plants growing beneath forest canopies or in certain tropical regions subjected to variable sunshine due to the monsoons.

4.1.2. Temperature. Photosynthesis and translocation through the plant are affected by air temperature but the influence of this factor on VA mycorrhizae has received little attention. Furlan and Fortin [60] and Hayman [87] showed that infection levels in onion roots increased with increasing ambient temperature up to $26\,^\circ$C and this was sometimes associated with increased plant growth response, especially with alternating day/night temperatures. However, in these studies it is difficult to distinguish the effects of air temperature from those of the soil since no effort was made to independently control the temperature of the latter.

4.2. Physical-chemical soil factors

4.2.1. Soil temperature. The common characteristic of tropical soils is the high temperatures found both in humid regions and in semi-arid regions. In the lowland humid tropics, soil temperature at the beginning of the growing season can be 45–50 °C at a depth of 5 cm [156]. Daily mean temperatures in the semi-arid zone of West Africa are also high (40 °C) and their seasonal variation is relatively small [39]. It is therefore necessary to consider the role of high temperatures rather than that of low temperatures when discussing the development and the eco-physiology of VA mycorrhizae in the tropics; very little information, however, is available on this subject.

Effect on VA infection. VA mycorrhiza establishment consists of three distinct phases: spore germination in soil, hyphal penetration of root cells and development within the root cortex. From the few studies made, it appears that optimum temperatures for spore germination can vary widely between different VA fungi. Certain *Gigaspora* species isolated from soils in Florida, in a subtropical zone, germinated best at a temperature (34 °C) which was considerably higher than that for optimum germination of *Glomus* species (20 °C) from cooler climates [49, 198]. Fungal penetration and development in roots are similarly sensitive to variations in soil temperature. Smith and Bowen [205] showed that infection by fungi in soil from a temperate climate in southern Australia increased with increasing soil temperature and reached a maximum at 16–25 °C, whilst Schenck and Schroder [197] reported that maximum infection by a *Gigaspora* species isolated from a Florida soil occurred at 30–33 °C.

Under field conditions precocious infections by indigenous VA fungi have been observed in peanuts sown during the hot season in Senegal, where infection reached 53% of roots on 16-day old seedlings [71], and similar observations have been also made on cowpea (Bertheau, personal communication).

These observations deserve further investigation for they suggest that VA fungi from different climates may be adapted to different soil temperatures, a point to be considered before introducing efficient VA fungi from temperate climates into tropical soils.

Effect on VA mycorrhizal function. For *Eupatorium odoratum*, *Guizotia abyssinica* and *Sorghum bicolor* better growth and enhanced P uptake by mycorrhizal plants occurs at 30 °C rather than at 25 °C [132, 152]. High temperature (ca. 35 °C) during the day is not harmful to the development and physiological activity of the mycorrhizae if night temperatures are only 25 or 30 °C [132]. The stimulatory effect of mycorrhizal infection only decreases at 40 °C. In the tropics, day-time soil temperatures are rarely below 15 °C or above 35 °C beneath plant canopy, although that of bare soil can reach 40–45 °C. According to Moawad's data, these temperatures are probably not a limiting factor for the activity of VA mycorrhizae.

Effect on survival of VA fungi. There is no report on the effect of high temperature on the survival of VA fungal structures in soil, although it may be assumed that high temperature existing in the bare soil after the host plant is harvested or has died could dramatically affect the survival of different VA fungal structures especially spores. Bowen [33] has suggested the necessity of selecting species or strains highly resistant to high temperatures. Temperature responses of different species or isolates of VA fungi, however, may also depend on other soil features, e.g. soil texture. Rhizobia, for example, are particularly sensitive to high temperatures in sandy soil but not to the same extent in clay soil because in the latter the clay envelope which surrounds the bacterial cells confers an increased resistance to high temperature and desiccation [122]. The role of clay soil in the storage of fungal cultures has been demonstrated [19], and it could be that high temperature tolerance of VA fungi is more marked in heavy textured soils than in sandy soils.

4.2.2. Soil water content. Tropical soils are quite different from one another in water content due to the wide range of soil textures and climates in the tropics.

Although VA mycorrhizae can occur in aquatic plants [15, 206], it is generally admitted that their development is adversely affected by waterlogged soil conditions [111, 124]. Vast regions in the tropics are governed by an arid or semi-arid climate, and water relations of VA mycorrhizae could be of particular importance here [33]. Knowledge of the ecophysiology of VA mycorrhizae in relation to soil water potential is essential for evaluation of the role of VA mycorrhizae in these regions.

Influence of drought on VA mycorrhiza development in soil. Many plants growing in desert and semi-arid regions are normally mycorrhizal showing that VA infections can develop under water-stressed conditions. The influence of drought conditions on the development of VA fungi inside and outside the root can be different. Daniels and Trappe [49] showed that spore germination was favoured in soil at or above field capacity. In experiments with *Khaya grandifolia* Redhead [174] found that whilst extra-matrical mycorrhizal spores and fungal development may be drastically affected by drought, mycorrhizal infection levels can be high, presumably because water content within the roots remains sufficiently high for continued fungal spread in the host tissues.

Role of VA mycorrhizae in plant growth under drought conditions. In dry climates plants are often subjected to relatively long periods of water stress, and an interesting question is whether soil water supplies to the plant can be improved by VA mycorrhizae. Although, as previously mentioned, many plants from desert and semi-arid regions are mycorrhizal, little is known of the significance of this association for growth and survival of these species.

Recently, Menge *et al.* [127] reported that mycorrhizal infection enabled avocado seedlings to resist transplant shock because VA mycorrhizae could improve the water absorption capacity of the host plant. Figure 5 illustrates the behaviour of mycorrhizal and non-mycorrhizal seedlings of *Acacia raddiana* grown in the same soil with the same water potential. In the middle of the day, when air humidity is low, leaflets of mycorrhizal seedlings remained open whereas those of the controls

Fig. 5. Effect of VA infection on the growth and the behaviour of *Acacia raddiana*; non-mycorrhizal seedling with closed leaflets (R) and mycorrhizal seedling with open leaflets (RM) observed when air humidity is low; 60-day old seedlings (photography by courtesy of F. Cornet).

are closed suggesting that evapotranspiration is higher on non-mycorrhizal plants than on mycorrhizal ones. According to Janos [102], not only do mycorrhizal perennial trees grow more rapidly than uninfected seedlings but also mycorrhizae improve the survival of some tree species. *Carica papaya* control seedlings died after wilting in hot, sunny days whereas inoculated plants survived.

It has been suggested that drought resistance of mycorrhizal plants may be due to a decreased resistance of roots to water flow and therefore an enhanced water transport into roots [190]. P-deficient plants are more susceptible to drought [7] and Safir *et al.* [191] reported that better P nutrition in mycorrhizal plants could enhance water transport in soybeans. Another possible mechanism would be the absorption of water by the extensive external mycelium of mycorrhizal roots. However, in either case water content in the soil would be quickly exhausted. A recent interesting finding on the relationship between soil water potential and mycorrhizal activity is that the amount of water used to produce 1 g dry matter was much lower in mycorrhizal than in non-mycorrhizal plants growing in dry soil fertilized with $Ca_5(PO_4)_3OH$ [202]. Consequently, the greater drought resistance of mycorrhizal plants may be simply due to a more economic utilization of water by plants growing in P-deficient soils [132].

How mycorrhizal plants could economize on water consumption remains to be determined. In recent work on citrus seedlings, stomatal conductance (which provides information on the resistance to water flow in the leaf—air interface) and transpiration flux density were slightly but not significantly influenced by mycorrhizae during a period of water stress whilst when plants were rewatered leaf conductance, transpiration flux density and photosynthesis were consistently higher in mycorrhizal plants [120]. In arid and semi-arid regions it seems therefore probable that mycorrhizae play an important role in the growth and drought resistance of a number of plants because of their ability to regulate uptake of soil nutrients and water.

4.2.3. Soil pH. VA fungi often show adaptation to soil pH and this can be a determining factor for endophyte efficiency. Both spore germination and mycorrhiza development by different fungal species can be significantly affected by variations in soil pH [49, 79, 137]. Hayman and Mosse [92] obtained infection and growth stimulation of *Coprosma robusta* by *Glomus mosseae* in two soils of pH 5.6 and 7.0 but not in acid soils of pH 3.3 to 4.4, whilst after liming to pH 6.5 infection occurred in all and growth stimulation in most of the soils. Similar results have also been obtained with *Paspalum notatum* growing in acid tropical soils [138] and with soybean which responded positively to both *Gigaspora gigantea* and *G. mosseae* in limed soil (pH. 6.2) but only to *Gigaspora gigantea* in acid soil (pH 5.1) even though infection levels with both fungi were comparable [204].

On the contrary, Cornet and Diem (unpublished data) found that *Acacia raddiana* responded positively to both *Glomus mosseae* and *Glomus* E_3 in an acid soil from Senegal (pH 4.5) but only to *Glomus mosseae* in the same soil after liming to pH 6.9.

The relationship between soil pH and VA mycorrhizal effect is complex and depends not only on the fungal species, soil type or forms of P but also on the plant species. In the case of *Guizotia abyssinica* VA mycorrhizae strongly depressed plant growth at pH 4.3 in the presence of different fertilizers whereas mycorrhizal *Tagetes minuta* was not affected at the same pH [77].

4.2.4. Organic matter and root residues. The important role of organic matter in the environment is evident, especially in the tropics where continuously high temperatures favour rapid rates of decay of plant residues in soil. Organic matter influences soil structure, pH, nutrient and water-holding capacity, all of which may directly and/or indirectly influence VA mycorrhizal development and efficiency. According to Sheikh *et al.* [201], endogonaceous spore population seems to be closely correlated with the level of organic matter content in Pakistan soils. Maximum spore numbers were recovered from soils containing 1–2% organic matter and spores were sparse in soils with below 0.5% organic matter. No such correlation has been observed in temperate soils with higher (2–13%) organic matter contents (Gianinazzi-Pearson, unpublished data) although organic manures often enhance mycorrhizal development in tropical soils [109].

An aspect of the study of organic matter which deserves attention is the impact of mycorrhizal root residues themselves on the ecology of VA fungi in soil. Numerous mycorrhizal plants are annuals and mycorrhizal root systems are thus continuously being incorporated into soils and degraded by soil microorganisms. Almost nothing is known regarding the fate of the endophyte mycelia outside and inside the root tissues.

Redhead [175] suggested that seasonal die-back of Sudan and Sahel savanna grasses could stimulate endogonaceous spore production thus increasing spore populations, as observed when arable crops such as maize, barley and wheat are harvested. Mycorrhizal root debris in soil can also be an important reservoir of inoculum. Daft *et al.* [45] claimed that most infections of *Endymion non-scriptus* in nature arose directly from the decaying old root systems through which the new roots grew and Warner and Mosse [221] indicated some saprophytic ability of VA fungi in soil that would enable them to establish a base (possibly in particles of organic material) from which they could infect a host plant.

Many authors have emphasized that spores are not important for maintaining infection when infected roots are present especially in the case of natural plant communities when VA fungi may be non-sporing or poorly-sporing types [25, 207]. Rives *et al.* [183] also suggested that in areas with low annual rainfall contact between infected root debris and roots of uninfected plants may constitute the most efficient mode of VA mycorrhizal spread.

4.2.5. Chemical soil factors. The relationship between VA mycorrhizae and chemical factors have been little studied in tropical soils. Sheikh *et al.* [210] reported that chemical factors and soil chemical treatments may directly influence the occurrence of VA fungi in Pakistan soils and that low P supply which limits plant growth favoured spore production. In Californian soils, P content in soil was inversely correlated with mycorrhizal spore numbers [126]. Studies in temperate climates have shown that high P concentrations in soils reduce VA infection, probably due to the high internal P concentrations developing within the host tissues [5, 130, 192].

The effect of N and P fertilizers on VA mycorrhizae have been extensively studied in temperate soils by Hayman [88]. He showed that N fertilizer (188 kg N ha^{-1} as 'Nitro-Chalk') had a large negative effect on the mycorrhizal population and plots not given N contained 2 to 7 times more endogonaceous spores and 2 to 4 times more VA infection than plots given N. Porter and Beute [162] also found that mycorrhizal peanuts growing in soils containing little N produced more spores than in soils containing much N. Hayman [88] found that the effects of N fertilizer were more marked than those of P fertilizer whilst Jensen and Jakobsen [105] observed that these two fertilizers generally affected mycorrhiza formation equally.

In spite of the observed effects of fertilizers, VA fungi can be very abundant in fertile soils [93] (Trouvelot, personal communication) and this is probably because VA populations are influenced not only by fertilizers but also by different crop plants, soils, management practices etc. [115].

As well as the extensive use of chemical fertilizers for plant production, there has been increased development and use of soil pesticides for crop protection. The majority of these chemicals, and especially fungicides, can greatly reduce both the development and sporulation of VA fungi [57, 129, 149]. The nematicide DBCP*, however, frequently increases VA infection and/or sporulation [30, 71].

It is not known whether VA mycorrhizae can affect chemical characteristics of soils. Bowen [33] hypothesized that VA fungi could counteract some forms of

*1-2 dibromo-3-chloropropane

soil toxicity by absorbing elements harmful to plants or assisting plant tolerance of high alkalinity or high salinity in tropical soils. These questions deserve further investigations.

4.3. Biological factors: interactions with soil microorganisms

Large populations of microorganisms live in the soil and an intense microbial activity exists around plant roots. Saprophytic soil-inhabiting microorganisms in the rhizosphere interact with the plant and more specialized parasitic or symbiotic organisms infect the living roots. There are, however, major gaps in our knowledge of the interactions that VA mycorrhizae may have with these other soil microorganisms, especially in tropical soils.

4.3.1. Phosphate-solubilizing bacteria. Many experiments (quoted by Mosse *et al.* [146]) have shown that rock phosphates only become available to plants in acid soils but not in neutral or alkaline soils, and VA infections did not alter these relationships VA mycorrhizae do not mobilize insoluble soil phosphate but only increase the absorption of that which is already available to roots. Experiments with ^{32}P confirm that both mycorrhizal and non-mycorrhizal plants utilize the same pool of available soil P [74, 145, 161, 193]. The possible processes involved in P absorption by VA mycorrhizae have been reviewed by Tinker [214]. Mosse [139] discussed different mechanisms involved in the solubilization of organic and inorganic phosphate in the soil and in the rhizosphere. It is well known that two groups of bacteria, chemo-organotrophs such as some *Pseudomonas* and *Bacillus* species and chemo-lithotrophs such as thiobacilli, are able to solubilize insoluble phosphate.

The possibility of synergistic interactions between VA fungi and phosphate-solubilizing chemo-organotroph bacteria has been investigated in alkaline soils by Barea *et al.* [21] and Azcon *et al.* [8] using lavender and maize. In some cases the combined inoculation of a *Glomus* species (E_3 type) with these bacteria significantly increased plant growth above that achieved with either microorganism separately. However, the effect of *G. mosseae* on plant growth was not enhanced by the same bacteria. Azcon *et al.* [8] attributed this difference to an improvement in the efficiency of *Glomus* E_3 due to a lowering of soil pH by the introduced bacteria. In fact, *Glomus* E_3 is more tolerant of acid than alkaline soils whereas *G. mosseae* is not favoured by a lowered pH. It was also suggested that some solubilization of rock phosphate by the bacteria did occur and that inoculation with VA fungi favoured the early establishment of phosphate-solubilizing bacteria in the rhizosphere, although after 2 months of plant growth bacterial populations declined in the usual way.

Recently, the influence of inoculation with *Glomus fasciculatus* and *Bacillus circulans* on growth of finger millet and P uptake from ^{32}P-labelled tricalcium phosphate was studied [171]. In the treatment receiving both inocula a synergistic effect was also recorded with increased P uptake and growth in mycorrhizal plants.

Thiobacilli are well known for their ability to solubilize insoluble rock phosphate in soil when they are introduced together with S; they oxidize S to H_2SO_4 which in turn partially dissolves the rock phosphate. Swaby [210] claimed that this principle could be exploited in agriculture and that it had given promising results in field tests. The ability of thiobacilli to solubilize rock phosphate in the presence of S is used commercially in the Lipman process. Hayman [89] found that this process was as cheap or as efficient as the use of standard P fertilizers only in warm, wet soils and Swaby [210] in fact considers that this method to solubilize rock phosphate could be used economically in the wetter tropics.

Table 1 shows how inoculation of a soil from Senegal with thiobacilli can improve growth of *Vigna unguiculata* in the presence of a supplement of rock phosphate. The effect of an interaction between the thiobacilli and VA fungi on growth and nodulation seems to vary in terms of the species of fungus used. A synergistic effect was only recorded with the combined inoculum *Glomus* E_3 and thiobacilli as far as nodule dry weight is concerned.

Edaphic factors in humid tropical regions or in irrigated tropical soils are certainly very favourable to the activity of thiobacilli. With the continuously increasing cost of P fertilizers, the possible use of phosphate-solubilizing thiobacilli alone or in combination with appropriate VA fungi deserves more attention. P-deficient tropical soils often harbour high populations of indigenous VA fungi and exploitation of combined activities of VA mycorrhizae and thiobacilli may contribute to a better use of rock phosphate in agriculture, especially in developing countries where rock phosphate is a natural resource.

4.3.2. N_2-fixing microorganisms

Free-living bacteria. There are few publications concerning interactions between VA mycorrhizae and free-living N_2-fixing bacteria. From three studies that have dealt with *Azotobacter* [14, 22, 34] and one with *Azospirillum* [28] the effect of interactions between VA mycorrhizae and these free-living bacteria can be summarized under four main aspects:

(1) *Populations of N_2-fixing bacteria and total soil microflora.* The common finding of studies on *Azotobacter* is that VA infections favourably affect bacterial populations in the rhizosphere of host plants [14, 22]. *Azotobacter* numbers decrease more slowly in the rhizosphere of mycorrhizal plants than in the rhizosphere of non-mycorrhizal plants and total soil bacteria populations have been found to increase in the presence of *Azotobacter* and VA mycorrhiza together.

(2) *VA infection.* VA infection is increased by the presence of *Azotobacter* in dual inoculation experiments [14, 34] and there is some evidence that this is due to the production of growth-promoting substances by the bacterium [9]. *Azospirillum*, on the contrary, does not appear to have any appreciable effect on VA infection levels [28].

Table 1. Effect of double inoculation with thiobacilli and VA fungi on growth and nodulation of *Vigna unguiculata* cv 58-185 in sterile soil. (Ollivier and Diem, unpublished data)

	Average shoot dry wt. (g plant^{-1})	Average nodule dry wt. (g plant^{-1})	Root segments infected (%)	Shoot N (%)	Shoot P (%)
Control	2.78[a]	46[a]	0	1.4	0.080
Thiobacilli	3.90[b]	64[b]	0	1.4	0.080
G. mosseae	5.88[c]	191[c]	100[a]	1.8	0.095
Glomus E$_3$	4.98[bc]	148[d]	99[a]	1.6	0.105
G. mosseae + Thiobacilli	5.68[c]	189[c]	99[a]	1.7	0.090
Glomus E$_3$ + Thiobacilli	5.22[c]	178[c]	99[a]	1.7	0.095

60-day old plants all inoculated with *Rhizobium* strains CB 756

Soil: pH (KCl): 6.2; total P (ppm): 79; available P, Olsen (ppm): 10 and supplemented with rock phosphate (0.25 g kg^{-1} soil)

Values followed by the same letters in columns are not significantly different ($P = 0.05$)

(3) *N₂ fixation.* There is no conclusive information concerning the effect of VA mycorrhizae on N_2 fixation by free-living bacteria.

(4) *Plant growth.* Growth stimulation is greater in plants inoculated with both *Azotobacter* and VA fungi than either microorganism alone [14, 34]. Bertheau [28] also found some synergistic effects of dual inoculation of wheat with *G. mosseae* and an *Azospirillum* species; although these were negative for a first crop, they greatly improved growth of a consecutive crop on the same soil.

Symbiotic bacteria: Rhizobium. With the exception of certain *Lupinus* species which are immune to VA infection [135, 216], legumes are relatively poor foragers of P and generally very responsive to VA infection. Much of the information regarding mycorrhizal effects on legume growth has been obtained from studies on soybeans cultivated in more temperate climates (see for example Ross and Harper [188], Schenck and Hinson [194], Bagyaraj *et al.* [16], Carling and Brown [37] and Asimi *et al.* [5]).

Crush [41] speculated that apart from effects on the host's P supply VA mycorrhizae could influence the legume–*Rhizobium* by altering the root and/or rhizosphere environment for rhizobia. There are many reports concerning the nutritional effects of VA mycorrhizae in legumes (see Munns and Mosse [147]), but virtually nothing is known of direct interactions between VA mycorrhizae and rhizobia. VA fungi do not appear to penetrate nodule tissues [5, 41] and Carling *et al.* [38] concluded, from observations that P fertilizer produces similar plant growth responses to mycorrhiza in soybean, that the VA fungi probably have no direct effect on the symbiotic bacterium. However, the more detailed investigations by Asimi *et al.* [5] indicate that nodulation and nitrogenase activities of *R. japonicum* can still be significantly enhanced by mycorrhiza at high P fertilizer levels and this subject deserves further attention.

Tropical legumes associated with VA mycorrhizae. Annual and perennial tropical legumes can be strongly mycotrophic and although much information comes from soybean studies the influence of VA mycorrhizae has been investigated in some strictly tropical plant species: peanut [43, 71, 78, 162], *Stylosanthes guyanensis, Centrosema pubescens* [41, 146], *Pueraria phaseolides* [220] and *Vigna unguiculata* [12, 76, 80, 99, 227] (see also Tables 1, 3, 5, 6). Among legume trees of economic importance, species belonging to only two genera have been studied: *Leucaena leucocephala* [147, 227] and *Acacia senegal* [53], *A. farnesiana* [108], *A. holosericea* and *A. raddiana* (Cornet and Diem, p. 224).

Effect of VA mycorrhizae on N₂ fixation and growth of tropical legumes. Asai [4] first demonstrated that several legumes grew poorly and failed to nodulate in sterilized soil if they were not mycorrhizal. Many investigations have since confirmed this early finding (see Munns and Mosse [147]), and considerable attention is now being given to the tripartite association between plants, rhizobia and VA fungi. As an adequate P supply is necessary not only for plant growth but also for a satisfactory nodulation and N_2 fixation, VA mycorrhizae, in increasing P uptake by the plant, are obviously an important factor in the tripartite association.

There is a vast amount of literature concerning the effect of VA mycorrhizae on nodulation, N_2 fixation and growth of legumes; this section has been confined to a few examples related to strictly tropical legume species. Interactions between VA mycorrhizae and symbiotic N_2-fixing bacteria have been extensively studied in *Centrosema pubescens* and *Stylosanthes guyanensis*, an important forage legume in the tropics, in a large number of both temperate and tropical soils [41, 141, 146]. VA mycorrhizae exerted stimulatory effects on nodulation and N_2 fixation of these two legumes in all the P-deficient soils tested and nodulation and growth were improved in both sterile and non-sterile soils. Similar mycorrhizal effects have been obtained with *Vigna unguiculata* in Nigeria [99] and Senegal (Ollivier and Diem, unpublished data), and with *Pueraria phaseolides* and *Stylosanthes guyanensis* in Sri Lanka [220]. It is highly probable that the most important factor involved in the mycorrhizal responses of tropical legumes growing in P-deficient soils is the improved P nutrition of the host plant.

Three main factors can determine the extent to which VA mycorrhizae assist P uptake and therefore their effect on legumes production: plant species, mycorrhizal fungal species and available soil P.

Influence of the host species. Legumes can differ widely in their growth responses to VA infection. Growth and nodulation of *Astragalus sinensis Glycine max*, and *Ornithopus* sp. are greatly improved by VA mycorrhiza whereas in *Vicia sativa* and *Lupinus luteus* this is less marked [4]; *Stylosanthes* sp. and *Centrosema* sp. are much more dependent on VA mycorrhizae than temperature legumes such as *Trifolium* and *Lotus* sp. [41]. The rapidity with which legumes respond to VA infection can also vary between species of the same genus; *A. holosericea* responds to VA infection 5 weeks after inoculation whereas 8 weeks are necessary before mycorrhizal effects are visible in *A. raddiana* growing under the same conditions and inoculated with the same VA fungus (Cornet and Diem, unpublished data). These variations in mycorrhizal dependency are probably related to the ability of a given legume species to forage for P in the soil. Differences in mycorrhiza development and effects on plant growth can even differ between cultivars of the same host species and this topic is fully discussed in Section 4.4.

Influence of the VA fungal species. VA fungi are not host specific but the efficiency of a mycorrhizal system will partly depend on the physiological characteristics of the fungal symbiont (ability to translocate and transfer nutrients), the amount and distribution of the soil mycelium and on interactions between the fungal species and the environment [141]. Comparisons between *Glomus* and *Gigaspora* species in soybean [37], *Glomus* and *Acaulospora* species in alfalfa [155] and *Glomus*, *Gigaspora* and *Acaulospora* species in cowpea [99] have confirmed that VA fungi, and even isolates of the same fungal species [1], often vary in their effects on growth of the host plant even when infection levels are similar. However, such comparison between effectiveness of different VA fungi are only valuable in a given soil and should not be generalized to another environment because effectiveness of a fungal strain can greatly vary in terms of soil characteristics and fertility [37] (see also Table 2).

Table 2. Effect of soluble phosphate additions to soil on yield responses of soybeans infected by two VA fungi in sterile soil (Gianinazzi-Pearson and Gianinazzi, unpublished data)

	ppm P added to soil			
	0	57	114	228
Control	1.2[a]	4.2[a]	7.6[a]	7.0[a]
G. mosseae	2.9[b]	5.4[b]	8.3[a]	7.5[a]
G. fasciculatus	5.0[c]	6.3[b]	9.5[b]	10.1[b]

P added in form of KH_2PO_4
Results expressed as pod dry wt, g plant^{-1}
Values followed by the same letters in columns are not significantly different ($P = 0.05$)
Soil: pH (H_2O): 7.1; total P (ppm): 698; available, P, Olsen (ppm): 56

Influence of added soluble phosphate and insoluble rock phosphate. Applications of increasing amounts of soluble phosphate to soil considerably influence the incidence and the efficiency of the VA infection [5], although the extent of this effect also depends on the fungal species involved (Table 2). However, in very poor soils addition of soluble phosphate up to an appropriate level can enhance mycorrhizal effects on plant growth [97, 168] (Table 3). Increased utilization of soluble and sparingly soluble phosphate fertilizers by mycorrhizal plants [143] will depend not only on the amount of available P already in the soil but also on the soil's P-fixing capacity and water content. The diffusion coefficient of phosphate decreases with decreasing soil humidity and available P content [3] so that in many arid and semi-arid P-deficient tropical soils, inoculation with VA fungi alone may be insufficient to improve yields greatly and should be accompanied by irrigation and/ or applications of appropriate levels of P fertilizers in order to obtain optimum plant productivity.

Interactions between VA fungi, nodulation and rock phosphate have been studied in several tropical legumes [99, 141, 146, 220]. Utilization of rock phosphate by mycorrhizal legumes for growth and nodulation depends on soil pH (see Section 4.3.1). In neutral or alkaline soils, added rock phosphate remains unavailable to both mycorrhizal and non-mycorrhizal plants.

Symbiotic actinomycetes: Frankia. Many VA fungi have been observed in close association with different non-leguminous N_2-fixing plants [56, 185, 186, 224] but there are few reports that VA mycorrhizae influence N_2 fixation in these actinomycete-nodulated plants although they could exert a similar influence to that observed in legumes. Amongst actinomycete-nodulated plants growing in tropical climates, *Casuarina* is of great economic interest because of its use for afforestation in semi-arid and nutrient-poor soils. Apart from the presence of proteoid roots usually observed on seedlings raised in nurseries [53], roots of *Casuarina* can be heavily infected by VA fungi [53, 56, 185] and preliminary experiments have shown that double inoculation of *C. equisetifolia* with *G. mosseae* and crushed nodules significantly improves plant growth and nodulation (Table 4). In *Ceanothus velutinus*, Rose and Youngberg [187] also found that plant dry weight, number

Table 3. Growth responses of *Vigna unguiculata* cv 58-185 to triple superphosphate, rock phosphate and inoculation with *Glomus mosseae* in sterile soil (Ollivier and Diem, unpublished data)

	Average shoot dry wt. (g plant⁻¹)	Average nodule dry wt. (mg plant⁻¹)	Root segments infected (%)	Shoot N (%)	Shoot P (%)
Control	3.14[a]	54[a]	0	2.12	0.055
Triple superphosphate	4.27[b]	119[b]	0	2.46	0.055
Rock phosphate	2.68[a]	51[a]	0	2.54	0.082
G. mosseae	4.30[b]	107[b]	100[a]	2.67	0.085
Triple superphosphate + G. mosseae	5.10[c]	147[c]	99[a]	3.38	0.142
Rock phosphate + G. mosseae	5.03[c]	109[b]	95[a]	2.67	0.085

Triple superphosphate added: 10 ppm P; rock phosphate added: 40 ppm P
54-day old plants, all inoculated with *Rhizobium* strain CB 756
Soil: pH (KCl): 6.2; total P (ppm): 79; available P, Olsen (ppm): 10
Values followed by the same letters in columns are not significantly different ($P = 0.05$)

Table 4. Effect of soluble phosphate, crushed nodules and *Glomus mosseae* on growth and nodulation of *Casuarina equisetifolia* (Diem and Gauthier, unpublished data)

	Average shoot dry wt. (g plant^{-1})	Average nodule dry wt. (mg plant^{-1})	Shoot N (%)	Root segments infected (%)
Control	2.69^a	0^a	0.82	0
Crushed nodules	4.23^b	57.2^b	1.20	0
Crushed nodules + *G. mosseae*	7.69^c	132.2^c	1.25	47
Crushed nodules + soluble phosphate	7.49^c	106.6^c	1.61	0

Soluble phosphate added as K$_2$HPO$_4$, [0.5 g kg^{-1} soil]
6-month old plants
Soil: pH (KCl): 6.2; total P (ppm): 79; available P, Olsen (ppm): 10
Values followed by the same letters in columns are not significantly different ($P = 0.05$)

and weight of nodules and N and P content were greater in mycorrhizal nodulated plants than in nodulated-only plants. The recent isolation of many strains of *Frankia* from nodules of *Comptonia* [36], *Alnus* [18, 27, 116], *Elaeagnus* [18] *Casuarina* [64] and *Hippophaë* (Gauthier, Diem and Dommergues, unpublished data) will no doubt contribute to the development of studies on the effect of VA mycorrhizae on *Frankia*-nodulated plants.

4.3.3. Phytohormone-producing bacteria. Gunze and Hennessy [80] suggested that indole-3-acetic acid application could influence arbuscule formation in VA mycorrhiza. Phytohormones synthesized by certain bacteria (*Rhizobium, Azotobacter, Pseudomonas*) can significantly increase VA infection [9]. A large proportion of rhizosphere bacteria are able to produce phytohormones but how and to what extent they influence VA infection is not known.

4.3.4. Soil-borne phytopathogenic fungi. Biological control by VA mycorrhizae. Many reports in the literature indicate that plants previously inoculated with VA fungi show an increased resistance to certain fungal root diseases, for example *Fusarium* wilts, and root rots [195]. The mechanism of this mycorrhizal effect on pathogens is not known in most instances and several hypotheses have been proposed to explain the observed protection. As Schönbeck [200] points out, VA fungi are not thought to act directly on the pathogen but rather by causing changes in the host tissues. For example, they may stimulate lignification or the development of callosities or lignitubers in the host cells thus creating a physical barrier to penetration of the pathogen (Becker, quoted by Schenck and Kellam [195]). VA infections can also induce biochemical changes in host tissues which could render them unfavourable for pathogen development (Dehne and Schönbeck, quoted by Schönbeck [200]). Alternatively, prior occupation of the root tissues by VA fungi as first colonizers could simply physically exclude a pathogen competing for the same infection sites.

The influence of VA mycorrhizae on fungal root pathogens however appears to vary greatly with the disease complex studied. There are reports in which VA fungi do not show any stimulating effect on plant resistance to attack by fungal pathogens and in some cases the presence of pathogens reduces the beneficial effects of mycorrhizae, or disease is more severe in the mycorrhizal plants than in non-mycorrhizal plants [50, 51, 126, 195].

It is now evident that interactions between VA fungi, fungal root pathogens and host plants are complex and each combination should be considered individually; the effectiveness of VA mycorrhizae in protecting plants varies according to the species, strain or variety of the VA fungus and plant involved [23, 195]. However, as Wilhelm [223] pointed out, the evaluation of VA mycorrhizae as biocontrol agents remains one of the most challenging areas in plant pathology. In fact, if prior root colonization by VA fungi can reduce certain root diseases then this would be of great interest for many tropical plants which must be grown in nurseries before transplanting. Inoculation of these plants with selected VA fungi when growing under nursery conditions could provide a good method for protecting plants against pathogen attacks which risk to subsequently occur in the field. Unfortunately, studies on this aspect of VA mycorrhizae are still lacking.

4.3.5. Plant parasitic nematodes. Most investigations of nematode–VA mycorrhizae interactions are related to plant species naturally cultivated under tropical or sub-tropical climates: tobacco [20, 58, 180], cotton [98, 184], citrus [153, 154], soybean [110, 196, 199] and cowpea (Table 5).

Effect of nematodes on VA infection. Data published by Fox and Spasoff [58] suggest that there is a competitive interaction between *Heterodera solanaceanum* and *Gigaspora gigantea* in tobacco. Each organism adversely affects the reproduction of the other and this seems to be due to a competition for both space and food supply in the root system. Table 5 shows a slight but significant reduction in mycorrhizal infection of *V. unguiculata* by *Pratylenchus sefaensis* and O'Bannon and Nemec [153] also found less vesicle formation and mycelium growth by *Glomus etunicatus* in nematode-infected citrus roots. The effect of nematodes on the sporulation of VA fungi seems to be variable; it may be detrimental [196, 199], indifferent [6, 98] or stimulative [17].

Effect of VA infection on nematodes. There are several reports of significantly reduced development of root-knot nematodes or reduced formation of root galls in different plants pre-inoculated with *G. mosseae, G. fasciculatus* and *G. macrocarpus* (respectively Sikora and Schönbeck [203]; Bagyaraj *et al.* [17]; Kellam and Schenck [110]).

Reduction in numbers of migratory endoparasitic nematodes by *Gigaspora margarita* in cotton roots have also been observed [98, 199]. Schenck *et al.* [199] indicated that nematode–VA mycorrhizae interactions can differ with the host cultivar and other workers have reported greater populations of root-knot larvae from mycorrhizal plants than from non-mycorrhizal plants [6, 179]. Thus, as in the case of protection against fungal pathogens, interactions between VA fungi, nematodes and plant roots appear complex and seem to vary with each combination [195].

Effect of nematode–VA mycorrhizae interactions on plant growth. In studies where parasitic nematodes that reduce plant growth have been used, it has generally been found that plants inoculated with both nematodes and VA fungi have intermediate yields between those inoculated with either microorganism alone, indicating that the beneficial effect of VA fungi does not completely compensate for the damage caused by the nematodes [6, 153, 154]. Such observations have, for example, been made on tropically grown *V. unguiculata* inoculated with both *G. mosseae* and *P. sefaensis* (Table 5). Although Germani *et al.* [72] found that the harmful effect of *Scutellonema cavenessi* on soybean can be totally suppressed by *G. mosseae*, growth of mycorrhizal plants infected with nematodes must generally be governed by an equilibrium between an inhibitory (nematodes) and a stimulative (VA fungi) factor, the state of this equilibrium depending on the host plant, the nematode and the VA fungus involved as well as the soil environment.

230

Table 5. Effect of single and combined inoculations with *Glomus mosseae* and *Pratylenchus sefaensis* on growth and nodulation of *Vigna unguiculata* cv TVX 7 – 5H in sterile soil (Ollivier, Almeida and Diem, unpublished data)

	Average pod dry wt. (g plant⁻¹)	Average nodule dry wt. (mg plant⁻¹)	Average total dry wt. (g plant⁻¹)	Root segments infected by VA fungus (%)	Shoot N (%)	Shoot P (%)
Control	5.75^a	175^a	11.09^a	0	2.05	0.06
P. sefaensis	4.66^b	181^b	9.85^b	0	2.40	0.06
G. mosseae	6.50^c	248^b	15.22^c	93^a	2.40	0.09
P. sefaensis + G. mosseae	5.90^a	258^b	12.92^d	82^b	2.40	0.12

60-day old plants, all inoculated with *Rhizobium* strain CB 756
Values followed by the same letters in columns are not significantly different (*P* = 0.05)
Soil: pH (KCl): 6.9; total P (ppm): 172; available P, Olsen (ppm): 102

4.3.6. Hyperparasitic fungi. Spores of certain VA fungi can often be parasitized by such hyperparasitic fungi as *Anquillospora pseudolongissimi, Humicola fuscoatra* and a species of *Phlyctochytrium* [46, 189]. The presence of hyperparasitic fungi in soil can cause a decrease in the population density of VA fungal species and affect the physiological function of the mycorrhizae [189].

VA fungi seem to differ in their susceptibility to hyperparasities; for example, *Glomus macrocarpus* is more susceptible to parasitism than *Gigaspora gigantea* [189], whilst *Gigaspora constrictus* is more resistant than *Gigaspora margarita* to attack [46]. It has been suggested that hyperparasitic fungi may play a role in controlling VA fungal flora in the soil and, according to Daniels and Menge [46], the use of hyperparasitized VA mycorrhizal inoculum may explain some of the erratic results obtained in tests with VA fungi.

4.4. Host genotype

VA mycorrhizae are not always beneficial associations and this depends not only on the culture conditions but also on the species of fungus and host plant involved. Whilst VA fungi are known to vary in their ability to infect and transfer P to the plant, very little is known of the role of the host genotype in the expression of VA mycorrhizae. The efficiency of a same VA fungus can vary markedly between different species of host plant so that certain host—fungus associations are more effective than others [117, 169]. Skipper and Smith [204] suggested that the specific cultivar-fungal response was dependent on soil pH. Menge *et al.* [128] attributed these variations in mycorrhizal dependency to the differing ability of plants to absorb P from low P soils but other characteristics inherent to plants may also be determinant (Section 3.1.3). Response to VA mycorrhizae can also vary within a plant species; in screening experiments Bertheau *et al.* [29] obtained positive, negative or no response to VA infection in wheat depending on the host cultivar and irrespective of infection levels. Similar observations have been made on maize lines [83] and on *Vigna unguiculata* cultivars; Table 6 shows the effect of inoculation of three VA fungi on growth of two cultivars. Whilst growth is stimulated by all three fungi in one cultivar (58-185) the other only responds to infection by *Glomus* E_3 and *Glomus mosseae* but not *Glomus epigaeus*. O'Bannon *et al.* [155], on the contrary, did not find such varietal effects in alfalfa, despite the differences in dormancy, hardiness and area of adaptation of the different cultivars.

5. Agricultural significance of VA mycorrhizae

5.1. Inoculation experiments

Experimental studies on the effect of VA mycorrhizae on plant growth only began in the 1960's. These were generally confined to pot experiments in a small volume

Table 6. Growth responses of two cultivars of *Vigna unguiculata* to inoculation with *Glomus mosseae, Glomus epigaeus* and *Glomus* E_3 in sterile soil (Ollivier, Bertheau, Diem and Gianinazzi-Pearson, unpublished data)

	cv 58-185		cv Bambey 30	
	Average shoot dry wt. (g plant^{-1})	Root segments infected (%)	Average shoot dry wt. (g plant^{-1})	Root segments infected (%)
Control	0.75^a	0	1.10^a	0
G. epigaeus	1.47^{ab}	40^a	1.28^a	32^a
Glomus E_3	2.40^c	91^b	2.24^b	89^b
G. mosseae	1.71^{cb}	81^b	1.99^b	85^b

50-day old plants, all inoculated with *Rhizobium* strain CB 756
Soil: pH (KCl): 6.2; total P (ppm): 79; available P, Olsen (ppm): 10
Values followed by the same letters in columns are not significantly different ($P = 0.05$)

of sand or sterilized soil, so that during a decade little was known about the effect of an introduced VA fungus on plant growth in the presence of a natural soil microflora and competition from indigenous VA fungi. Knowledge is still very limited, in the absence of suitable techniques, of the competitive ability of introduced preselected VA fungi in natural, non-sterilized soils.

The success of VA inoculation under natural conditions depends on many factors including establishment of introduced VA mycorrhizae, crop management, production of inoculum and up to now greater success has been obtained with nursery-raised perennials, which are often produced in disinfected soils, than in field-sown annuals.

5.1.1. Nursery-raised perennial plants. A number of perennial plants particularly citrus and forest tree species cannot apparently develop normally unless the roots become mycorrhizal and it has been suggested that a large number of species of hardwood seedlings have an obligate dependence on VA mycorrhizae for normal growth. Stunting of citrus seedlings in fumigated nurseries soils can be corrected by VA inoculation [113, 125, 213] and some VA inoculation experiments have also been applied to improve growth of important forest tree species [35, 40, 108, 114, 119].

These plant species are normally raised in fumigated or steamed soil in nurseries, so that VA inoculation at this stage can ensure good infection before transplanting the seedlings in the field. Under these conditions, VA inoculations should have a great possibility to succeed. The importance of VA mycorrhizae for nursery production of seedlings has been reviewed by Menge *et al.* [125] for citrus and by Kormanik *et al.* [114] for hardwood species.

5.1.2. Field-sown annual plants. More than twenty VA inoculation experiments have been attempted up to now to study the effects of artificial inoculation on growth of annual crops growing in non-sterile soils. Mosse *et al.* [144] first demonstrated that beneficial effects of VA mycorrhizae on pre-inoculated onions were maintained after transplanting into unsterilized field soil and almost simultaneously, increased onion growth and corn yield were reported after direct inoculation of unsterilized soil at the time of sowing [100, 142].

Important growth responses have also been obtained in wheat, maize and alfalfa seedlings inoculated with selected VA fungi before transplanting into field soils with low levels of available P [11, 112]. These and subsequent experiments raise the hope of practical applications; however, only limited extrapolations can be made of results from pot or transplant experiments to field conditions.

In spite of this and of the ubiquity of indigenous VA fungi in field soils, positive responses to inoculation of crops directly in the field have occasionally been obtained in non-disinfected soils. The increased growth and development of VA inoculated cotton obtained by Rich and Bird [179] is the first report of successful inoculation of a crop plant in the field. Black and Tinker [32] also increased potato yield in a field inoculation experiment using naturally infected soil which was applied in the furrows before planting.

Promising results have been obtained with soybeans in India [16], onion, barley and alfalfa in England [158] and alfalfa in Spain [10] using infected roots and soil as inoculum placed below seeds at the time of planting. Other successful field inoculation experiments have been reported with *Lotus pedunculatus* and clover in New Zealand using respectively soil pellets [85] or seeds pelleted with infested soil [165]. In each example indigenous VA fungal populations were either extremely low or very inefficient for plant growth. There is evidence that under natural conditions and in the presence of indigenous VA fungal flora, an introduced fungus can sometimes act as a regulator of plant productivity rather than as a stimulative factor for plant growth. Results obtained with inoculated field-grown soybeans in Senegal (Table 7) suggest a role of VA mycorrhizae in producing homogeneous rather than increased shoot and grain yields.

5.2. Mycorrhizal inoculum

One of the major obstacles to the establishment of pre-selected VA fungi in field-grown crops is to provide sufficient inoculum for large-scale operations. As long as

Table 7. Effect of field inoculation with *Glomus mosseae* on yield of soybean cv 44A-73 (Ganry and Diem, unpublished data)

	Yield (g m^{-2})		Coefficient of variation (%)		Root segments infected (%)[a]
	Grain	Shoot + pod	Grain	Shoot + pod	
Control	137	284	33	27	38
G. mosseae	138	339	30	14	48
Control + P	155	412	12	15	36
G. mosseae + P	185	416	4	8	42

[a] 55-day old plants

P added as triple superphosphate: 60 kg P_2O_5 ha^{-1}

isolated VA fungi cannot be grown on synthetic medium, VA mycorrhizal inoculum has to be prepared by multiplication of the selected fungi in roots of susceptible host plants growing in sterilized substrates or soil. Pots or large containers are currently used to produce large quantities of inoculum under greenhouse conditions with maximum protection against contamination with other VA fungal species, pathogenic or hyperparasitic fungi.

5.2.1. Infectivity of inoculum. Plant growth responses appear to be related not only to the ability of VA fungi to translocate and transfer nutrients to the plant but also to the rapidity with which various inocula of the same fungus infect roots. Hall [82] found than an inoculum of root segments caused more rapid plant response to VA infection than spores, whilst Manjunath and Bagyaraj [121] reported that, on the contrary, mycorrhizal establishment was better with spores. According to Powell [164] and Johnson [107] hyphae from root segments would infect more rapidly since they infect immediately without forming pre-infection structures as germinating spores do. Differences in the extent of external hyphae developing from either root segments or spores may account for the large differences in inoculum infectivity ratings between different mycorrhizal fungi [167].

5.2.2. Types of inoculum. Inoculum can consist of infested soil, mycorrhizal roots with external mycelium and/or spores obtained from pot cultures. For certain fungi like *Glomus mosseae* pure suspensions of hyphae and spores can be collected from infected roots and soil. Several methods exist for recovering spores from soil: wet-sieving and decantation [70], flotation-adhesion [209] and the flotation-bubbling method [61]. For physiological studies, axenic mycorrhizal spores can be subsequently obtained by different sterilization treatments [131].

Different types of inoculum have been studied in order to facilitate manipulations and to maintain fungal infectivity (survival) during storage. Infested soil can be a highly infective type of inoculum, suitable for large-scale inoculations [32, 170, 179]. VA fungi have also been introduced into soil as soil pellets containing a mixture of infected root fragments, infested soil and clay [84]. Recently, in Senegal, Ganry and Diem (Table 7) successfully infected field-grown soybeans with an inoculum consisting of spores and infected roots of *Glomus mosseae* entrapped in beads of alginate gel [54].

There is almost no information on methods for long-term storage of mycorrhizal inoculum. Mosse and Hayman [143] claim that air-dried sievings of soil mycelium and infected root fragments can retain infectivity up to three years when stored at low temperatures. Jackson *et al.* [100] successfully used lyophilized mycorrhizal roots to inoculate soybean growing in non-sterile soil, but in more extensive tests Crush and Pattison [42] found that lyophilization led to greatly reduced infectivity of VA fungi. Recently preservation of VA spores by L-drying was proposed for routine long-term storage of mycorrhizal inoculum [215].

5.2.3. Methods to improve inoculum production. Different approaches have been made to improve both qualitatively and quantitatively the production of inoculum:

Selection of the host plant. One way to massively produce inoculum is to grow a given VA fungus in roots of strongly mycotrophic plants. With this aim in mind, Bagyaraj and Manjunath [13] recently compared the mycotrophy of different host plants, and it appears from their screening that, under tropical conditions, *Panicum maximum* is a suitable host for culturing VA fungi in large quantities.

An original and potentially effective method for production of mycorrhizal inoculum has been recently suggested by Parke and Linderman [159] using the moss *Funaria hygrometrica* as a host for culturing VA fungi. The moss-mycorrhizal inoculum can easily be peeled off the soil surface so that the mycorrhizal pot culture may be left intact for successive crops of moss. Mosses have the advantage of supporting few pathogens that attack most higher plants and the merit of this method is to produce 'clean' rather than important amounts of mycorrhizal inoculum.

Host plant culture. Menge *et al.* [125] proposed a scheme for producing large amounts of inoculum for inoculation of citrus in nurseries. The pre-selected VA fungus, carefully maintained in pot culture in the greenhouse, is multipled on a selected healthy host plant grown in sterilized soil in large containers. For phytosanitary reasons this host plant should have no diseases in common with the crop for which the inoculum (infected roots and soil) is intended. Another technique for mass production of inoculum which has recently been devised (Mosse and Thompson, quoted by Mosse and Hayman [143]) consists of growing an infected host plant in a nutrient flow culture system and harvesting the external mycelium and spores that are produced.

Selection of the VA fungus. This must take into account the efficiency of the VA fungus, its ability to produce large amounts of inoculum and the facility with which the latter can be manipulated. A fungus which seems to meet these requirements is *Glomus epigaeus* recently described by Daniels and Trappe [48]. Its efficiency on a wide range of host plants, the ability to produce abundant sporocarps on the soil surface, and the ease with which these can be harvested and stored make it a commercially interesting fungus [47]. However, *Glomus epigaeus* appears to be adapted to cooler climates [49] and research is needed for fungi or similar potential in tropical soils.

5.3. Inoculation techniques

Since annual crops are generally sown directly whilst perennial crops are often transplanted into the field, inoculation techniques will vary: perennials can be pre-infected in nursery soils whilst for annuals inoculum must be introduced into soil either directly or by pelleting of seeds.

5.3.1. Pre-inoculated transplants. A number of perennial tropical crops like coffee, tea, cocoa, rubber, oil palm and citrus are normally raised in nurseries. Seedlings

can be pre-inoculated in disinfected soils or substrates and transplanted into the field when the roots are heavily infected [125]. Pre-inoculation at the nursery stage may also be usefully applied to tree species used in arboriculture or for afforestation (see Kormanik *et al.* [114]). The transplants could act as centers of infection for adjacent plants if the VA fungi are able to successfully compete with indigenous populations. As nothing is known of this aspect of VA mycorrhizae, the necessity for repeated inoculation after transplant should be investigated.

5.3.2. Pelleting seeds with inoculum. Selected VA fungi have been successfully introduced into soil with annual crops by pelleting seeds with mycorrhizal inoculum [84, 165, 211]. Inoculum, consisting of root pieces and/or spores, can be coated onto seeds using inert additives, such as perlite or methyl cellulose [63, 86, 211]. The pelleted seed method would be a convenient method of inoculation in practice but according to certain reports [100, 125] the effectiveness of the inoculum is reduced because it is not in an area of root proliferation. Possible harmful effects of pelleting on seed germination should also be taken into account [84].

5.3.3. Direct inoculation of soil. Inocula in the form of soil pellets [84], lyophilized infected roots [100], infested soil and infected roots [1, 16, 32, 158, 170] or a slurry of infected sievings [226] are introduced into the soil either with or below seeds. These methods favour early infection of the actively growing main host roots by the introduced VA fungus rather than by the more dispersed propagules of indigenous VA fungi. Comparing different methods of direct inoculation of two VA fungi, Hayman *et al.* [94] found that plants inoculated in furrows with crude or fluid-drilled inoculum applied with the seed were most infected. They suggested that fluid drilling could be a suitable technique for field inoculation because smaller quantities of inoculum are needed and inocula such as rhizobia can also be incorporated for legumes.

5.3.4. Concluding remarks. The problem of establishment of selected VA fungi in the field is not only one of introduction of inoculum but also one of its survival and spread in the face of competition from indigenous VA fungi.

Based on results obtained in greenhouse experiments, Powell [166] calculated that even a mycorrhizal fungus having a fast rate of spread would only move 65 m in 150 years which suggests that natural spread of VA fungi through soil is probably slow. According to Powell the movement of topsoil (and VA fungi) by soil erosion, worm activities or agricultural machinery can contribute to the spread of VA fungi. The method used by Rich and Bird [179] in their successful field inoculation trials is an example to illustrate the high effectivity of the displacement of infested soil to soil lacking VA fungi.

An alternative field inoculation technique could therefore consist of creating a few reservoirs for infection in the same field and then spreading the infested topsoil over the surrounding areas. This proposed '*in situ* multiplication—propagation' method is somewhat similar to that used for the multiplication of *Azolla* to fertilize rice fields (Watanabe, this volume).

The soil in the area to be infested can be previously fumigated to ensure good multiplication of the selected VA fungus, which would be cultured using a strongly mycotrophic host plant provided with an abundant fine root system (Sudan grass, Guinea grass . . .). When soil and roots are well infected, propagation of the fungus in the top soil can be carried out using agricultural machinery. The major advantages of this method are that *in situ* mass production eliminates problems of inoculum conservation and transport, inoculum should be highly infective due to the abundant freshly infected roots with extra-matrical hyphae and spores, and the adaptation of a given efficient VA fungus to characteristics of the field soil can be determined before its use.

6. Use of VA mycorrhizae as biofertilizers for tropical plants

The beneficial effect of VA mycorrhizae on plant growth is particularly spectacular in P-deficient soils and because of the generally low availability of P in tropical soils, the potential for the exploitation of VA mycorrhizae in agriculture seems to be much greater than in temperate soils. Cornet and Diem (unpublished data) found that seedlings of *Acacia raddiana* failed to grow in a sterilized soil from Senegal unless they were mycorrhizal or supplemented with a soluble phosphate source. In these conditions, they grew well even though they were not nodulated. This suggests that in some tropical soils P deficiency may be even more important than N deficiency as a factor limiting plant growth.

As Black [31] points out, VA mycorrhizae could be important factors for increasing plant productivity in developing countries for several reasons: the intrinsically low availability of P or high P-fixing ability of many tropical soils, the difficulties for locally manufacturing soluble phosphate fertilizers and the high cost of importation, and the mycotrophic habits of many tropical plant species of economic importance especially legumes. Jehne [104] examined some aspects of managing VA mycorrhizae for enhancing nutrition, growth and productivity of tropical pastures. However, as Hayman [91] emphasized, 'undue faith in mycorrhizae as an agriculture elixir should be tempered by a realistic appraisal of which aspects might best be exploited on a practical scale', the last sections of this review are therefore devoted to an assessment of the limitations, potentialities and research needs for applications of VA mycorrhizae in agricultural practices in the tropics.

6.1. Limitations

6.1.1. Lack of inoculum. Because of the difficulties in obtaining important quantities of inoculum, practical VA inoculation is at present provisionally restricted to the nursery level. Two major problems remain for annual crops: the massive production of high quality inoculum and inoculation techniques applicable on a field scale [91].

Mosse and Hayman [143] calculated that the amount of inoculum required to inoculate 1 ha (2,500–8,000 kg) is impractical to produce, handle or transport. Thus as long as VA fungi cannot be cultured *in vitro* or suitable techniques for field inoculation have not been developed (Section 5.3.4), the application of VA mycorrhizae for improved production of annual crops will be severely limited.

6.1.2. Competition with indigenous VA fungi. The success of field inoculation with a given VA fungus depends not only on the nutrient status of a soil (e.g. P levels) but also on an ensemble of factors controlling infection and establishment of the introduced efficient fungus. Inoculation with specific VA fungi has in fact been successful in field soils where indigenous VA fungal populations were either extremely low or very inefficient for plant growth (Section 5.1.2).

Density and competitivity of indigenous VA fungal populations in the soil are important limiting factors for the introduction of new species of VA fungi. In the situation that the indigenous fungi are efficient, their effect on plant growth would mask that induced by the introduced fungi, but even if the former are less or not efficient, their high population in the soil could be so important that they readily infect most of the host root system preventing subsequent infection by the more efficiently introduced fungi. Unfortunately very little is known concerning the competition between VA fungi for infection sites in roots.

In pot inoculation experiment with *Glomus monosporus* and *Glomus fasciculatus*, Abbott and Robson [1] found that there was no antagonistic effect by any of the introduced fungi on early stages of infection by the indigenous fungi and the VA fungi which were most efficient (*G. monosporus*) produced more mycorrhizae at an earlier stage in plant growth than the less efficient fungi. Generally, in the case of pot experiments with efficient VA fungi in non-sterile soil, it is probable that the quantity of inoculum introduced into the soil is so great that early infection of the root system occurs and its propagation through the host tissue is favoured by the pot-imposed restrictions on plant growth [99]. This is probably one of the reasons for which extrapolations of pot experiments to field trials can lead to discrepancies in the mycorrhizal effects obtained.

6.2. Potentialities

The potential benefits of VA inoculations concern a wide range of tropical plant species normally raised in nurseries before transplanting VA inoculations could be extended to industrial perennials of great importance in the tropics such as cocoa, coffee, tea, rubber or fruit trees. *In vitro* vegetative propagation techniques are being increasingly employed for the rapid reproduction of high quality, healthy plants of a wide variety of temperate and tropical species. The precocious inoculation of such plants with efficient VA fungi during their multiplication *in vitro* is a recent promising perspective for the use of VA mycorrhizae [134].

Other situations where inoculation of VA fungi would most probably be highly beneficial is in the reclamation of marginal soils or of eroded, degraded or unstable habitats. Such soil situations are frequent in the tropics under both semi-arid and humid climates and investigations have recently begun into the potential role of VA mycorrhizae in revegetation practices in such severely disturbed habitats of semi-arid regions, where reduction of active inoculum in disturbed marginal soils seems to be an important ecological factor determining the subsequent colonizing plant species [133, 178]. Soil structure could be also controlled by external hyphae of VA fungi which contribute to stabilize e.g. dunes by aggregating sand grains [150].

7. Research needs and conclusion

Since interactions between the host plant, VA fungus, soil and climate will determine the mycorrhizal effect on plant growth, research for practical applications of VA mycorrhizae should bear in mind all these factors and therefore be directed towards finding appropriate host—VA fungus combinations adapted to well-defined soil types and climatic conditions.

Recent research has shown that varieties within a plant species can differ greatly in their response to inoculations with VA fungi, although the reasons for this genetical variation is yet to be determined. Screening of the most responsive varieties and species of crop plants is of utmost importance for mycorrhiza research in the tropics. Similar variations can be observed in the ability of different species or isolates of VA fungi to improve plant growth. However, a large number of inoculation experiments with tropical plants have employed VA fungi isolated from soils in temperate regions and not necessarily adapted to conditions occurring in tropical soils. There is therefore an urgent need to isolate VA fungi from different types of tropical soils and to assess their efficiency for plant growth, their ecological limits and their ability to develop and survive under adverse environmental conditions. This research for efficient tropical VA fungal isolates should also take into account their ability to produce large amounts of inoculum which can be easily manipulated and stored.

There are still large gaps in our knowledge concerning certain aspects of VA mycorrhizae which are relevant to temperate as well as tropical zones and which limit the advances that can be made towards practical applications. Amongst these are the inability to successfully culture isolated VA fungi, the lack of information concerning competition between VA fungi in soils and for infection sites, the relationship between inoculum potential and infectivity of VA fungi, and the influence of nutrients other than phosphate on mycorrhizal responses.

In spite of this and the need for further research into the more fundamental aspects of VA mycorrhizae (fungal physiology, host—fungus interactions, . . .), the utilization of this symbiotic association for improving production of certain, especially perennial crop plants and forest trees is coming to age as a potential practical reality.

240

References

1. Abbott, L.K. and Robson, A.D. 1978 Growth of subterranean clover in relation to the formation of endomycorrhizas by introduced and indigenous fungi in a field soil. New Phytol. 81, 575–585.
2. Abbott, L.K. and Robson, A.D. 1979 A quantitative study of the spores and anatomy of mycorrhizas formed by a species of *Glomus* with reference to its anatomy. Aust. J. Bot. 27, 363–375.
3. Aboulroos, S.A., Ezzedin Ibrahim, M., Wassif, M. and El-Shall, M. 1980 Relationship between self-diffusion coefficient and uptake of phosphate. Z. Pflanzenernaehr. Bodenkd. 143, 659–665.
4. Asai, T. 1944 Über die Mykorrhizen-bildung der leguminosen Pflanzen. Jap. J. Bot. 13, 463–485.
5. Asimi, S., Gianinazzi-Pearson, V. and Gianinazzi, S. 1980 Influence of increasing soil phosphorus levels on interactions between vesicular-arbuscular mycorrhizae and *Rhizobium* in soybeans. Can. J. Bot. 58, 2200–2205.
6. Atilano, R.A., Rich, J.R., Ferris, H. and Menge, J.A. 1976 Effect of *Meloidogyne arenaria* on endomycorrhizal grape (*Vitis vinifera*) rootings. J. Nematol. 8, 278 (Abstract).
7. Atkinson, D. and Davison, A.W. 1973 The effects of phosphorus deficiency on water content and response to drought. New. Phytol. 72, 307–313.
8. Azcon, R., Barea, J.M. and Hayman, D.S. 1976 Utilization of rock phosphate in alkaline soils by plants inoculated with mycorrhizal fungi and phosphate-solubilizing bacteria. Soil Biol. Biochem. 8, 135–138.
9. Azcon, R., Azcon-G De Aguilar, C. and Barea, J.M. 1978 Effects of plant hormones present in bacterial cultures on the formation and responses to VA endomycorrhiza. New Phytol. 80, 359–364.
10. Azcon-Aguilar, C. and Barea, J.M. 1981 Field inoculation of Medicago with VA mycorrhiza and *Rhizobium* in a phosphate-fixing agricultural soil. Soil Biol. Biochem. 13, 19–22.
11. Azcon-G De Aguilar, C., Azcon, R. and Barea, J.M. 1979 Endomycorrhizal fungi and *Rhizobium* as biofertilizers for *Medicago sativa* in normal cultivation. Nature, London, 279, 325–327.
12. Bagyaraj, D.J. and Manjunath, A. 1980 Response of crop plants to VA mycorrhizal inoculation in an unsterile Indian soil. New Phytol. 85, 33–36.
13. Bagyaraj, D.J. and Manjunath, A. 1980 Selection of a suitable host for mass production of VA mycorrhizal inoculum. Plant and Soil 55, 494–498.
14. Bagyaraj, D.J. and Menge, J.A. 1978 Interaction between a VA mycorrhiza and *Azotobacter* and their effects on rhizosphere microflora and plant growth. New Phytol. 80, 567–573.
15. Bagyaraj, D.J., Manjunath, A. and Patil, R.B. 1979 Occurrence of vesicular-arbuscular mycorrhizas in some tropical aquatic plants. Trans. Br. mycol. Soc. 72, 164–169.
16. Bagyaraj, D.J., Manjunath, A. and Patil, R.B. 1979 Interaction between a vesicular-arbuscular mycorrhiza and *Rhizobium* and their effects on soybean in the field. New Phytol. 82, 141–146.
17. Bagyaraj, D.J., Manjunath, A. and Reddy, D.D.R. 1979 Interaction of vesicular-arbuscular mycorrhiza with root knot nematodes in tomato. Plant and Soil 51, 397–403.
18. Baker, D. and Torrey, J.G. 1979 The isolation and cultivation of actino-mycetous root nodule endophytes. In: Symbiotic nitrogen fixation in the management of temperate forests, pp. 38–56. Gordon, J.C., Wheeler, C.T. and Perry, D.A. (eds.), Oregon State University Press, Corvallis.

19. Bakerspiegel, A. 1953 Soil as a storage medium for fungi. Mycologia 45, 595–604.
20. Baltruschat, H., Sikora, R.A. and Schönbeck, F. 1973 Effect of VA mycorrhizae (*Endogone mosseae*) on the establishment of *Thielaviopsis basicola* and *Meloidogyne incognita* in tobacco. 2nd Intern. Congr. Plant Pathol. Abstr. No. 0661.St Paul.
21. Barea, J.M., Azcon, R. and Hayman, D.S. 1975 Possible synergistic interactions between *Endogone* and phosphate-solubilizing bacteria in low-phosphate soils. In: Endomycorrhizas, pp. 409–417. Sanders, F.E., Mosse, B. and Tinker, P.B. (eds.), Academic Press, London.
22. Barea, J.M., Brown, M.E. and Mosse, B. 1973 Association between VA mycorrhiza and *Azotobacter*. Rothamsted Exp. Stat. Report for 1972, pp. 81–82.
23. Bärtschi, H., Gianinazzi-Pearson, V. and Vegh, I. 1981 Vesicular-arbuscular mycorrhiza formation and root rot disease (*Phytophthora cinnamomi* Rands) development in *Chamaecyparis lawsoniana* (Murr.) Parl. Phytopathol. Z. 102, 213–218.
24. Baylis, G.T.S. 1959 The effect of vesicular-arbuscular mycorrhizas on growth of *Griselinia littoralis* (Cornaceae) New Phytol. 58, 274–280.
25. Baylis, G.T.S. 1969 Host treatment and spore production by Endogone. N.Z. J. Bot. 7, 173–174.
26. Baylis, G.T.S. 1975 The magnolioid mycorrhiza and mycotrophy in root systems derived from it. In: Endomycorrhizas, pp. 373–389. Sanders, F.E., Mosse, B. and Tinker, P.B. (eds.), Academic Press, London.
27. Berry, A. and Torrey, J.G. 1979 Isolation and characterization *in vivo* and *in vitro* of an actinomycetous endophyte from *Alnus rubra* Bong. In: Symbiotic nitrogen fixation in the management of temperate forests, pp. 69–83. Gordon, J.C., Wheeler, C.T. and Perry, D.A. (eds.), Oregon State University Press, Corvallis.
28. Bertheau, Y. 1979 Les mycorhizes du blé: effet variétal et quelques aspects biochimiques. Thèse de 3è cycle, Université de Dijon.
29. Bertheau, Y., Gianinazzi-Pearson, V. and Gianinazzi, S. 1980 Développement et expression de l'association endomycorhizienne chez le blé. I. Mise en évidence d'un effet variétal. Ann. Amélior. Plantes 30, 67–78.
30. Bird, G.W., Rich, J.R. and Glover, S.U. 1974 Increased endomycorrhizae of cotton roots in soil treated with nematicides. Phytopathology 64, 48–51.
31. Black, R. 1980 The role of mycorrhizal symbiosis in the nutrition of tropical plants. In: Tropical Mycorrhiza Research, pp. 191–202. Mikola, P. (ed.), Clarendon Press, Oxford.
32. Black, R.L.B. and Tinker, P.B. 1977 Interaction between effects of vesicular-arbuscular mycorrhizae and fertilizer phosphorus on yields of potatoes in the field. Nature. London 267, 510–511.
33. Bowen, G.D. 1980 Mycorrhizal roles in tropical plants and ecosystems. In: Tropical Mycorrhiza Research, pp. 165–190. Mikola, P. (ed.), Clarendon Press, Oxford.
34. Brown, M.E. and Carr, G.H. 1979 Effects on plant growth of mixed inocula of VA endophytes and root microorganisms. Rothamsted Exp. Stat. Report for 1979, Part 1, p. 187.
35. Bryan, W.C. and Kormanik, P.P. 1977 Mycorrhizae benefit survival and growth of sweetgum seedlings in the nursery South J. Appl. For. 1, 21–23.
36. Callaham, D., Del Tredici, P. and Torrey, J.G. 1978 Isolation and cultivation *in vitro* of the actinomycete causing root nodulation in *Comptonia*. Science 199, 899–902.

37. Carling, D.E. and Brown, M.F. 1980 Relative effect of vesicular-arbuscular mycorrhizal fungi on the growth and yield of soybeans. Soil Sci. Soc. Amer. J. 44, 528–531.
38. Carling, D.E., Riehle, W.G., Brown, M.F. and Johnson, D.R. 1978 Effects of a vesicular-arbuscular mycorrhizal fungus on nitrate reductase and nitrogenase activites in nodulating and non-nodulating soybean. Phytopathology 68, 1590–1596.
39. Charreau, C. 1974 Soils of tropical dry and dry-wet climatic areas of West Africa and their use and management: a series of lectures at the Department of Agronomy. Cornell University AID Grant csd 2834. Agronomy mimeo, 74–26.
40. Clark, F.B. 1969 Endotrophic mycorrhizal infection of tree seedlings with *Endogone* spores. Forest Sci. 15, 134–138.
41. Crush, J.R. 1974 Plant growth responses to vesicular-arbuscular mycorrhiza VII. Growth and nodulation of some herbage legumes. New Phytol. 73, 743–749.
42. Crush, J.R. and Pattison, A.C. 1975 Preliminary results on the production of vesicular-arbuscular mycorrhizal inoculum by freeze drying. In: Endomycorrhizas, pp. 485–493. Sanders, F.E., Mosse, B. and Tinker, P.B. (eds.), Academic Press, London.
43. Daft, M.J. and El-Giahmi, A.A. 1976 Studies on nodulated and mycorrhizal peanuts. Ann. Appl. Biol. 83, 273–276.
44. Daft, M.J. and El-Giahmi, A.A. 1978 Effect of vesicular-arbuscular mycorrhiza on plant growth VIII. Effects of defoliation and light on selected hosts. New Phytol. 80, 365–372.
45. Daft, M.J., Chilvers, M.T. and Nicolson, T.H. 1980 Mycorrhizas of the Liliiflorae I. Morphogenesis of *Endymion non-scriptus* (L.) Garcke and its mycorrhizas in nature. New Phytol. 85, 181–189.
46. Daniels, B.A. and Menge, J.A. 1980 Hyperparasitization of vesicular arbuscular mycorrhizal fungi. Phytopathology 70, 584–588.
47. Daniels, B.A. and Menge, J.A. 1981 Evaluation of the commercial potential of the vesicular-arbuscular mycorrhizal fungus *Glomus epigaeus*. New Phytol. 87, 345–354.
48. Daniels, B.A. and Trappe, J.M. 1979 *Glomus epigaeus* sp. nov. a useful fungus for vesicular-arbuscular mycorrhizal research. Can. J. Bot. 57, 539–542.
49. Daniels, B.A. and Trappe, J.M. 1980 Factors affecting spore germination of the vesicular-arbuscular mycorrhizal fungus, *Glomus epigaeus*. Mycologia 72, 457–471.
50. Davis, R.M., Menge, J.A. and Zentmyer, G.A. 1978 Influence of vesicular-arbuscular mycorrhizae on *Phytophthora* root rot of three crop plants. Phytopathology 68, 1614–1617.
51. Davis, R.M., Menge, J.A. and Erwin, D.C. 1979 Influence of *Glomus fasciculatus* and soil phosphorus on *Verticillium* wilt of cotton. Phytopathology 69, 453–456.
52. De Alwis, D.P. and Abeynayake, K. 1980 A survey of mycorrhizae in some forest trees of Sri-Lanka. In: Tropical Mycorrhiza Research, pp. 146–153. Mikola, P. (ed.), Clarendon Press, Oxford.
53. Diem, H.G., Gueye, I., Gianinazzi-Pearson, V., Fortin, J.A. and Dommergues, Y.R. 1981 Ecology of VA mycorrhizae in the tropics: the semi-arid zone of Senegal. Acta Oecologica. Oecol. Plant. 2, 53–62.
54. Diem, H.G., Jung, G., Mugnier, J., Ganry, F. and Dommergues, Y. 1981 Alginate-entrapped *Glomus mosseae* for crop inoculation. In: Proc. 5th Nacom, Université Laval, Quebec, Canada, August 16–21, 1981 (abstract).

55. Dommergues, Y.R., Belser, L.W. and Schmidt E.L. 1978 Limiting factors for microbial growth and activity in soil. In: Adv. Microbiol. Ecol. Vol. 2, pp. 49–104. Alexander, M. (ed.), Plenum Press, New York and London.
56. El-Giahmi, A.A., Nicolson, T.H. and Daft, M.J. 1976 Endomycorrhizal fungi from Libyan soils. Trans. Br. mycol. Soc. 67, 164–169.
57. El-Giahmi, A.A., Nicolson, T.H. and Daft, M.J. 1976 Effects of fungal toxicants on mycorrhizal maize. Trans. Br. mycol. Soc. 67, 172–173.
58. Fox, J.A. and Spasoff, L. 1972 Interaction of *Heterodera solanacearum* and *Endogone gigantea* on tobacco. J. Nematol. 4, 224–225.
59. Frank, A.B. 1885 Über die auf Wurzelsymbiose beruhende Ernährung gewisser Bäume durch Unterirdische Pilze. Ber. Dtsch. Bot. Ges. 3, 128–145.
60. Furlan, V. and Fortin, J.A. 1973 Formation of endomycorrhizae by *Endogone calospora* on *Allium cepa* under three temperature regimes. Naturaliste Can. 100, 467–477.
61. Furlan, V. and Fortin, J.A. 1975 A flotation-bubbling system for collecting Endogonaceae spores from sieved soil. Naturaliste Can. 102, 663–667.
62. Furlan, V. and Fortin, J.A. 1977 Effects of light intensity on the formation of vesicular-arbuscular endomycorrhizas on *Allium cepa* by *Gigaspora calospora*. New Phytol. 79, 335–340.
63. Gaunt, R.E. 1978 Inoculation of vesicular-arbuscular mycorrhizal fungi on onion and tomato seeds. N.Z. J. Bot. 16, 69–71.
64. Gauthier, D., Diem, H.G. and Dommergues, Y. 1981 *In vitro* nitrogen fixation by two actinomycete strains isolated from *Casuarina* nodules. Appl. Envir. Microbiol. 41, 306–308.
65. Gerdemann, J.W. 1955 Relation of a large soil-borne spore to phycomycetous mycorrhizal infections. Mycologia 47, 619–632.
66. Gerdemann, J.W. 1964 The effect of mycorrhiza on the growth of maize. Mycologia 56, 342–349.
67. Gerdemann, J.W. 1965 Vesicular-arbuscular mycorrhizae formed on maize and tuliptree by *Endogone fasciculata*. Mycologia 57, 562–575.
68. Gerdemann, J.W. 1968 Vesicular-arbuscular mycorrhiza and plant growth. Ann. Rev. Phytopathol. 6, 397–418.
69. Gerdemann, J.W. and Bakshi, B.K. 1976 Endogonaceae of India: two new species. Trans. Br. mycol. Soc. 66, 340–343.
70. Gerdemann, J.W. and Nicolson, T.H. 1963 Spores of mycorrhizal *Endogone* species extracted from soil by wet sieving and decanting. Trans. Br. mycol. Soc. 46, 235–244.
71. Germani, G., Diem, H.G. and Dommergues, Y.R. 1980 Influence of 1,2-dibromo-3-chloropropane (DBCP) fumigation on mycorrhizal infection of field-grown groundnut. In: Tropical Mycorrhiza Research, pp. 245–246. Mikola, P. (ed.), Clarendon Press, Oxford.
72. Germani, G., Ollivier, B. and Diem, H.G. 1982 Interaction of *Scutellonema cavanessi* and *Glomus mosseae* on growth and N$_2$-fixation of soybean. Rev. Nematol. 4, 277–280.
73. Gianinazzi-Pearson, V. and Gianinazzi, S. 1981 The role of endomycorrhizal fungi in phosphorus cycling in the soil. In: The fungal community, its organization and role in the ecosystem, pp. 637–652. Wicklow, D.T. and Carrol, G.C. (eds.), Marcel Dekker, New York.
74. Gianinazzi-Pearson, V., Fardeau, J.C., Asimi, S. and Gianinazzi, S. 1981 Source of additional phosphorus absorbed from soil by vesicular-arbuscular mycorrhizal soybeans. Physiol. vég. 19, 33–43.
75. Ginsburg, O. and Avizohar-Hershenson, Z. 1965 Observations on vesicular-

arbuscular mycorrhiza associated with avocado roots in Israel. Trans. Br. mycol. Soc. 48, 101—104.

76. Godse, D.B., Wani, S.P., Patil, R.B. and Bagyaraj, D.J. 1978 Response of cowpea (*Vigna unguiculata* (L.) Walp.) to *Rhizobium* VA mycorrhiza dual inoculation. Current Sci. (India) 47, 784—785.

77. Graw, D. 1979 The influence of soil pH on the efficiency of vesicular-arbuscular mycorrhiza. New Phytol. 82, 687—695.

78. Graw, D. and Rehm, S. 1977 Vesikular-arbuskuläre Mykorrhiza in den Fruchtträgern von *Arachis hypogea* L. Z. Acker-und Pflanzenbau 144, 75—78.

79. Green, N.E., Graham, S.O. and Schenck, N.C. 1976 The influence of pH on the germination of vesicular-arbuscular mycorrhizal spores. Mycologia 68, 929—933.

80. Gunze, C.M.B. and Hennessy, C.M.R. 1980 Effect of host applied auxin on development of endomycorrhiza in cowpeas. Trans. Br. mycol. Soc. 74, 247—251.

81. Hall, I.R. 1975 Endomycorrhizas of *Metrosideros umbellata* and *Weimannia racemosa*. N.Z. J. Bot. 13, 463—472.

82. Hall, I.R. 1976 Response of *Coprosma robusta* to different forms of endomycorrhizal inoculum. Trans. Br. mycol. Soc. 67, 409—411.

83. Hall, I.R. 1978 Vesicular-arbuscular mycorrhizas on two varieties of maize and one of sweetcorn. N.Z. J. Agric. Res. 21, 517—519.

84. Hall, I.R. 1979 Soil pellets to introduce vesicular-arbuscular mycorrhizal fungi into soil. Soil Biol. Biochem. 11, 85—86.

85. Hall, I.R. 1980 Growth of *Lotus pedunculatus* Cav. in an eroded soil containing soil pellets infested with endomycorrhizal fungi. N.Z. J. Agric. Res. 23, 103—105.

86. Hattingh, M.J. and Gerdemann, J.W. 1975 Inoculation of Brazilian sour orange seed with an endomycorrhizal fungus. Phytopathology 65, 1013—1016.

87. Hayman, D.S. 1974 Plant growth responses to vesicular-arbuscular mycorrhiza. VI. Effect of light and temperature. New Phytol. 73, 71—80.

88. Hayman, D.S. 1975 The occurrence of mycorrhiza in crops as affected by soil fertility. In: Endomycorrhizas, pp. 495—509. Sanders, F.E., Mosse, B. and Tinker, P.B. (eds.), Academic Press, London.

89. Hayman, D.S. 1975 Phosphorus cycling by soil microorganisms and plant roots. In: Soil Microbiology, pp. 67—91. Walker, N. (ed.), Butterworths, London.

90. Hayman, D.S. 1978 Endomycorrhizae. In: Interactions between non-pathogenic soil microorganisms and plants, pp. 401—442. Dommergues, Y.R. and Krupa, S.W. (eds.), Elsevier, Amsterdam.

91. Hayman, D.S. 1980 Mycorrhiza and crop production. Nature, London 287, 487—488.

92. Hayman, D.S. and Mosse, B. 1971 Plant growth responses to vesicular-arbuscular mycorrhiza. I. Growth of *Endogone*—inoculated plants in phosphate-deficient soils. New Phytol. 70, 19—27.

93. Hayman, D.S., Barea, J.M. and Azcon, R. 1976 Vesicular-arbuscular mycorrhiza in southern Spain: its distribution in crops growing in soil of different fertility. Phytopathol. Mediterranea 15, 1—6.

94. Hayman, D.S., Morris, E.J. and Page, R. 1979 Inoculation techniques. Rothamsted Exp. Stat. Report for 1979, Part 1, p. 187.

95. Herrera, R.A. and Ferrer, R.L. 1980 Vesicular-arbuscular mycorrhiza in Cuba. In: Tropical Mycorrhiza Research, pp. 156—162. Mikola, P. (ed.), Clarendon Press, Oxford.

96. Hirrel, M.C., Mehraveran, H. and Gerdemann, J.W. 1978 Vesicular-arbuscular mycorrhizae in the Chenopodiaceae and Cruciferae: do they occur? Can. J. Bot. 56, 2813–2817.

97. Howeler, R.H. 1980 The effect of mycorrhizal inoculation on the phosphorus nutrition of Cassava. In: Cassava cultural practice, pp. 131–137. Weber, E.J., Toro, J.C. and Graham, M. (eds.), International development Research Center, Ottawa.

98. Hussey, R.S. and Roncadori, R.W. 1978 Interaction of *Pratylenchus brachyurus* and *Gigaspora margarita* on cotton J. Nematol. 10, 16–20.

99. Islam, R., Ayanaba, A. and Sanders F.E. 1980 Response of cowpea (*Vigna unguiculata*) to inoculation with VA mycorrhizal fungi and to rock phosphate fertilization in some unsterilized Nigerian soils. Plant and Soil 54, 107–117.

100. Jackson, N.E., Franklin, R.E. and Miller, R.H. 1972 Effect of vesicular-arbuscular mycorrhizae on growth and phosphorus content of three agronomic crops. Soil Sci. Soc. Amer. Proc. 36, 64–67.

101. Janos, D.P. 1975 Effects of vesicular-arbuscular mycorrhizae on lowland tropical rainforest trees. In: Endomycorrhizas, pp. 437–446. Sanders, F.E., Mosse, B. and Tinker, P.B. (eds.), Academic Press, London.

102. Janos, D.P. 1980 Vesicular-arbuscular mycorrhizae affect lowland tropical rain forest plant growth. Ecology 61, 151–162.

103. Janse, J.M. 1896 Les endophytes radicaux de quelques plantes javanaises. Ann. Jard. Bot. Buitenz. 14, 53–212.

104. Jehne, W. 1980 Endomycorrhizas and the productivity of tropical pastures: the potential for improvement and its practical realization. Tropical Grasslands 14, 202–209.

105. Jensen, A. and Jakobsen, I. 1980 The occurrence of vesicular-arbuscular mycorrhiza in barley and wheat grown in some Danish soils with different fertilizer treatments. Plant and Soil 55, 403–414.

106. Johnson, P.N. 1976 Effects of soil phosphate level and shade on plant growth and mycorrhizas. N.Z. J. Bot. 14, 333–340.

107. Johnson, P.N. 1977 Mycorrhizal Endogonaceae in a New Zealand forest. New Phytol. 78, 161–170.

108. Johnson, C.R. and Michelini, S. 1974 Effect of mycorrhizae on container grown *Acacia*. Proc. of the Florida State Hort. Soc. 87, 520–522.

109. Johnston, A. 1949 Vesicular-arbuscular mycorrhiza in Sea Island cotton and other tropical plants. Tropic. Agric. Trin. 26, 118–121.

110. Kellam, M.K. and Schenck, N.C. 1980 Interactions between a vesicular-arbuscular mycorrhizal fungus and root-knot nematode on soybean. Phytopathology 70, 293–296.

111. Khan, A.G. 1974 The occurrence of mycorrhizas in halophytes, hydrophytes and xerophytes and of *Endogone* spores in adjacent soils. J. Gen. Microbiol. 81, 7–14.

112. Khan, A.G. 1975 Growth effects of VA mycorrhiza on crops in the field. In: Endomycorrhizas, pp. 419–435. Sanders, F.E., Mosse, B. and Tinker, P.B. (eds.), Academic Press, London.

113. Kleinschmidt, G.D. and Gerdemann, J.W. 1972 Stunting of citrus seedlings in fumigated nursery soils related to the absence of endomycorrhizae. Phytopathology 62, 1447–1453.

114. Kormanik, P.P., Bryan, W.C. and Schultz, R. 1976 Endomycorrhizae; their importance in nursery production of hardwood seedlings. Proc. 1976. S E Area Nurserymen's Conf. Eastern Session. Charleston, SC August 3–5, 1976, pp 16–21.

115. Kruckelman, H.W. 1975 Effects of fertilizers, soils, soil tillage and plant

species on frequency of *Endogone* chlamydospores and mycorrhizal infection in arable soils. In: Endomycorrhizas, pp. 511–525. Sanders, F.E., Mosse, B. and Tinker, P.B. (eds.), Academic Press, London.

116. Lalonde, M., Calvert, H.E. and Pine, S. 1981 Isolation and use of *Frankia* strains in actinorrhizae formation. In: Current perspectives in nitrogen fixation, pp. 296–299. Gibson, A.H. and Newton, W.E. (eds.), Australian Academy of Sciences, Canberra.

117. Lambert, D.H., Cole Jr. H. and Baker, D.E. 1980 Variation in the response of Alfalfa clone and cultivars to mycorrhizae and phosphorus. Crop Sci. 20, 615–618.

118. Laycock, D.H. 1945 Preliminary investigations into the function of the endotrophic mycorrhiza of *Theobroma cacao* L. Trop. Agric. Trin. 22, 77–80.

119. Le Tacon, F. 1978 Le rôle des ectomycorrhizes et des endomycorhizes dans la nutrition minérale et le comportement des arbres forestiers. 103é Congr. Nat. Soc. Savantes Nancy. Sciences 1, 281–293.

120. Levy, Y. and Krikun, J. 1980 Effect of vesicular-arbuscular mycorrhiza on *Citrus jambhiri* water relations. New Phytol. 85, 25–31.

121. Manjunath, A. and Bagyaraj, D.J. 1981 Components of VA mycorrhizal inoculum and their effects on growth of onion. New Phytol. 87, 355–361.

122. Marshall, K.C. 1967 Methods of study and ecological significance of *Rhizobium*-clay interactions. In: Methods of study in soil ecology, pp. 107–110. Phillipson, J. (ed.), Proc. UNESCO/IBP Symposium, Paris.

123. Marx, D.H., Bryan, W.C. and Campbell, W.A. 1971 Effect of endomycorrhizae formed by *Endogone mosseae* on growth of citrus. Mycologia 63, 1222–1226.

124. Mejstrik, J. 1965 Study of the development of endotrophic mycorrhiza in association with *Cladietum marisci*. In: Plant microbe relationships, pp. 283–290. Macura, J. and Vancura, V. (eds.), Czech. Acad. Sci., Prague.

125. Menge, J.A., Lembright, H. and Johnson, E.L.V. 1977 Utilization of mycorrhizal fungi in citrus nurseries. Proc. Int. Soc. Citriculture 1, 129–132.

126. Menge, J.A., Nemec, S., Davis, R.M. and Minassian, V. 1977 Mycorrhizal fungi associated with *Citrus* and their possible interactions with pathogens. Proc. Int. Soc. Citriculture 3, 872–876.

127. Menge, J.A., Davis, R.M., Johnson, E.L. and Zentmyer, G.A. 1978 Mycorrhizal fungi increase growth and reduce transplant injury in avocado. Calif. Agric. April, 6–7.

128. Menge, J.A., Johnson, E.L.V. and Platt, R.G. 1978 Mycorrhizal dependency of several citrus cultivars under three nutrient regimes. New Phytol. 81, 553–559.

129. Menge, J.A., Munnecke, D.E., Johnson, E.L.V. and Carnas, D.W. 1978 Dosage response of the vesicular-arbuscular mycorrhizal fungi *Glomus fasciculatus* and *G. constrictus* to methyl bromide. Phytopathology 68, 1368–1372.

130. Menge, J.A., Steirle, O., Bagyaraj, D.J., Johnson, E.L.V. and Leonard, R.T. 1978 Phosphorus concentrations in plants responsible for inhibition of mycorrhizal infection. New Phytol. 80, 575–578.

131. Mertz, Jr., S.M., Heithaus III J.J. and Bush, R.L. 1979 Mass production of axenic spores of the endomycorrhizal fungus *Gigaspora margarita*. Trans Br. mycol. Soc. 72, 167–169.

132. Moawad, M. 1979 Ecophysiology of vesicular-arbuscular mycorrhiza in the tropics. In: The soil-root interface, pp. 197–209. Harley, J.L. and Scott Russell, R. (eds.), Academic Press, London.

133. Moorman, T. and Reeves, F.B. 1979 The role of endomycorrhizae in revegetation practices in the semi-arid West II. A bioassay to determine the effect of

land disturbance on endomycorrhizal populations. Amer. J. Bot. 66, 14—18.

134. Morandi, D., Gianinazzi, S. and Gianinazzi-Pearson, V. 1979 Intérêt de l'endomycorrhization dans la reprise et la croissance du Framboisier issu de multiplication végétative *in vitro*. Ann. Amélior. Plantes 29, 623—630.

135. Morley, C.D. and Mosse, B. 1976 Abnormal vesicular-arbuscular mycorrhizal infections in white clover induced by lupin. Trans. Br. mycol. Soc. 67, 510—513.

136. Mosse, B. 1953 Fructification associated with mycorrhizal strawberry roots. Nature, London, 171, 974.

137. Mosse, B. 1972 The influence of soil type and *Endogone* strains on the growth of mycorrhizal plants in phosphate deficient soils. Rev. Ecol. Biol. Sol 9, 529—537.

138. Mosse, B. 1972 Effects of different *Endogone* strains on the growth of *Paspalum notatum*. Nature, London, 239, 221—223.

139. Mosse, B. 1973 The role of mycorrhiza in phosphorus solubilization. Global Impacts of Applied Microbiology. 4th Intern. Conf. Sao Paulo. Brazil 1973, pp. 543—561.

140. Mosse, B. 1975 Specificity in VA mycorrhizas. In: Endomycorrhizas, pp. 469—484. Sanders, F.E., Mosse, B. and Tinker, P.B. (eds.), Academic Press, London.

141. Mosse, B. 1977 The role of mycorrhiza in legume nutrition on marginal soils. In: Exploiting the legume-*Rhizobium* symbiosis in tropical agriculture, pp. 275—292. Vincent, J.M., Whitney, A.S. and Bose, J. (eds.), College of Tropical Agriculture, University of Hawaii Misc. Publ. 145.

142. Mosse, B. and Hayman, D.S. 1971 Plant growth responses to vesicular-arbuscular mycorrhiza. II. In unsterilized field soils. New Phytol. 70, 29—34.

143. Mosse, B. and Hayman, D.S. 1980 Mycorrhiza in agricultural plants. In: Tropical Mycorrhiza Research, pp. 213—230. Mikola, P. (ed.), Clarendon Press, Oxford.

144. Mosse, B., Hayman, D.S. and Ide, G.J. 1969 Growth responses of plants in unsterilized soil to inoculation with vesicular arbuscular mycorrhiza. Nature, London, 224, 1031.

145. Mosse, B., Hayman, D.S. and Arnold, D.J. 1973 Plant growth responses to vesicular-arbuscular mycorrhiza. V. Phosphate uptake by three plant species from P-deficient soils labelled with ^{32}P. New Phytol. 72, 809—815.

146. Mosse, B., Powell, C.Ll. and Hayman, D.S. 1976 Plant growth responses to vesicular-arbuscular mycorrhiza. IX. Interactions between VA mycorrhiza, rock phosphate and symbiotic nitrogen fixation. New Phytol. 76, 331—342.

147. Munns, D.N. and Mosse, B. 1980 Mineral nutrition of legume crops. In: Advances in legume science, pp. 115—125. Summerfield, R.J. and Bunting, A.H. (eds.), Royal Botanical Gardens, Kew.

148. Nadarajah, P. 1980 Species of Endogonaceae and mycorrhizal association of *Elaeis guineensis* and *Theobroma cacao*. In: Tropical Mycorrhiza Research, pp. 232—237. Mikola, P. (ed.), Clarendon Press, Oxford.

149. Nemec, S. and O'Bannon, J.H. 1979 Response of *Citrus aurantium* to *Glomus etunicatus* and *G. mosseae* after soil treatment with selected fumigants. Plant and Soil 53, 351—359.

150. Nicolson, T.H. and Johnston, C. 1979 Mycorrhiza in the Gramineae. III. *Glomus fasciculatus* as the endophyte of pioneer grasses in a maritime sand dune. Trans. Br. mycol. Soc. 72, 261—268.

151. Nicolson, T.H. and Schenck, N.C. 1979 Endogonaceous mycorrhizal endophytes in Florida. Mycologia 71, 178—198.

152. Nyabyenda, P. 1977 Einfluss der Bodentemperatur und organischer Stoffe

248

im Boden auf die Wirkung der vesikulär arbuskulären Mykorrhiza. Dissertation, Göttingen.

153. O'Bannon, J.H. and Nemec, S. 1979 The response of *Citrus limon* seedlings to a symbiont, *Glomus etunicatus*, and a pathogen, *Radopholus similis*. J. Nematol. 11, 270–275.

154. O'Bannon, J.H., Inserra, R.N., Nemec, S. and Vovlas, N. 1979 The influence of *Glomus mosseae* on *Tylenchulus semipenetrans*-infected and uninfected *Citrus limon* seedlings. J. Nematol. 11, 247–250.

155. O'Bannon, J.H., Evans, D.W. and Peaden, R.N. 1980 Alfalfa varietal response to seven isolates of vesicular-arbuscular mycorrhizal fungi. Can J. Plant Sci. 60, 859–864.

156. Okigbo, B.N. and Lal, R. 1979 Soil fertility maintenance and conservation for improved agroforestry systems in the lowland humid tropics. In: Soils Research in Agroforestry, pp. 41–77. Mongi, H.O. and Huxley, P.A. (eds.), ICRAF, Nairobi.

157. Old, K.M., Nicolson, T.H. and Redhead, J.F. 1973 A new species of mycorrhizal *Endogone* from Nigeria with a distinctive spore wall. New Phytol. 72, 817–823.

158. Owusu-Bennoah, E. and Mosse, B. 1979 Plant growth responses to vesicular-arbuscular mycorrhiza XI. Field inoculation responses in barley, lucern and onion. New Phytol. 83, 671–679.

159. Parke, J.L. and Linderman, R.G. 1980 Association of vesicular-arbuscular mycorrhizal fungi with the moss *Funaria hygrometrica*. Can. J. Bot. 58, 1898–1904.

160. Philips, J.M. and Hayman, D.S. 1970 Improved procedures for clearing roots and staining parasitic and vesicular-arbuscular mycorrhizal fungi for rapid assessment of infection. Trans. Br. Mycol. Soc. 55, 158–161.

161. Pichot, J. and Binh, T. 1976 Action des endomycorrhizes sur la croissance et la nutrition phosphatée de l'agrostis en vases de végétation et sur le phosphate isotopiquement diluable du sol. Agron. Trop. 31, 375–378.

162. Porter, D.M. and Beute, M.K. 1972 *Endogone* species in roots of Virginia type peanuts. Phytopathology 62, 783 (Abstract).

163. Possingham, J.V., Groot Obbink, J. and Jones, R.K. 1971 Tropical legumes and vesicular-arbuscular mycorrhiza. J. Aust. Inst. Agric. Sci. 37, 160–161.

164. Powell, C.Ll. 1976 Development of mycorrhizal infections from Endogone spores and infected root segments. Trans. Br. Mycol. Soc. 66, 439–445.

165. Powell, C.Ll. 1979 Inoculation of white clover and ryegrass seed with mycorrhizal fungi. New Phytol. 83, 81–85.

166. Powell, C.Ll. 1979 Spread of mycorrhizal fungi through soil. N.Z. J. Agric. Res. 22, 335–339.

167. Powell, C.Ll. 1980 Effect of phosphate fertilizers on the production of mycorrhizal inoculum in soil. N.Z. J. Agric. Res. 23, 219–223.

168. Powell, C.Ll. and Daniel, J. 1978 Mycorrhizal fungi stimulate uptake of soluble and insoluble phosphate fertilizer from a phosphate-deficient soil. New Phytol. 80, 351–358.

169. Powell, C.Ll. and Sithamparanatham, J. 1977 Mycorrhizas in hill country soils IV. Infection rate in grass and legume species by indigenous mycorrhizal fungi under field conditions. N.Z. J. Agric. Res. 20, 489–494.

170. Powell, C.Ll. Groters, M. and Metcalfe, D. 1980 Mycorrhizal inoculation of a barley crop in the field. N.Z. J. Agric. Res. 23, 107–109.

171. Raj, J., Bagyaraj, D.J. and Manjunath, A. 1981 Influence of soil inoculation with vesicular-arbuscular mycorrhiza and a phosphate-dissolving bacterium on plant growth and ^{32}P uptake. Soil Biol. Biochem. 13, 105–108.

172. Ramirez, B.N., Mitchell, D.J. and Schenck, N.C. 1975 Establishment and growth effects of three vesicular-arbuscular mycorrhizal fungi on papaya. Mycologia 57, 1039–1041.
173. Redhead, J.F. 1968 Mycorrhizal associations in some Nigerian forest trees. Trans. Br. Mycol. Soc. 51, 377–387.
174. Redhead, J.F. 1975 Endotrophic mycorrhizas in Nigeria: some aspects of the ecology of the endotrophic mycorrhizal association of *Khaya grandifolia* C. DC. In: Endomycorrhizas, pp. 447–459. Sanders, F.E., Mosse, B. and Tinker, P.B. (eds.), Academic Press, London.
175. Redhead, J.F. 1977 Endotrophic mycorrhizas in Nigeria: species of the Endogonaceae and their distribution. Trans. Br. Mycol. Soc. 69, 275–280.
176. Redhead, J.F. 1979 Soil mycorrhiza in relation to soil fertility and productivity. In: Soils Research in Agroforestry, pp. 175–204. Mongi, H.O. and Huxley, P.A. (eds.), ICRAF, Nairobi.
177. Redhead, J.F. 1980 Mycorrhiza in natural tropical forests. In: Tropical Mycorrhiza Research, pp. 127–142. Mikola, P. (ed.), Clarendon Press, Oxford.
178. Reeves, F.B., Wagner, D., Moorman, T. and Kiel, J. 1979 The role of endomycorrhizae in revegetation practices in the semiarid West. I. A comparison of incidence of mycorrhizae in severely disturbed vs. natural environments. Amer. J. Bot. 66, 6–13.
179. Rich, J.R. and Bird, G.W. 1974 Association of early-season vesicular-arbuscular mycorrhizae with increased growth and development of cotton. Phytopathology 64, 1421–1425.
180. Rich, J.R. and Schenck, N.C. 1979 Survey of North Florida flue-cured tobacco fields for root-knot nematodes and vesicular-arbuscular mycorrhizal fungi. Plant Dis. Reptr. 63, 952–955.
181. Richards, P.W. 1952 The tropical rain forest. Cambridge University Press, London.
182. Riess, S. and Rambelli, A. 1980 Preliminary notes on mycorrhizae in a natural tropical rain forest. In: Tropical Mycorrhiza Research, pp. 143–145. Mikola, P. (ed.), Clarendon Press, Oxford.
183. Rives, C.S., Baswa, M.I. and Liberta, A.E. 1980 Effects of topsoil storage during surface mining on the viability of VA mycorrhiza. Soil Sci. 129, 253–257.
184. Roncadori, R.W. and Hussey, R.S. 1977 Interaction of the endomycorrhizal fungus *Gigaspora margarita* and root-knot nematode on cotton. Phytopathology 67, 1507–1511.
185. Rose, S.L. 1980 Mycorrhizal associations of some actinomycete nodulated nitrogen-fixing plants. Can. J. Bot. 58, 1449–1454.
186. Rose, S.L. and Trappe, J.M. 1980 Three new endomycorrhizal *Glomus* spp. associated with actinorrhizal shrubs. Mycotaxon 10, 413–420.
187. Rose, S.L. and Youngberg, C.T. 1981 Tripartite associations in snowbrush (*Ceanothus velutinus*): effect of vesicular-arbuscular mycorrhizae on growth, nodulation and nitrogen fixation. Can. J. Bot. 59, 34–39.
188. Ross, J.P. and Harper, J.A. 1970 Effect of *Endogone* mycorrhiza on soybean yields. Phytopathology 60, 1552–1556.
189. Ross, J.P. and Ruttencutter, R. 1977 Population dynamics of two vesicular-arbuscular endomycorrhizal fungi and the role of hyperparasitic fungi. Phytopathology 67, 490–496.
190. Safir, G.R., Boyer, J.S. and Gerdemann, J.W. 1971 Mycorrhizal enhancement of water transport in soybean. Science 172, 581–583.

191. Safir, G.R., Boyer, J.S. and Gerdemann, J.W. 1972 Nutrient status and mycorrhizal enhancement of water transport in soybean. Plant Physiol. 49, 700–703.
192. Sanders, F.E. 1975 The effect of foliar-applied phosphate on the mycorrhizal infections of onion roots. In: Endomycorrhizas, pp. 261–276. Sanders, F.E., Mosse, B. and Tinker, P.B. (eds.), Academic Press, London.
193. Sanders, F.E. and Tinker, P.B. 1971 Mechanism of absorption of phosphate from soil by *Endogone* mycorrhizas. Nature, London, 232, 278–279.
194. Schenck, N.C. and Hinson, K. 1973 Response of nodulating and non-nodulating soybeans to a species of *Endogone* mycorrhiza. Agron. J. 65, 849–850.
195. Schenck, N.C. and Kellam, M.K. 1978 The influence of vesicular-arbuscular mycorrhizae on disease development. Flo. Agric. Exp. Stn. Tech. Bull. 798.
196. Schenck, N.C. and Kinloch, R.A. 1974 Pathogenic fungi, parasitic nematodes and endomycorrhizal fungi associated with soybean roots in Florida. Plant Dis. Reptr. 58, 169–173.
197. Schenck, N.C. and Schroder, V.N. 1974 Temperature response of *Endogone* mycorrhiza on soybean roots. Mycologia 66, 600–605.
198. Schenck, N.C., Graham, S.O. and Green, N.E. 1975 Temperature and light effect on contamination and spore germination of vesicular-arbuscular mycorrhizal fungi. Mycologia 67, 1189–1192.
199. Schenck, N.C., Kinloch, R.A. and Dickson, D.W. 1975 Interaction of endomycorrhizal fungi and root-knot nematode on soybean. In: Endomycorrhizas, pp. 602–617. Sanders, F.E., Mosse, B. and Tinker, P.B. (eds.), Academic Press, London.
200. Schönbeck, F. 1979 Endomycorrhiza in relation to plant diseases. In: Soilborne plant pathogens. pp. 271–280. Schippers, B. and Gams, W. (eds.), Academic Press, London.
201. Sheikh, N.A., Saif, S.R. and Khan, A.G. 1975 Ecology of *Endogone*. II. Relationship of Endogone spore population with chemical soil factors. Islamabad J. Sci. 2, 6–9.
202. Sieverding, E. 1979 Einfluss der Bodenfeuchte auf die Effektivitat der VA Mykorrhiza. Angew. Botanik 53, 91–98.
203. Sikora, R.A. and Schönbeck, F. 1975 Effect of vesicular-arbuscular mycorrhiza (*Endogone mosseae*) on the population dynamics of the root-knot nematode (*Meloidogyne incognita* and *Meloidogyne hapla*). 8th Intern. Cong. Plant Protect. 5, 158–166.
204. Skipper, H.D. and Smith, G.W. 1979 Influence of soil pH on the soybean-endomycorrhiza symbiosis. Plant and Soil 53, 559–563.
205. Smith, S.E. and Bowen, G.D. 1979 Soil temperature, mycorrhizal infection and nodulation of *Medicago truncatula* and *Trifolium subterraneum*. Soil Biol. Biochem. 11, 469–473.
206. Sondergaard, M. and Laegaard, S. 1977 Vesicular-arbuscular mycorrhiza in some aquatic vascular plants. Nature, London 268, 232–233.
207. Sparling, G.P. and Tinker, P.B. 1978 Mycorrhizal infection in Pennine Grassland. I. Levels of infection in the field. J. Appl. Ecol. 15, 943–950.
208. St John, T.V. 1980 Root size, root hairs and mycorrhizal infection: a reexamination of Baylis's hypothesis with tropical trees. New Phytol. 84, 483–487.
209. Sutton, J.C. and Barron, G.L. 1972 Population dynamics of *Endogone* spores in soil. Can. J. Bot. 50, 1909–1914.
210. Swaby, R.J. 1975 Biosuper-biological superphosphate. In: Sulfur in

Australian agriculture, pp. 213–222. McLachlen, K.D. (ed.), CSIRO, Glen Osmond.

211. Swaminathan, K. and Verma, B.C. 1979 Responses of three crop species to vesicular-arbuscular mycorrhizal infection on zinc-deficient Indian soils. New Phytol. 82, 481–487.

212. Thomazini, L.I. 1974 Mycorrhiza in plants of the 'Cerrado'. Plant and Soil 41, 707–711.

213. Timmer, L.W. and Leyden, R.F. 1978 Stunting of citrus seedlings in fumigated soils in Texas and its correction by phosphorus fertilization and inoculation with mycorrhizal fungi. J. Amer. Soc. Hort. Sci. 103, 533–537.

214. Tinker, P.B. 1978 Effects of vesicular-arbuscular mycorrhizas on plant nutrition and plant growth. Physiol. vég. 16, 743–752.

215. Tommerup, I.C. and Kidby, D.K. 1979 Preservation of spores of vesicular-arbuscular endophytes by L-drying. Appl. Envir. Microbiol. 37, 831–835.

216. Trinick, M.J. 1977 Vesicular-arbuscular infection and soil phosphorus utilization in Lupinus spp. New Phytol. 78, 297–304.

217. Tupas, G.L. and Sajise, R.E. 1976 Mycorrhizal associations in some savanna and reforestation trees. Kalikasan 5, 235–240.

218. Vander Zaag, P., Fox, R.L., De la Pena R.S. and Yost, R.S. 1979 P nutrition of cassava, including mycorrhizal effects on P, K, S, Zn and Ca uptake. Field Crops Research 2, 253–263.

219. Waidyanatha, U.P. de S. 1980 Mycorrhizae of Hevea and leguminous ground covers in rubber plantations. In: Tropical Mycorrhiza Research, pp. 238–241. Mikola, P. (ed.), Clarendon Press, Oxford.

220. Waidyanatha, U.P. de S., Yogaratnam, N. and Ariyaratne, W.A. 1979 Mycorrhizal infection on growth and nitrogen fixation of Pueraria and Stylosanthes and uptake of phosphorus from two rock phosphates. New Phytol. 82, 147–152.

221. Warner, A. and Mosse, B. 1980 Independent spread of vesicular-arbuscular mycorrhizal fungi in soil. Trans. Br. Mycol. Soc. 74, 407–410.

222. Wastie, R.L. 1965 The occurrence of an Endogone type of endotrophic mycorrhiza in Hevea brasiliensis. Trans. Br. mycol. Soc. 48, 167–178.

223. Wilhelm, S. 1973 Principles of biological control of soil-borne plant disease. Soil Biol. Biochem. 5, 729–737.

224. Williams, S. 1979 Vesicular-arbuscular mycorrhizae associated with actinomycete-nodulated shrubs, Cercocarpus montanus Raf. and Purshia tridentata (Pursh) D.C. Bot. Gaz. 140, S115–S119.

225. Williams, C.N. 1975 The agronomy of the major tropical crops. Oxford University Press, London.

226. Witty, J.F. and Hayman, D.S. 1978 Slurry-inoculation of VA mycorrhiza. Rothamsted Exp. Stat. Report for 1977, Part 1, pp. 239–240.

227. Yost, R.S. and Fox, R.L. 1979 Contribution of mycorrhizae to P nutrition of crops growing on an oxisol. Agron. J. 71, 903–908.

9. Ectomycorrhizae in the tropics

J.F. REDHEAD

1. Introduction

Ectomycorrhizae are symbiotic associations between fungi and plant roots in which the fungus forms a sheath around all or some of the fine absorbing rootlets. Hyphae penetrate between the root cells and occasionally enter the cells but they never penetrate beyond the cortex and any intracellular hyphae do not cause destruction of the host cell.

The association was first described on tree roots by Frank [19] who realised the association was a naturally occurring, non-pathogenic relationship. Other workers, including the distinguished forester Hartig [31], believed the fungus was a parasite. At that time work was largely descriptive and speculation on the nature of the relationship continued for many years. Gradually experimental evidence accumulated confirming that the ectomycorrhizal association had a beneficial effect and under certain conditions was essential for the growth of both partners [26, 27, 32, 50, 51, 52].

In contrast to the ubiquitous endomycorrhizae, ectomycorrhizae are not common in the tropics. They are, however, very important in the afforestation of poor lands and members of the genus *Pinus* will not grow unless they develop ectomycorrhizae. Their occurrence in the families Caesalpiniaceae and Dipterocarpaceae may help to explain why these families dominate two of the world's most extensive plant formations in the tropics, the miombo woodland of Central Africa and the tropical rain forest of Malaysia and South East Asia.

2. Occurrence

Ectomycorrhizae were not described in the tropics until the early 1960's despite extensive surveys of mycorrhizal associations [35, 37]. The association was first recorded in the West African lowland rain forest on several species of the family Caesalpiniaceae [15, 64].

Records of the ectotrophic association in natural ecosystems are still few: eleven species of Caesalpiniaceae [14, 15, 36, 64, 65, 79], 30 species of Dipterocarpaceae [12, 33, 71], one species of Euphorbiaceae [67], one species of Fagaceae [71] and one species of Myrtaceae [79]. It is also reported that ectomycorrhizae occur in the Sapindaceae [2].

Y.R. Dommergues and H.G. Diem (eds.), Microbiology of Tropical Soils
and Plant Productivity. ISBN 978-94-009-7531-6.
© 1982 Martinus Nijhoff/Dr W. Junk Publishers, The Hague/Boston/London.

More recently Mr. P. Högberg, of the Swedish University of Agricultural Sciences, has found ectomycorrhizae on several miombo woodland species in Tanzania (Table 1).

Table 1. Some ectomycorrhizal species in the miombo woodland of Tanzania

Caesalpiniaceae	*Brachystegia boehmii*
	B. bussei
	B. longifolia
	B. microphylla
	B. spiciformis
	Julbernardia globiflora
Dipterocarpaceae	*Monotes elegans*
Euphorbiaceae	*Uapaca kirkiana*

Dr. M.H. Ivory of the Commonwealth Forestry Institute, Oxford, has recently made extensive collections in natural stands of *Pinus caribaea* and *P. oocarpa* in Central America and the Bahamas. He confirmed that many of these fungi form ectomycorrhizae on pine under laboratory conditions (Table 2).

Table 2. Fungi forming mycorrhizal associations on pines in the laboratory

Fungus	Source
Gyroporus castaneus	Mixed *P. caribaea* and *P. oocarpa* forests
Pisolithus tinctorius	Both *P. caribaea* and *P. oocarpa* forests
Rhizopogon nigrescens	*P. caribaea* and *P. oocarpa* hosts and from the alkaline soils of the Bahamas
Suillus cothurnatus	*P. caribaea* on alkaline soils
Suillus cf. *Holoeacus*	*P. caribaea* forest
Scleroderma (geaster)?	*P. caribaea* and mixed pine forests
Tylopilus gracilis	*P. caribaea* and *P. oocarpa* forests

Sporocarps of fungi, well known to be mycorrhizal in temperate regions, are regularly found in stands of exotic pines (Table 3).

The identity of fungi-forming ectomycorrhizae on tropical angiosperms is not known. Hong [33] has associated six genera of Basidiomycetes with a range of dipterocarp species and there is circumstantial evidence that a species of *Inocybe* forms ectomycorrhizae on *Afzelia bella* [66]. Attempts to culture fungi from ectomycorrhizae of *Afzelia africana*, *A. bella*, *A. bipindensis* and *Brachystegia eurycoma* were unsuccessful. The varied hyphal structure indicated that several different fungi could form associations on these species; some bearing clamp connections were Basidiomycetes. Sterile-raised plants of these species exposed to the open air readily formed ectomycorrhizae with a range of different fungi from air-borne propagules and from mycorrhizal root fragments. These species, however, would not form ectomycorrhizae when inoculated with ectomycorrhizae formed by *Rhizopogon luteolus* on *Pinus caribaea.*

Table 3. Fungal sporocarps commonly found associated with exotic pines in the tropics

Fungal sporocarp	Country	References	
Amanita muscaria	Malawi		[62]
Boletus spp.	Malawi		[62]
Corticia cinnamomea	West Malaysia		[34]
Hebeloma crustiliniforme	Kenya		[21]
Lycoperdon perlatum	Tanzania	Coll. Maghembe 1980, unpublished	
Pisolithus tinctorius	Ghana	Coll. Ivory 1978, unpublished	
Rhizopogon luteolus	Nigeria		[60]
Rhizopogon sp.	Sri Lanka	Coll. Redhead 1976, unpublished	
	Tanzania	Coll. Maghembe 1979, unpublished	
Scleroderma bovista	Kenya		[21]
Scleroderma sp.	Malawi		[62]
Scleroderma dictyospotum	Tanzania	Coll. Maghembe 1980, unpublished	
Suillus granulatus	Kenya		[63]
	Tanzania		[63]
	West Malaysia		[34]
	Uganda		[8]
	Zaire		[78]
S. luteus (Boletus luteus)	Kenya		[21]
	Tanzania		[63]
S. sibiricus	Tanzania		[63]
Thelephora terrestis	Malawi		[62]
	Tanzania		[38]

3. Development and structure

An ectomycorrhiza develops when a rootlet comes into contact with hyphae spreading from ectomycorrhizae on an adjacent rootlet, from residual ectomycorrhizal root fragments, or from a germinating spore. Growth of the hyphae on the root surface is stimulated by root exudates such as the M-factor [53] and a dense sheath or mantle is formed over the root surface. Other substances produced by the root appear to keep a balance between the parasitic relationship of the host and mycorrhizal fungus [24, 54]. In turn the fungus produces growth regulating substances which cause changes in the rootlet morphology. This is particularly marked in pines and some ectomycorrhizae develop very complex coralloid structures [73]. The morphology is generally changed less in Angiosperms, e.g. in *Brachystegia eurycoma* ectomycorrhizal rootlets are usually shorter and stouter than non-mycorrhizal rootlets due to the thick fungal sheath but in *Afzelia bella* ectomycorrhizae are merely stouter due to the fungal sheath but little affected in length.

In section the mantle may be made up of regularly sized pseudo-parenchymatous 'cells', of compressed hyphae, or have a 2-layered structure. The hyphae penetrating between the rootlet cells are able to dissolve the pectin of the middle lamella [61] and mechanically separate the cells [17, 18]. Hyphae may only penetrate between the cells of the piliferous layer or they may extend between the cortical cells as far as the endodermis to form the so-called Hartig network.

A very high proportion of absorbing rootlets are mycorrhizal in the case of pine [59] and also with *Afzelia* spp. and *Brachystegia eurycoma* [67]. Dissection of beech (*Fagus sylvatica*) mycorrhizae indicated that 40 percent of the total dry weight consisted of mantle. On the assumption that the feeding roots represent 10 percent of the total root mass, Harley [28] concluded that something like 4 percent of this would consist of fungus. Calculations based on ratios of root to shoot dry weights of heavily mycorrhizal and non-mycorrhizal *Brachystegia* seedlings indicated that 23 percent of the seedling root mass consisted of fungus (data from Redhead [68].

Ectomycorrhizal rootlets do not develop root hairs and, although some ectomycorrhizae may appear to be smooth, hyphae radiate out from the mantle surface, often as extensively developed hyphal strands. In *Brachystegia* mycorrhizae were formed by two fungi with distinctly different hyphae which had diameters of 2–3 μm for the hyphae of a white mycorrhizal fungus and 3–4 μm for those of a brown fungus. The extent of mycelium in the soil is difficult to ascertain because it breaks off on examination. In *Brachystegia* mycelial strands were frequently over 3 cm from the mantle, branching frequently to form dense wefts of hyphae. Such hyphae are able to penetrate the fine interstices of the soil whereas rootlets must depend on root hairs for this and, being so short, root hairs can only tap a limited soil volume.

The colour of the mantle, its surface texture, hyphal thickness, and the presence or absence of clamp connections vary with the fungal species and are features which may be used to characterize ectomycorrhizae. All these features may be used in the construction of keys [13, 84]. Unfortunately the morphology also varies with the age of the association and to a lesser extent with the nature of the soil environment. This makes the use of keys to identify mycorrhizal fungi rather difficult and they must be used with caution.

4. Ecological advantage in tropical ecosystems

4.1. Significance for growth

It is common experience that without an appropriate mycorrhizal fungus pines stagnate and eventually die [56]. Briscoe [7] in Puerto Rico found that inoculated seedlings of *Pinus elliottii* var. *elliottii* grew only 12 cm in 4 years after planting out whereas inoculated seedlings grew 149 cm. Few other species have been studied in the tropics but Redhead [68] found that two different mycorrhizal fungi significantly increased the dry weight of *Brachystegia eurycoma* (Tables 4 and 5).

Table 4. Mean stem heights and dry weights of *Brachystegia eurycoma* grown for 13 months at low levels of nutrients after inoculation with two mycorrhizal fungi [68]

Treatment	Stem height (cm)	Mean dry weight (g)			
		Leaves	Stem	Root	Total
Inoculation Sources					
All fertilizer levels combined					
No inoculation	30.5	0.76	1.94	1.60	4.30
B. eurycoma White mycorrhiza	30.5	0.87	2.10[a]	2.14[b]	5.11[b]
B. eurycoma Brown mycorrhiza	32.9	0.86	2.58[a]	2.33[b]	5.77[b]
Fertilizer treatment					
All inocula combined					
No added nutrient	27.3	0.66	1.61	1.58	3.85
6 ppm N + 1 ppm P	31.9	0.79	2.06	1.75	4.60
6 ppm N + 5 ppm P	29.8	0.77	1.94	2.09	4.80
6 ppm N + 25 ppm P	36.0[b]	1.07[b]	3.05[b]	2.55[b]	6.67[b]

[a] Difference significant at the 5% level
[b] Difference significant at the 1% level

Table 5. Mean stem heights, dry weights and nitrogen content of *Brachystegia eurycoma* grown for 5 months at three levels of nitrogen and two levels of potassium after inoculation with two mycorrhizal fungi [68]

Treatment	Stem height (cm)	Mean dry weight (g)				Nitrogen content dry matter %		
		Leaves	Stem	Roots	Total	Leaves	Stem	Roots
Inoculation sources								
All fertilizer levels combined								
No inoculation	22.4	0.74	1.07	1.10	2.91	2.2	0.8	0.9
White mycorrhiza	26.3[a]	1.23[b]	1.65[b]	2.05[b]	4.93[b]	2.1	0.8	0.9
Brown mycorrhiza	28.6[a]	1.54[b]	1.94[b]	2.08[b]	5.56[b]	2.2	0.7	0.8
Fertilizer treatments								
Total daily amount per plant. All fungal inocula combined	[b]	[b]	[b]	[b]	[b]	[b]	[b]	[b]
No added nutrient	18.3	0.58	1.00	1.21	2.79	1.7	0.5	0.6
0.0016g N	23.3	0.98	1.24	1.70	3.92	1.9	0.7	0.7
0.0024g N + 0.0011g K	28.2	1.41	2.41	2.74	6.56	2.1	0.7	0.7
0.0040g N + 0.0011g K	30.5	1.46	2.16	2.75	6.37	1.9	0.5	0.8
0.0112g N	23.0	1.01	0.91	0.79	2.71	2.6	1.1	1.2
0.0120 g N + 0.0011g K	31.1	1.57	1.62	1.29	4.48	2.7	1.0	1.3

[a] Difference significant at the 5% level
[b] Difference significant at the 1% level

This beneficial effect on growth is linked to the physiology of the ectomycorrhizal association. The research involved has been carried out in temperate regions but an understanding of the principles is especially relevant for tropical research, especially that related to the afforestation of difficult sites.

In addition to the beneficial effect on growth, ectomycorrhizae also confer disease resistance against some of the most serious nursery pathogens in tropical nurseries.

4.2. Mechanisms of nutrition

The physiology of ectomycorrhizae and the metabolic exchanges between host and fungus have been reviewed by Harley [27] and Marks and Kozlowski [45]. All ectomycorrhizal fungi depend wholly on their host for sugars and the amounts used can be very high [28]. The fungus is an important sink for carbohydrate in the system and there are indications that the fungal share is increased through inhibition of root respiration by the fungus [81]. In turn the root receives water and mineral salts through the fungal component of the system.

With pines a high content of soluble carbohydrate in the rootlet and a low to moderate amount of available nitrogen appear to favour mycorrhizal development. A high light intensity favours a high sugar content [1] and this may partially explain why pines are light-demanding species. The inter-relationships between the sugar content and available nitrogen may be more complicated than Bjorkman suggests and more recent work on this aspect is discussed by Slankis [72].

Hatch [32] found four elements to be important for the development of ectomycorrhizae of pines: nitrogen, phosphorus, potassium and calcium. A certain minimum amount is necessary; if the supply of these elements was in abundance mycorrhizae were few but if there was an inbalance in the availability of one or more of these elements then mycorrhizae were plentiful.

Experience with Angiosperms has been different both in temperate and tropical species. Meyer [55] found in beech that adding nitrogen and phosphorus did not preclude formation of mycorrhizae in soils of high fertility. Redhead [68] obtained similar results with *Brachystegia* (Tables 4 and 5). Abundant ectomycorrhizae developed both at very low and very high levels of nitrogen.

It has been suggested that mycorrhizal hyphae are capable of bringing relatively insoluble minerals into solution. This has not been demonstrated experimentally but, there is evidence that mycorrhizal rootlets are several times more efficient than non-mycorrhizal roots at absorbing ions from the soil solution. Harley and McCready [29] showed that excised mycorrhizae of beech had 2.3 to 8.9 times the uptake of phosphorus than did non-mycorrhizal roots. Bowen and Theodorou [4] found similar results with p,ne. Ions of K, Na, Rb and Zn are also absorbed more rapidly by mycorrhizae [6, 30, 82].

Mycorrhizal hyphae are able to absorb mineral ions from very dilute soil solutions and their place is made up by more ions entering solution from minerals, e.g. rock phosphate. This is of special significance in the tropics where phosphorus is usually a limiting factor in degraded soils and it enables foresters to use the more cheaply available rock phosphate rather than the more expensive superphosphate fertilizers.

The extensive mycelial phase and the ability for rapid uptake give ectomycorrhizae a competitive advantage in the litter layer. Although the fungi involved cannot break down complex organic molecules such as cellulose [61] in the litter, once this has been done by other organisms the mycorrhizal hyphae can absorb the available solutes more quickly than most other soil microflora, even to the extent of starving such decomposing organisms of nutrients [20]. This rapid absorption is especially important in the tropics where nutrient cycling and leaching takes place so quickly.

4.3. N_2 fixation

Some of the early speculation on the function of ectomycorrhizae include claims that they could fix atmospheric N_2. Some pine plantations have shown evidence of N_2 fixation but there is no experimental evidence that the fungal component of the symbiosis was responsible [69, 70]. It appears that bacteria and blue-green algae, which develop in intimate contact with the fungal mantle, are likely to be responsible for any fixation of atmospheric N_2 by ectomycorrhizae. Foster and Marks [18] published electron micrographs which clearly show bacteria in association with ectomycorrhizae and Katznelson et al. [39] recorded up to ten times the number of bacteria associated with ectomycorrhizae as compared to the rhizosphere of non-mycorrhizal roots.

Recently the N_2-fixing bacterium Azotobacter has been successfully incorporated into the hyphae of Rhizopogon and used to synthesise mycorrhizae on Pinus radiata [22, 23]. This is a development with exciting possibilities for the afforestation of degraded sites in the tropics.

4.4. Water balance

Ectomycorrhizae do not thrive in heavy waterlogged soils but they are efficient at procuring water when it is in short supply. Theodorou and Bowen [76] found that vigorous mycorrhizal Pinus radiata seedlings showed fewer deaths from summer drought in the first planting season than did uninoculated seedlings. This is the experience in the tropics and there appear to be two reasons. Firstly the mycelial strands can be so dense and extensive that they help to bind the soil together in a ball when the seedling is lifted and the seedling quickly adjusts to its new site. Secondly some fungal hyphae are able to tolerate much lower water potential than root cells [75] and continue to absorb moisture when root hairs would have collapsed.

This aspect is very important when planting out in the sandy soils of the drier tropical regions as death from drought is the commonest cause of failure. As fungi vary in their ability to grow at low humidities [83] it would be an advantage to ensure that the most suitable mycorrhizal fungus was available.

4.5. Heat tolerance

Fungi vary in their growth at various temperatures and, for most, growth ceases above 35 °C. Lamb and Richards [42] studied the survival of several mycorrhizal fungi and found that the hyphae of six of these fungi were killed by a 48-hour exposure to temperatures between 28 ° and 38 °C and that of *Pisolithus tinctorius* by 45 °C. This can be a guide to the expected behaviour of ectomycorrhizae in the field although Hacskaylo *et al.* [25] showed that growth of mycorrhizal fungi was greatly influenced by the laboratory medium, and Theodorou and Bowen [77] consider growth in a laboratory medium to be a poor indication of their growth in the rhizosphere.

Marx and Bryan [48] compared the growth of mycorrhizal and non-mycorrhizal *Pinus taeda* at 40 °C and found that 55 percent of the non-mycorrhizal plants died and were generally smaller than the mycorrhizal pines. Marx [47] found that pines with *Pisolithus tinctorius* as the mycorrhizal partner survived on adverse sites, such as exposed coal spoils, better than pines which were mycorrhizal with other fungi. He considers that *Pisolithus* has special merit as a mycorrhizal partner for tropical pines.

Adaptation of a mycorrhiza to heat stress is likely to vary with the fungal partner and most of the known mycorrhizal fungi of pine in tropical plantations (Table 3) are temperate species. This may present little problem for highland areas above 1,000 m. Lowland areas planted with species such as *Pinus caribaea* and *P. oocarpa* present more of a problem. In lowland nurseries in Nigeria using black polypots, and planting out in bare land, it is not uncommon for the surface soil to reach 50 °C for considerable periods of the day. The mycorrhizae formed with *Rhizopogon luteolus* often died and nursery beds had to be sheltered with low walls of straw and planting carried out quickly in order to overcome this problem. During the long, hot, dry season most of the mycorrhizae in the surface layers died but grew again during the following rainy season. It is likely that a more heat tolerant fungus would be a more suitable mycorrhizal partner for these lowland sites. There is the danger that otherwise suitable introductions of species and provenances could be discarded through lack of an appropriate mycorrhizal partner.

4.6. Disease resistance

The developing feeder root attacts or stimulates certain pathogenic fungi in the group commonly known as 'damping-off' fungi, *Fusarium, Phytophthora, Pythium* and *Rhizoctonia*. These fungi are stimulated by root exudates in a similar way to that in which a mycorrhizal hypha is stimulated.

The cell's reaction to penetration by the hypha of a mycorrhizal fungus is in several ways similar to its reaction to invasion by a pathogen. However, the mycorrhizal fungus is kept in balance and does not cause breakdown of the root cells. The cell nucleus and cytoplasm grow in volume, respiration increases and starch reserves are mobilized [43]. The mycorrhizal fungus also develops to form a symbiotic mycorrhizal association. In contrast the pathogenic fungus invades the succulent tissues of the primary cortex and destroys the rootlet. It depends which fungus reaches the rootlet first, for if a mantle forms before the pathogen arrives the pathogen has to possess the physical and chemical ability to penetrate the fungal mantle and the Hartig net tissues before it can attack the root cells [46].

In addition to the physical barrier afforded by the mantle, at least some mycorrhizal fungi also secrete antibiotics inhibitory to pathogens. This has been demonstrated in the case of the mycorrhizal fungi *Cenococcum graniforme* [40] and *Leucopaxillus cerealis* var. *piceina* and *Suillus luteus* [49].

Damping-off diseases caused by the above pathogens are the most serious diseases in tropical forest nurseries. Soil drenches using Zineb or partial sterilization using methyl bromide are the normal methods of control but the use of an appropriate ectomycorrhizal fungus may offer a potential alternative.

5. Special problems of afforestation

5.1. Afforestation of new lands

An indigenous species will have evolved a relationship with a mycorrhial fungus either endo- or ectomycorrhizal but an exotic species, introduced for the first time, has either to form an association with an indigenous fungus or to be inoculated with an introduced fungus.

There is more information on exotic pines than any other tropical ectomycorrhizal species. If inoculum is present in the soil pines normally take it up readily and become mycorrhizal [80] which perhaps explains why many pines in their native habitats are excellent pioneers and are rapid, aggressive colonizers of open base-poor soils. Introduction into tropical Africa, and other areas where pines were not previously grown, consistently failed until it was realized that it was necessary to inoculate with a suitable mycorrhizal fungus. Mikola [58] gives the following examples: Kenya [21], Malawi [11], Nigeria [44], Puerto Rico [7], Trinidad [41, 56]. It is believed that settlers unwittingly introduced mycorrhizal fungi into Australia, New Zealand, South Africa and South America on the roots of living plants carried from Europe. These have been the origins of most mycorrhizal inoculum in the tropics and for this reason most of the fungi so far identified are themselves exotics (Table 3). In some instances in tropical Africa pines have appeared to have become spontaneously mycorrhizal. The transport of pine soil or living plants as an inoculum is often prohibited by quarantine regulations and there have been cases where foresters have smuggled inoculum. As Mikola [58] comments, 'such cases, of course, have not been well documented'.

Many species of the genus *Eucalyptus* are widely planted as exotics in tropical countries. Several of these are known to be mycorrhizal in their native Australia [9. 10] but their mycorrhizal status has not been studied in the tropics. Eucalypts are key species for the provision of fuel and poles in village afforestation schemes throughout the semi-arid areas of tropical Africa. These are often degraded, overgrazed sites with a long dry season; considerable research is being carried out to select the most suitable species and provenances for planting. There is need to undertake parallel studies of the mycorrhizal associations of the selected species under these exacting conditions.

5.2. Use of soil sterilants

It has become standard practice in several countries to partially sterilize nursery soils with methyl bromide to kill weed seeds and damping-off fungi. In long-established pine nurseries mycorrhizal associations develop both from residual root fragments and air-borne spores [38] and re-inoculation is not essential. Kalaghe demonstrated that if the planting medium is inoculated after sterilization the mycorrhizal association develops much more rapidly and uniformly than if left to nature. It becomes a question of economics as to whether the saving in raising seedlings in six months instead of eight is worth the cost of inoculating the potting medium.

5.3. Provision of inoculum

Marx [47] considers that inoculation of pine with the most appropriate fungus at the nursery stage is very worthwhile. This may be true when it is required to introduce a fungus especially adapted to certain ecological conditions. If the nursery is raising plants for routine planting on favourable sites then purpose-raised inoculum is probably not worthwhile. Unless the new fungus has special survival attributes, it will be replaced on new feeder roots by the fungi already well adapted to the area. The sequences of fungal succession in the formation of ectomycorrhizae is an area requiring further research.

Marx [47] describes the large-scale production of inoculum using a peat moss and vermiculite medium. Prepared inocula have long been used in South America [74] but are still at the experimental stage in tropical Africa. Mikola [57] has reviewed the early work in producing inocula, and Marx [47] discusses present developments which include the incorporation of fungal spores in the coating of pelleted seeds.

6. Research priorities

6.1. Surveys in natural ecosystems

It is not sufficient to find a sporocarp near to a plant and to assume that the plant is mycorrhizal. Careful excavation of the feeder roots is necessary, followed by microscopic examination to be sure that a plant is mycorrhizal. Little has been done in this respect: in tropical Africa less than 5 percent of tree species have been examined.

It is important to carry out this type of survey apart from the ecological interest. Some of the species may prove to be useful afforestation species for special sites and others may be trees of special importance to the village community. For example, Högberg (Table 1) found that *Uapaca kirkiana* is ectomycorrhizal. This tree is a wild fruit tree esteemed throughout large parts of Tanzania and Zambia yet entirely unimproved by selection and breeding. A knowledge of the ectomycorrhizae of such a potentially important tree is obviously important.

6.2. The relationship between ectomycorrhizae and N_2 fixation

The relationship between ectomycorrhizae and associated bacteria, or blue-green algae, requires further study, to ascertain their precise interdependence and the optimum conditions favouring N_2 fixation.

Some members of the Caesalpiniaceae are both ectomycorrhizal and nodulated by N_2 fixing bacteria. Preliminary studies have been carried out on these by Högberg (personal communication) on Tanzanian miombo species and it is hoped this work will continue.

6.3. Identification and culture of ectomycorrhizal fungi

In some cases mycorrhizal fungi, isolated from ectomycorrhizae, can be grown in culture and subsequently matched with cultures from identified sporocarps. Zak [85] discusses the problems involved. Once this has been done it is possible to inoculate plants and to use these plants to select ectomycorrhizal combinations most suited to special sites such as hot, dry lowland areas, areas of unfavourable pH or other poor lands.

Current work in Tanzania indicates that under unfavourable conditions a fungus may grow vegetatively as an ectomycorrhiza but may not fruit. It is therefore difficult to identify except by matching it with cultures from elsewhere. This aspect of mycorrhizal research may eventually help the forester to match his species to site more effectively.

6.4. The ectomycorrhizae of eucalypts

This has already been mentioned as a neglected field. Eucalypts are the most widely planted exotics in the tropics and merit a level of research commensurate with their afforestation importance.

This brief review of some aspects of ectomycorrhizae related to tropical forestry illustrate how important a knowledge of ectomycorrhizae is to efficient use of land in the tropics. The forester has to make a greater impact on the marginal lands which are usually hot and dry, and degraded through neglect and overgrazing.

Most of our present knowledge derives from research in temperate regions. An intelligent application of this experience can extend knowledge to the solution of the many and varied problems of the tropics. These studies have both great ecological interest and the satisfaction that they are essential to meet the material needs for fuel and industrial wood of a rapidly increasing population.

References

1. Björkman, E. 1942 Über die Bedingungen der Mykorrhizabildung bei Kiefer und Fichte. Symb. Bot. Upsal. 6, 1–190.
2. Black, R. 1980 The role of mycorrhizal symbiosis in the nutrition of tropical plants. In: Tropical Mycorrhiza Research, pp. 191–202. Mikola, P. (ed.), Clarendon Press, Oxford.
3. Bowen, G.D. 1973 Mineral nutrition of ectomycorrhizae. In: Ectomycorrhizae, Their Ecology and Physiology, pp. 151–205. Marks, G.C. and Kozlowski, T.T. (eds.), Academic Press, New York.
4. Bowen, G.D. and Theodorou, C. 1967 Studies on phosphate uptake by mycorrhizas. Proc. Int. Union For. Res. Organ., 14th, 1967, Vol. 5, p. 116.
5. Bowen, G.D. and Theodorou, C. 1973 Fungal growth around seeds and roots. In: Ectomycorrhizae, Their Ecology and Physiology, pp. 107–150. Marks, G.C. and Kozlowski, T.T. (eds.), Academic Press, New York.
6. Bowen, G.D., Skinner, M.F. and Bevege, D.I. 1974 Zinc uptake by mycorrhizal and uninfected roots of *Pinus radiata* and *Araucaria cunninghamii*. Soil Biol. Biochem. 6, 141–144.
7. Briscoe, C.B. 1959 Early results of mycorrhizal inoculation in Puerto Rico. Caribbean Forester 20, 73–77.
8. Chaudhry, M.A. 1980 Ectomycorrhiza of *Pinus caribaea* in Uganda. In: Tropical Mycorrhiza Research, pp. 88–89. Mikola, P. (ed.), Clarendon Press, Oxford.
9. Chilvers, G.A. 1968 Some distinctive types of eucalypt mycorrhiza. Aust. J. Bot. 16, 49–70.
10. Chilvers, G.A. and Pryor, L.D. 1965 The structure of eucalypt mycorrhizas. Aust. J. Bot. 13, 245–259.
11. Clements, J.B. 1941 The introduction of pines into Nyasaland. Nyasaland Agric. Quart. J. 1, 5–15.
12. de Alwis, D.P. and Abeynayake, K. 1980 A survey of mycorrhizae in some

forest trees in Sri Lanka. In: Tropical Mycorrhiza Research, pp. 146–153. Mikola, P. (ed.), Clarendon Press, Oxford.

13. Dominik, T. 1959 Synopsis of a new classification of the ectotrophic mycorrhizae established on morphological and anatomical characteristics. Mycopathologia 11, 359–367.

14. Fassi, B. 1963 The distribution of ectotrophic mycorrhiza in the litter and upper soil layers of the *Gilbertiodendron dewevrei* (Caesalpiniaceae) forest in the Congo. In: mycorrhiza, pp. 297–302. Proc. Mykorrhiza Intern. Mykorrhizas-symposium, Weimar, 1960. Rawald, W. and Lyr, H. (eds.), G. Fischer, Jena.

15. Fassi, B. and Fontana, A. 1961 Le micorize ectotrofiche di *Julbernardia seretii*, Caesalpiniacea del Congo. Alliona 7, 131–151.

16. Fassi, B. and Fontana, A. 1962 Micorize ectotrofiche di *Brachystegia laurentii* e di alcune altre Cesalpiniaceae minori del Congo. Alliona 8, 121–131.

17. Foster, R.C. and Marks, G.C. 1966 The fine structure of the mycorrhizas of *Pinus radiata* D. Don. Aust. J. Biol. Sci. 19, 1027–1038.

18. Foster, R.C. and Marks, G.C. 1967 Observations on the mycorrhizas of forest trees. II. The rhizosphere of *Pinus radiata* D. Don. Aust. J. Biol. Sci. 20, 915–926.

19. Frank, A.B. 1885 Über die auf Wurzelsymbiose beruhende Ernährung gewisser Bäume durch Unterirdische Pilze. Ber. Dtsch. bot. Ges. 3, 128–145.

20. Gadgil, R.L. and Gadgil, P.D. 1975 Suppression of litter decomposition by mycorrhizal roots of *Pinus radiata*. New Zealand J. Forest Sc. 5, 33–41.

21. Gibson, I.A.S. 1963 Eine Mitteilung über die Kiefernmykorrhiza in den Waldern Kenias. In: Mykorrhiza. pp. 49–51. Proc. Mykorrhiza Intern. Mykorrhizas – Symposium, Weimar, 1960. Rawald, W. and Lyr, H. (eds.), G. Fischer, Jena.

22. Giles, K.L. and Whitehead, H.C.M. 1976 Uptake and continued metabolic activity of *Azotobacter* within fungal protoplasts. Science 193, 1125–1126.

23. Giles, K.L. and Whitehead, H.C.M. 1977 Reassociation of a modified mycorrhiza with the host plant roots (*Pinus radiata*) and the transfer of acetylene reduction activity. Plant and Soil 48, 143–152.

24. Hacskaylo, E. 1969 Metabolic exchanges in ectomycorrhizae. In: Mycorrhizae, pp. 175–182. Hacskaylo, E., (ed.), Proc. 1st N. American Conf. on Mycorrhizae, April 1969. USDA, Washington, D.C.

25. Hacskaylo, E., Palmer, J.G. and Vozzo, J.A. 1965 Effect of temperature on growth and respiration of ectotrophic mycorrhizal fungi. Mycologia 57, 748–756.

26. Harley, J.L. 1968 Fungal symbiosis. Trans. Br. mycol. Soc. 51, 1–11.

27. Harley, J.L. 1969 The Biology of Mycorrhiza. 2nd Ed. Leonard Hill, London.

28. Harley, J.L. 1971 Fungi in ecosystems. J. Ecol. 59, 635–668.

29. Harley, J.L. and McCready, C.C. 1950 Uptake of phosphate by excised mycorrhizal roots of the beech. New Phytol. 49, 388–397.

30. Harley, J.L. and Wilson, J.M. 1959 The absorption of potassium by beech mycorrhizas. New Phytol. 58, 281–298.

31. Hartig, R. 1888 Die pflanzlichen Wurzelparasiten. Allg. Forst. und Jagdztg. 64, 118–123.

32. Hatch, A.B. 1937 The physical basis of mycotrophy in the genus *Pinus*. Black Rock Forest Bull. 6, 1–168.

33. Hong, L.T. 1979 A note on dipterocarp mycorrhizal fungi. Malaysian Forester 42, 280–283.

34. Ivory, M.H. 1975 Mycorrhizal studies on exotic conifers in West Malaysia. Malaysian Forester 38, 149–152.

35. Janse, J.M. 1896 Les endophytes radicaux de quelques plantes Javanaises. Ann. Jard. Bot. Buitenz. 14, 53–212.
36. Jenik, J. and Mensah, K.O.A. 1967 Root system of tropical trees. I. Ectotrophic mycorrhizae of *Afzelia africana* Sm. Preslia (Praha) 39, 59–65.
37. Johnston, A. 1949 Vesicular-arbuscular mycorrhiza in Sea Island Cotton and other tropical plants. Trop. Agric. Trinidad 26, 118–121.
38. Kalaghe, A.G. 1980 Studies on the fungi forming ectotrophic mycorrhizal associations on pines at Sao Hill. M.Sc. thesis, University of Dar es Salaam.
39. Katznelson, H., Rouatt, J.W. and Peterson, E.A. 1962 The rhizosphere effect of mycorrhizal and non-mycorrhizal roots of yellow birch seedlings. Can. J. Bot. 40, 377–382.
40. Krywalop, G.N., Grand, L.F. and Casida, L.E. 1964 The natural occurrence of an antibiotic in the mycorrhizal fungus *Cenococcum graniforme*. Can. J. Microbiol. 10, 323–328.
41. Lamb, A.F.A. 1956 Exotic forest trees in Trinidad and Tobago. Govt. Printer, Trinidad and Tobago.
42. Lamb, R.J. and Richards, B.N. 1971 Unpublished data. Cited by Bowen, G.D. and Theodorou, C. 1973.
43. Lewis, D.H. 1974 Micro-organisms and plants: the evolution of parasitism and mutualism. Symp. Soc. gen. Microbiol. 24, 367–392.
44. Madu, M. 1967 The biology of ectotrophic mycorrhiza with reference to the growth of pines in Nigeria. Obeche, J. of the Tree Club, Univ. of Ibadan 3, 9–18.
45. Marks, G.C. and Kozlowski, T.T. 1973 Ectomycorrhizae, their ecology and physiology. Academic Press, New York.
46. Marx, D.H. 1973 Mycorrhizae and feeder root diseases. In: Ectomycorrhizae, Their Ecology and Physiology, pp. 351–382. Marks, G.C. and Kozlowski, T.T. (eds.), Academic Press, New York.
47. Marx, D.H. 1980 Ectomycorrhizal fungus inoculations: A tool for improving forestation practices. In: Tropical Mycorrhiza Research, pp. 13–71. Mikola, P. (ed.), Clarendon Press, Oxford.
48. Marx, D.H. and Bryan, W.C. 1971 Influence of ectomycorrhizae on survival and growth of aseptic seedlings of loblolly pine at high temperature. Forest Sci. 17, 37–41.
49. Marx, D.H. and Davey, C.B. 1969 The influence of ectotrophic mycorrhizal fungi on the resistance of pine roots to pathogenic infections. III. Resistance of aseptically formed mycorrhizae to infection by *Phytophthora cinnamomi*. Phytopathology 59, 549–558.
50. Melin, E. 1923 Experimentelle Untersuchungen über die Konstitution und Ökologie der Mykorrhizen von *Pinus sylvestris* und *Picea abies*. Mykol. Unters. u. Ber. 2, 73–331.
51. Melin, E. 1925 Untersuchungen über die Bedeutung der Baummykorrhiza. G. Fischer, Jena.
52. Melin, E. 1936 Methoden der experimentellen Untersuchungen mykotropher Pflanzen. Handb. d. biol. Arbeitsmeth. II 4, 1015–1108.
53. Melin, E. 1954 Growth factor requirements of mycorrhizal fungi of forest trees. Svensk. Bot. Tidskr. 48, 86–94.
54. Melin, E. 1955 Nyare undersökningar över skogsträdens mykorrhizasvampar och det fysiologiska växelspelet mellan dem och trädens rötter. Uppsala Univ. Arsskr. 3, 3–29.
55. Meyer, F.H. 1962 Die Buchen und Fichtenmykorrhiza in verschiedenen Bodentypen, ihre Beeinflussung durch Mineraldünger sowie für die Mykor-

rhizabildung wichtige Faktoren. Mitt. Bundesforschungsant. Forst-Holzwirt. 54, 1−73.

56. Mikola, P. 1970 Mycorrhizal inoculation in afforestation. Int. Rev. For. Res. 3, 123−196.

57. Mikola, P. 1973 Mycorrhizal symbiosis in forestry practice. In: Ectomycorrhizae, Their Ecology and Physiology, pp. 383−411. Marks, G.C. and Kozlowski, T.T. (eds.), Academic Press, New York.

58. Mikola, P. 1980 Mycorrhizae across the frontiers. In: Tropical Mycorrhiza Research, pp. 3−10. Mikola, P. (ed.), Clarendon Press, Oxford.

59. Mikola, P., Hahl, J. and Torniainen, E. 1966 Vertical distribution of mycorrhizae in pine forests with spruce undergrowth. An. bot. Fen. 3, 406−409.

60. Momoh, Z.O. 1972 The problem of mycorrhizal establishment in the savanna zone of Nigeria. In: The Development of Forest Resources in the Economic Advancement of Nigeria, pp. 408−415. Onochie, C.F.A. and Adeyoju, S.K. (eds.), Proc. Inaug. Conf. Forestry Assoc. of Nigeria, Ibadan, 1970.

61. Palmer, J.G. and Hacskaylo, E. 1970 Ectomycorrhizal fungi in pure culture. I. Growth on single carbon sources. Physiol. Plant. 23, 1187−1197.

62. Pawsey, R.G. 1980 A review of mycorrhizal inoculation practice in Malawi. In: Tropical Mycorrhiza Research, pp. 90−92. Mikola, P. (ed.), Clarendon Press, Oxford.

63. Pegler, D.N. 1977 A Preliminary Agaric Flora of East Africa. Kew Bulletin Additional Series VI. H.M.S.O., London.

64. Redhead, J.F. 1960 A study of mycorrhizal associations in some trees of Western Nigeria. Diploma in Forestry Thesis, University of Oxford.

65. Redhead, J.F. 1968 Mycorrhizal associations in some Nigerian forest trees. Trans. Brit. Mycol. Soc. 51, 377−387.

66. Redhead, J.F. 1968 *Inocybe* sp. associated with ectotrophic mycorrhiza on *Afzelia bella* in Nigeria. Comm. Forest. Rev. 47, 63−65.

67. Redhead, J.F. 1974 Aspects of the biology of mycorrhizal associations occurring on tree species in Nigeria. Ph.D. Thesis, University of Ibadan.

68. Redhead, J.F. 1980 Mycorrhiza in natural tropical forest. In: Tropical Mycorrhiza Research, pp. 127−142. Mikola, P. (ed.), Clarendon Press, Oxford.

69. Richards, B.N. 1964 Fixation of atmospheric nitrogen in coniferous forests. Aust. Forest. 28, 68−74.

70. Richards, B.N. and Voigt, G.K. 1965 Nitrogen accretion in coniferous forest ecosystems. In: Proc. Second N. American Forest Soils Conf., pp. 105−116. Youngberg, C.T. (ed.), Oregon State Univ. Press, Corvallis.

71. Singh, K.G. 1966 Ectotrophic mycorrhiza in equatorial rain forest. Malayan Forester 39, 13−19.

72. Slankis, V. 1969 Formation of ectomycorrhizae of forest trees in relation to light, carbohydrates and auxins. In: Mycorrhizae, pp. 151−167. Hacskaylo, E. (ed.), Proc. 1st N. American Conf. on Mycorrhizae, April 1969. USDA, Washington, D.C.

73. Slankis, V. 1973 Hormonal relationships in mycorrhizal development. In: Ectomycorrhizae, their ecology and physiology, pp. 231−298. Marks, G.C. and Kozlowski, T.T. (eds.), Academic Press, New York.

74. Tackacs, E.A. 1967 Producción de cultivos puros de hongos micorrhizógenos en el Centro Nacional de Investigaciones Agropecuarias, Castelar. IDIA Suppl. For. 4, 83−87.

75. Theodorou, C. 1978 Soil moisture and the mycorrhizal association of *Pinus radiata* D. Don. Soil Biol. Biochem. 10, 33−37.

76. Theodorou, C. and Bowen, G.D. 1970 Mycorrhizal responses of radiata pine in experiments with different fungi. Aust. Forest. 34, 183—191.

77. Theodorou, C. and Bowen, G.D. 1971 Influence of temperature on the mycorrhizal associations of *Pinus radiata* D. Don. Aust. J. Bot. 19, 13—20.

78. Thoen, D. 1974 Preliminary data on mycorrhizas and mycorrhizal fungi in plantations of exotics of Upper Shaba (Republic of Zaire). Bulletin des Recherches Agronomiques de Gembloux 9, 215—227.

79. Thomazini, L.I. 1974 Mycorrhiza in plants of the "Cerrado". Plant and Soil 41, 707—711.

80. Watling, R. 1980 Mycorrhizal fungi. Bull. Brit. Mycol. Soc. 14, 59—62.

81. Wedding, R.T. and Harley, J.L. 1976 Fungal polyol metabolism in control of carbohydrate metabolism of mycorrhizal roots of beech. New Phytol. 77, 675—688.

82. Wilson, J.M. 1957 A study of the factors affecting the uptake of potassium by the mycorrhiza of beech. Ph.D. thesis, University of Oxford.

83. Worley, J.F. and Hacskaylo, E. 1959 The effect of available soil moisture on the mycorrhizal association of Virginia pine. Forest Sci. 5, 267—268.

84. Zak, B. 1969 Characterization and identification of Douglas fir mycorrhiza. In: Mycorrhizae, pp. 38—53. Hacskaylo, E. (ed.), Proc. 1st N. American Conf. on Mycorrhizae, April 1969. USDA, Washington, D.C.

85. Zak, B. 1973 Classification of ectomycorrhizae. In: Ectomycorrhizae, Their Ecology and Physiology, pp. 43—78. Marks, G.C. and Kozlowski, T.T. (eds.), Academic Press, New York.

10. The significance of the biological sulfur cycle in rice production

J.R. FRENEY, V.A. JACQ and J.F. BALDENSPERGER

1. Introduction

Rice is one of the world's most important food crops, forming nearly 20 percent of the world's food grain production [226]. It is grown on 137×10^6 ha throughout the world but more than 90 percent of all rice grain produced is grown in Asia where it is the dominant food crop.

Rice is grown under many different climatic conditions from the hot tropical conditions at the equator to the cold temperate conditions in Hungary at $\sim 48°N$ and from sea level to the tops of mountains ($> 2,500$ m). It is also cultivated in many different hydrological regimes, (a) as an upland crop with rain as the only source of water, (b) under flooded conditions with full control of irrigation water, (c) under intermittently flooded conditions with rain as the main source of water and (d) as deepwater or floating rice in water as deep as 5 m. Even when grown under flooded conditions the field may be drained before harvest and then reflooded before the next crop, so the same field may range in moisture status from air dry to saturated a number of times during the year.

Even in a permanently flooded soil a range of conditions exists. For example, aerobic and anaerobic zones occur in both planted and unplanted soils. An oxidized layer develops in the upper part of the flooded horizon of unplanted and planted soils when oxygen supply from the atmosphere and from photosynthesis by algae and aquatic weeds exceeds oxygen consumption by soil microorganisms [196, 300]. In flooded soils planted to rice another oxidized zone develops in the rhizosphere in addition to the oxidized surface layer. Rice plants have the capacity to transmit oxygen absorbed via the stomates of leaf blades and leaf sheaths, or produced during photosynthesis, through air passages in the leaves, stems, nodes and roots to the surrounding soil or soil solution [10, 29, 58, 148, 269]. As rice roots can occupy a large volume of the planted soil a significant fraction of the planted soil can be aerobic and the soil solution can be maintained at a high redox potential [238].

The presence of these oxidized zones in flooded soils allows the growth and metabolism of aerobic organisms and thus the processes taking place in these zones are similar to those occurring in well-drained, aerated soils. Outside the oxidized zones the processes occurring involve facultative and true anaerobic organisms [193].

Rice is grown on a wide range of soils, from those low in organic matter, e.g. Lithosols, to highly organic Histosols, from dry Yermosols to wet Planosols, from

Y.R. Dommergues and H.G. Diem (eds.), Microbiology of Tropical Soils and Plant Productivity. ISBN 978-94-009-7531-6.
© *1982 Martinus Nijhoff/Dr W. Junk Publishers, The Hague/Boston/London.*

acid Thionic Fluvisols to alkaline Solonetz [257], and thus with sulfur status which ranges from deficiency conditions to excess (e.g. in acid sulfate soils).

Thus a wide range of conditions can occur in rice soils; from very hot to extreme cold, flooded to air-dry, aerobic to anaerobic, deficiency to an excess of sulfur, little organic matter to an abundance in mucks and peats, and from acid to alkaline pHs. Consequently, many different redox potentials and microorganisms can exist, sulfur can occur through the full range of oxidation states, from sulfate at +6 through elemental sulfur, 0, to sulfide at −2 [228], many different sulfur transformations can take place and the reaction rates can vary considerably.

This article presents a discussion of the chemistry and microbiology of rice soils as affected by some of these conditions.

2. Sulfur in rice fields

Rice plants will continue to grow well only if they receive an adequate supply of sulfur and other nutrients throughout much of their growing period. This sulfur may come from the soil, irrigation or rainwater, by absorption from the atmosphere or from applied manures or fertilizers. Rice plants remove between 8 and 17 kg S ha^{-1} from a paddy soil, depending on variety, to yield between 4 and 9 tonnes of grain per hectare [278].

2.1. Soils

Rice is grown on a wide variety of soils, usually on clayey, impervious soils and often in permanently or temporarily waterlogged soils unsuitable for other crops. It is also grown, however, on well drained soils more suited for other upland grain crops, such as wheat and maize, and is cultivated on sandy soils and even on volcanic ash soils.

It is grown on nearly all the soils described in any soil classification system [173]. The relative importance of different soils, classified according to soil taxonomy [258], in a number of rice growing countries is shown in Table 1. It can be seen from this table that Alfisols, Entisols, Inceptisols and Vertisols are commonly used for rice growing in these countries. This observation is reinforced by the information given in Table 2 [173], which shows that few orders in this classification [258] are of major importance and only a limited number of sub-orders are of significance in rice growing.

The soils included within each of these important sub-groups are also varied and thus it is to be expected that the total sulfur concentration in soil within these groupings will also vary greatly. This can be readily seen from two examples of the Aquepts sub-order, (1) the surface horizons of three rice soils from the lower Amazon Basin, classified as Tropaquepts ranged in total sulfur from 34 to 139 ppm [280], (2) the surface layers of rice soils from the Telok series in Malaysia classed

Table 1. Major rice growing soils in different countries

Country or continent	Order[a]									
	Alfisols	Aridisols	Entisols	Histosols	Inceptisols	Mollisols	Oxisols	Spodosols	Ultisols	Vertisols
Japan [164]	x		x	x	x				x	
Korea [299]	x		x		x					
Philippines [209]	x		x	x	x				x	x
Indonesia [223]	x		x		x				x	x
Malaysia [190]	x		x		x	x			x	x
Thailand [212]	x		x		x	?			x	x
Burma [298]	?		?			?				?
Bangladesh [54]			x	x	x	x			x	
India [178]	x	x	x	x	x	x			x	x
Sri Lanka [189]	x	x	x	x	x	x			x	x
Pakistan [69]		x	x		x					x
Egypt & Near East Countries [73]		x	x		x					x
European Countries [163]			x		x	x				x
USA [79]	x				x	x				x

[a] Soil taxonomy [258]

Table 2. Importance of different soil types [258] for rice growing [173]

Order	Suborders		
	Major importance	Local importance	Minor importance
Alfisols	Aqualfs, Ustalfs[a]	Udalfs[a]	Xeralfs[b]
Aridisols			Orthids[b], Argids[b]
Entisols	Aquents	Fluvents[a]	Orthents, Psamments
Histosols			Hemists, Saprists
Inceptisols	Aquepts, Ochrepts[a] Tropepts[a]		Andepts
Mollisols		Aquolls	Udolls
Oxisols			Orthox, Ustox
Spodosols			Aquods
Ultisols	Aquults, Udults	Humults	Usfults[a]
Vertisols		Uderts, Usterts	Torrerts[b], Xererts[b]

[a] Mainly aquic subgroup
[b] Exclusively under irrigation

as Sulfaquepts [190] had total sulfur concentrations which varied from 1600 to 4300 ppm [70].

Very few systematic analyses have been published for rice soils. Apart from Venkateswarlu et al. [270] who surveyed the sulfur distribution in selected rice soils in India, only isolated analyses have been reported in the literature. Even fewer authors have classified the soils analyzed according to some internationally recognized scheme [257, 258].

Venkateswarlu et al. [270] found that the total sulfur concentration in the surface, 0–15 cm, layer of Indian rice soils varied from 113 to 275 ppm S, organic sulfur ranged from 49 to 99 ppm S and sodium bicarbonate extractable sulfur from 26 to 144 ppm S. Total sulfur and organic sulfur in these soils tended to decrease with depth, but sodium bicarbonate extractable sulfur (which they considered to be mainly sulfate) did not change uniformly with depth. Organic sulfur formed 14 to 84 percent of the total sulfur and 'sulfate' made up 7 to 62 percent and non-sulfate inorganic sulfur varied from 0 to 57 percent. Total sulfur and organic sulfur in upland soils usually decrease with depth in the same way [286], but the distribution of sulfur between the organic and inorganic fractions is markedly different from that obtained for these Indian rice soils; sulfur in the upper horizons of upland soils is mainly in the organic form and there is very little sulfate or non-sulfate inorganic sulfur [81, 237, 286].

Bhan and Tripathi [36] analyzed Tarai soils, (Haplaquolls) which are used for rice growing in India [178] using different analytical techniques from those used by Venkateswarlu et al. [270] and found a similar distribution of organic sulfur, sulfate and non-sulfate sulfur, i.e. rather low values for organic sulfur and high values for sulfate and non-sulfate sulfur. For example, in the soil from Deoria, the total, organic, sulfate and non-sulfate sulfur concentrations were 121, 30, 18 and 73 ppm S respectively.

Other unpublished analyses [192] for 30 different types of Indian rice soils showed a large range in values for total sulfur, from 99 ppm S for a lateritic soil to 8794 ppm S for acid sulfate soils. Extractable sulfur from these rice soils in the dry state also varied widely and the value obtained depended on the reagent used. These workers used Morgan's solution, ammonium acetate at pH 4.6, Olsen's reagent and Bray's No. 1 reagent and the respective values obtained were in the ranges, 10–1733, 3–1646, 34–1833 and 7–1569 ppm S. Flooding these soils for 20 days increased the extractable sulfur [192]. In view of the strong adsorption of sulfate under acid conditions the choice of acid extractants to study soluble sulfate is surprising.

Current knowledge of the nature of organic sulfur in flooded or upland soils is far from complete. Trace amounts of free cystine and methionine occur in soils and larger amounts of these and related sulfur containing amino acids occur in combined forms [83, 208, 229].

Apart from the amino acid fraction we know very little about the nature of the organic sulfur and can only group the sulfur compounds on the basis of their chemical bonding or reactivity with certain reducing agents. The known groupings at present are:

(a) Organic sulfur which can be reduced to hydrogen sulfide by reagents containing hydriodic acid. This sulfur is not bonded directly to carbon and is believed to be mainly ester sulfate [81, 108].

(b) Organic sulfur which is bonded directly to carbon. This sulfur is not reduced by hydriodic acid [82].

(c) Organic sulfur which is reduced to inorganic sulfide by Raney nickel. This forms a substantial proportion of the carbon-bonded sulfur and may be amino acid sulfur [82].

In upland soils the hydriodic acid-reducible sulfur accounts for between 30–70 percent of the organic sulfur. Very few analyses are available for flooded soils but some unpublished results by Furusaka and Freney (Table 3) suggest that the distribution of organic sulfur is similar to that in upland soils.

Further information on the sulfur concentrations in rice soils is given in Table 4. It is apparent from the limited information available that the range of values is as great as the range of soils on which rice is grown.

2.2. Acid sulfate soils

Large areas of acid sulfate soils (classified as Sulfaquents, Sulfaquepts, Sulfic Tropaquepts [258], or Thionic Fluvisols [257]) which can be used for rice growing when kept in the flooded state, are found in coastal areas in the tropics. Many of the acid sulfate or potentially acid sulfate soils used for rice growing are derived from mangrove areas or from estuarine deposits. Their distribution has been reviewed by Kawalec [139]. Approximately 6.6 million hectares of acid sulfate soils are found in West Africa [35], 5 million hectares in Asia [267], of which 2–3 million hectares

Table 3. Distribution of sulfur in paddy soils from Japan[a] (ppm S)

Soil	Sulfur fraction					
	Total S	Organic S reducible by hydriodic acid	Carbon bonded S	Raney Ni-reducible S	Inorganic sulfate	Monosulfidic S
1. Alluvial Soil (Iwanuma)	1182	534	588	351	60	0
2. Muck Soil (Iwanuma)	1959	876	1004	580	76	2.8
3. Peat (Iwanuma)	1591	690	859	477	41	0.5
4. Clay Soil (Kashmadai)	597	309	251	179	37	0
5. Sandy Soil (Yamamoto)	280	148	117	84	14	0.5
% of total S	100	45.6	50.3	29.8	4.6	0.1

[a] Furusaka and Freney (unpublished data)

Table 4. Sulfur in the surface layer of rice soils

Country	Soil description or location	Total sulfur (ppm)	Sulfate (ppm)
Indonesia [117]	Muara, Bogor	1320	10
	Citayam, Bogor	1350	11
	Singamerta, Serang	200	14
	Cihea, Cianjur	1440	27
	Magelang	810	11
	Meguwoharjo, Yogyakarta	560	38
	Ngale, Ngawi	480	14
	Pacet	1400	163
	Pusakanegara	1490	143
Malaysia [224]	Kuala Kedah		960
	Tebangau		200
	Chengai		660
	Rotan		0
	Sedaka		150
	Kuala Perlis		6330
	Telok		530
	Guar		740
Pakistan [69]	Inceptisols		
	(Marghazar series)		307
	(Kanju series)		10
Philippines [155]	Alfisol (Quingua clay loam)	125	30
	Vertisol (Bantog loam)	125	51

are found in southern Kalimantan [46] and about 1 million hectares are found in the Mekong Delta [267], and other areas occur in Africa, South America and Australia. Large areas of estuarine land at the mouth of the Amazon and in the Orinoco delta may also be acid sulfate soils [46].

Acid sulfate soils have a pH below 4 within the top 50 cm which is caused by sulfuric acid formed by oxidation of pyrite (FeS_2) or other reduced sulfur compounds. Potentially acid sulfate soils contain pyrite which will oxidize to sulfuric acid when the soil is drained [267].

These soils may contain 2–10% pyrite which was formed by microbial reduction of sulfate from sea water. Detailed information on the sulfur distribution in the profile of acid sulfate soils is given in a paper by Chow and Ng [70]. All fractions studied (water-soluble sulfate, acid-soluble sulfate, oxidisable sulfur and total sulfur) increased with depth [70].

2.3. Irrigation and rainwater

Rice plants also obtain sulfur for growth and metabolism from irrigation and rainwater. In some areas irrigation water can be a major source of sulfur for flooded rice, especially when brackish water must be used. In the traditional rice fields of West Africa the irrigation water may be partly of tidal origin or it may be water that has previously been used for irrigating acid sulfate soils. Yoshida and Chaudhry [302] studied the relative importance of irrigation water and soil as sources of available sulfur and obtained results which suggested that irrigation water was twice as effective as soil in supplying sulfur for growth. They also found that only 54 percent of the sulfur supplied in irrigation water for flooded rice was recovered by the plant compared with 93 percent under upland conditions. Some of the sulfate would be converted to plant unavailable forms under both conditions; some would be incorporated into organic matter by soil microorganisms under flooded and upland conditions, and some reduced to sulfide under flooded conditions.

Ishizuka and Tanaka [114] found that 1.7 ppm S was required in water culture to achieve satisfactory growth of flooded rice, but Blair *et al.* [43] obtained results which indicated that this level of supply was insufficient to meet the sulfur requirements of flooded rice in the field. They conducted field experiments and obtained responses to sulfur in all of the major river basins of South Sulawesi. They found that the Maros tap and well water contained 1.7 and 2.8 ppm S respectively and that those levels of soluble sulfur were insufficient to meet the sulfur requirements of rice at two field sites [43].

Wang [279] calculated the uptake of sulfur by rice from irrigation waters of different sulfur contents assuming that lowland rice recovered 54 percent of the sulfur in the water, that the water consumption was 300 g water per g dry matter [302], and that the grain straw ratio was equal to 1. His calculations suggest that irrigation water containing 6.4 ppm S should result in a sulfur uptake of 10 kg S ha^{-1} which should be sufficient to produce 5 tons of rice grain per hectare.

The sulfur content of river waters varies widely, from 0.2 ppm S in the Jari River, Brazil [280] to 20.2 ppm S at Pacet in Indonesia [116, 117], but according to Takahashi [240] most river waters in the major rice growing areas contain relatively high sulfate concentrations. Some published figures for the sulfate concentration in waters of rice growing areas along with the average concentrations for the rivers of the various continents is given in Table 5. These figures suggest that irrigation waters in Asia, Australia and South America would probably not supply sufficient sulfur for flooded rice.

Few data are available on the amount of sulfur supplied in rainwater in tropical areas and even fewer for the input to rice fields, but indications are that input from this source is low. Khemani and Ramana Murty [143] studied the sulfate concentration in rainwater at Delhi, India and found that the concentration varied with the amount of rainfall, the type of cloud from which the rain fell and the season in which the rain fell. Significantly less sulfur was found in rainwater in the monsoon season, e.g. the respective concentrations for summer, winter and monsoon seasons

Table 5. Sulfate concentration in irrigation and river waters

Location	Concentration (ppm S)
Jari River, Brazil [280]	0.2
225 Japanese Rivers [240]	3.5
Indonesia [116, 117]	1.3–20.2
Indonesia [42]	0.9–2.8
Continents	
Africa [154]	4.5
Asia [154]	2.8
Australia [154]	0.9
Europe [154]	8
North America [154]	6.7
South America [154]	1.6
South America [87]	1.4

were 2.0, 2.1 and 1.0 ppm S. Similar results were found for other years [135]. Kapoor and Paul [136] analyzed snow samples at Gulmarg and found a mean concentration of 1 ppm S in these samples.

Probert [207] measured the sulfur input in rain at a number of sites near Townsville, Australia and found that it varied with distance from the sea; 9 km from the sea the input in 1973 was 6.2 kg S ha^{-1} year^{-1} compared with 2.7 kg ha^{-1} year^{-1} when measured 40 km inland. The mean annual input in rainfall in Nigeria was even lower and was estimated to be 1.14 kg ha^{-1} year^{-1} [59].

3. Redox processes and sulfur species in flooded soils

Sulfur occurs in various oxidation states in nature ranging from +6 in sulfates to −2 in sulfides. Sulfate is the stable species under oxidized conditions but under waterlogged or reduced conditions sulfide is the principal stable form [228].

Oxygen diffuses 10,000 times slower through a water phase than a gas phase [92] and thus the supply of oxygen is drastically reduced when a soil is saturated. When the rate of oxygen supply falls below its rate of metabolism the system becomes anaerobic and any sulfate present in the aerobic soil is subsequently reduced to sulfide.

This reduction is biologically mediated as the strict chemical reduction of sulfate by organic matter cannot occur at normal soil temperatures and pressures [88]. Sulfate is not biologically reduced immediately after oxygen removal; when a soil is initially flooded the various components are reduced in a more or less sequential order [195, 241, 242, 244, 245]. Soon after the soil is flooded, oxygen disappears and this is followed by the reduction of nitrate, nitrite, manganic compounds, ferric compounds, sulfate, etc. Reduction of one component does not need to be complete before reduction of the next one can commence, although oxygen and nitrate must be removed before ferric iron is reduced and sulfate is not reduced to sulfide if oxygen and nitrate are present. Consequently, addition of oxidants such as nitrate to the soil will delay the reduction of sulfate to sulfide [72, 74, 293]. The stage at which sulfate is reduced is indicated by the oxidation — reduction status of the soil and this is not generally reached until most of the reducible iron is in ferrous form. This ensures that there is usually sufficient ferrous iron to react with and precipitate any sulfide formed [194].

Both oxidized and reduced zones can occur in flooded rice fields, often in close proximity (e.g. near the soil-water interface and near the rhizosphere), and the redox status of these zones can change with water regime and stage of growth. Thus the pools of oxidized and reduced sulfur in these zones can be interconverted, by both chemical and microbial reactions, and in these redox reactions the reducing agent (electron donor) is oxidized and the oxidizing agent (electron acceptor) is reduced.

The driving force for these reactions in which electrons are transferred from the reduced substrate to the acceptor is the tendency for the free energy of the system to decrease until, at equilibrium, the sum of the free energies of the products equals that of the remaining reactants [202]. The change in free energy, ΔG, for the reduction (1)

$$\text{Oxidant} + ne \rightleftarrows \text{Reductant} \tag{1}$$

is given by equation (2)

$$\Delta G = \Delta G_0 + RT \ln \frac{(\text{Reductant})}{(\text{Oxidant})} \tag{2}$$

where (Reductant) and (Oxidant) are the activities of the reduced and oxidized species and ΔG_0 is the free energy change when the activities are unity. Using the relationship

$$E = \frac{-\Delta G}{nF} , \tag{3}$$

where n is the number of electrons involved in the reaction and F is the Faraday constant (23.063 calories/electron volt), we obtain an expression (E) for the voltage of the reaction (equation (4))

$$E = E_0 + \frac{RT}{nF} \ln \frac{(\text{Oxidant})}{(\text{Reductant})} \tag{4}$$

E_0 is the voltage when (Oxidant) and (Reductant) are each unity. If E is measured against the standard hydrogen electrode it is denoted by Eh and we have an expression (equation (5)) for the redox potential [202]

$$Eh = E_0 + \frac{RT}{nF} \ln \frac{(\text{Oxidant})}{(\text{Reductant})} \tag{5}$$

It is also possible to define an electron activity

$$pE = -\log(E)$$
(6)

which is a convenient measure of the oxidizing intensity of a system and this is related to the redox potential by equation (7) [231]

$$pE = E_0/(2.3\ RTF^{-1}) + \frac{1}{n} \log \frac{(Oxidant)}{(Reductant)}$$
(7)

If oxidation-reduction reactions are arranged one below the other in descending order of E_0 (as in [202], Table 1) then theoretically, under standard conditions, any one system can oxidize the system below it. The sequence of reactions which occurs when a soil is first flooded (see above) proceeds roughly in the same sequence [202, 282]. Changes in pH and activities can alter the sequence [202].

The flooded rice field is a dynamic system into which energy is fed either by photoautotrophic organisms synthetizing organic molecules, or by direct input of such molecules from outside, e.g. in irrigation water or manure. The activity of fermentative, organolithotrophic and chemolithotrophic microorganisms tends to restore the system to its thermodynamic equilibrium by dissipating this input of energy. As free energy concepts (redox potentials) can only describe a thermo-dynamically stable state, the significance of redox potentials is highly questionable in such a system [16, 39, 202, 231].

In submerged rice fields, a diurnal cycle of oxygen production by photo-synthetic organisms, similar to that observed in lakes [130] exists in the water layer and the first few millimeters of soil [24, 205, 217]. Consequently short-term fluctuations and steep gradients in oxygen and sulfide concentrations can be found with depth in the flooded field. Similar gradients are possible in the rice rhizo-sphere where fluctuations in the oxygen supply may occur due to the diurnal opening of stomata [9, 268]. In such sites it may be assumed that the redox system is not in equilibrium and redox potentials become meaningless. Although micro-organisms act as catalysts, redox processes take a long time to reach equilibrium and do not couple with one another; therefore it is frequently possible to have different apparent oxidation-reduction levels at the same site.

For these and other reasons the direct measurement of Eh in different sites of a flooded rice field by the traditional platinum electrode provides relative infor-mation only. For example, the measured redox potential in the flood water of a rice field in Camargue (south of France) was always lower than that calculated from the oxygen concentration, showing that equilibrium had not been reached between oxygen supply and the oxidation of reduced compounds released from the water-soil interface [205]. However, observations concerning the distribution of chemical species in environments displaying redox gradients [22, 38, 51, 55, 78, 89, 193] show that equilibrium is almost attained within the subsystem formed by the main redox couples yielding energy for microbial growth (i.e. O_2/H_2O, Mn^{3+}/Mn^{2+}, NO_3^-/N_2, Fe^{3+}/Fe^{2+}, SO_4^{2-}/S^{2-}, CO_2/CH_4). In the most favourable case, measure-ments of Eh can be related to a particular redox system or systems in partial

282

equilibrium [174]. The redox system must be electrochemically reversible at the surface of the platinum electrode at a rate which is rapid when compared with the electron supply or removal by the measuring electrode [47, 51]. A more complete discussion of the concept of redox potentials and the problems involved in their measurement is given in the review by Ponnamperuma [202].

As a redox equilibrium exists for each redox couple it is theoretically possible to calculate the concentration of a particular sulfur compound from equations relating Eh, pH and concentrations of the other components of a system [50, 200].

From thermodynamic data Boulègue and Michard [51] constructed stability diagrams which showed the fields of dominance of certain sulfur species. Direct laboratory and field measurements of sulfur species, pH and redox potential using platinum and silver/silver sulfide electrodes were in accordance with expected values from these diagrams. Examples of these equilibrium diagrams for the H_2S-S_8-H_2O and the H_2S-S_8-O_2-H_2O systems [48, 50, 51] for a total sulfur concentration of $2.15 \times 10^{-3} M$ are given in Fig. 1.

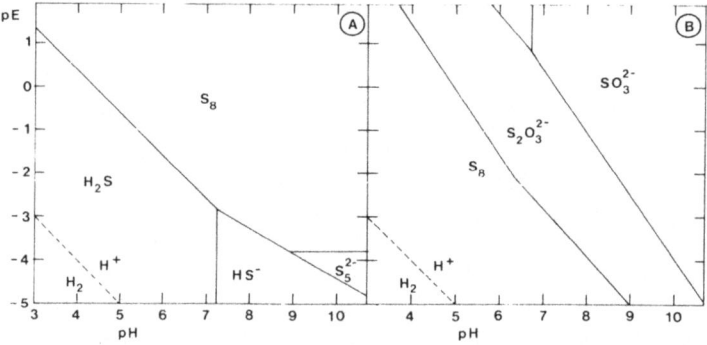

Fig. 1. pE versus pH diagrams for the H_2S—colloidal S_8—H_2O system (A), and H_2S—colloidal S_8—O_2—H_2O system (B) [48, 50, 51].

Pitts *et al.* [200] constructed a phase diagram for an aqueous iron-sulfur system from thermodynamic data, an initial hydrogen sulfide and iron concentrations of 100 ppm, and predicted that the principal precipitate in the Eh—pH range of Louisiana rice paddys would be pyrite (FeS_2). Analyses of precipitates from simulated systems in the laboratory showed that both FeS_2 and FeS were present and that more FeS than was predicted precipitated in the FeS_2 field of dominance. The relative time required for the precipitation of the iron sulfides is apparently an important factor in the composition of the precipitate at any Eh—pH level [200]. They were, however, able to predict hydrogen sulfide concentrations which were in agreement with measured values.

In certain reducing environments several factors such as the slow diffusion of oxygen and the presence of ferric minerals and organic matter may result in the incomplete oxidation of hydrogen sulfide. Under these conditions polysulfide ions ($S_n^=$) and thiosulfate can be formed [67, 91].

Redox conditions also have significant effects on the types and amounts of sulfur compounds emitted from soils [28, 76] and on the incorporation of sulfur into soil organic matter [49, 182]. According to Boulègue [49] hydrogen sulfide produced during dissimilatory sulfate or sulfur reduction at low redox potentials reacted with organic matter and polysulfide chains were incorporated into humic acids.

Unlike fermentative metabolism, corresponding to an internal reorganization of organic matter without modifications of the gross redox level, respiratory metabolism consists in the exoenergetic oxidation of a reduced substrate at the expense of an electron acceptor. A necessary condition for the existence of such a process is that the mean free energy change per electron transfer must be negative under standard conditions, although all metabolism predicted possible by thermodynamic data do not exist. The free energy change depends on the availability of the oxidant and therefore on the Eh of the medium. This relationship, represented in Fig. 2, shows that within the Eh range encountered in natural media, oxidation of organic compounds (i.e. organolithotrophic metabolism) is always exoenergetic, although the energy yield is lower at lower Eh values [39].

Considering the sulfur compounds available as electron acceptors in the rice paddy for the oxidation of organic matter, Widdel and Pfennig [285] have shown that the free energy change was more favourable for the growth of sulfate-reducing than for sulfur-reducing bacteria on the same substrate, acetate. However, both organisms have been found to be active at similar concentrations in flooded rice soils of West Africa, indicating that sulfate-sulfide and sulfur-sulfide reactions were competitive for the decomposition of low molecular weight organic molecules in the reduced sites of that ecosystem (Jacq, unpublished).

Oxidation of mineral compounds (i.e. chemolithotrophic metabolism), however, is not exoenergetic over the whole range of Eh, and a critical value exists, below which the oxidation of a given mineral substrate is impossible. Baas-Becking and Ferguson Wood [15] have shown that sulfur oxidizing bacteria of the *Thiobacillus* genus were only active above the upper limits of stability of the reduced sulfur compounds metabolized. From thermodynamic data is it therefore unlikely that the non-phototrophic sulfur-oxidizing bacteria, oxidising inorganic sulfur compounds (H_2S, S^0, $S_2O_3^{2-}$, SO_3^{2-}) at the expense of either oxygen or nitrate, coexist in the same microsites as sulfide-producing bacteria; diffusion of either oxidants or reductants could allow coupling between these organisms.

However, phototrophic bacteria, such as the purple and green sulfur bacteria, have the ability to oxidize sulfide to sulfur and sulfate in the light under strict anaerobic conditions in natural environments [15, 134, 162]. Electron transfer from sulfide (or sulfur) to carbon dioxide is driven by ATP generation by the photosynthetic system [197] and therefore growth is possible in the same habitat in which sulfide is produced.

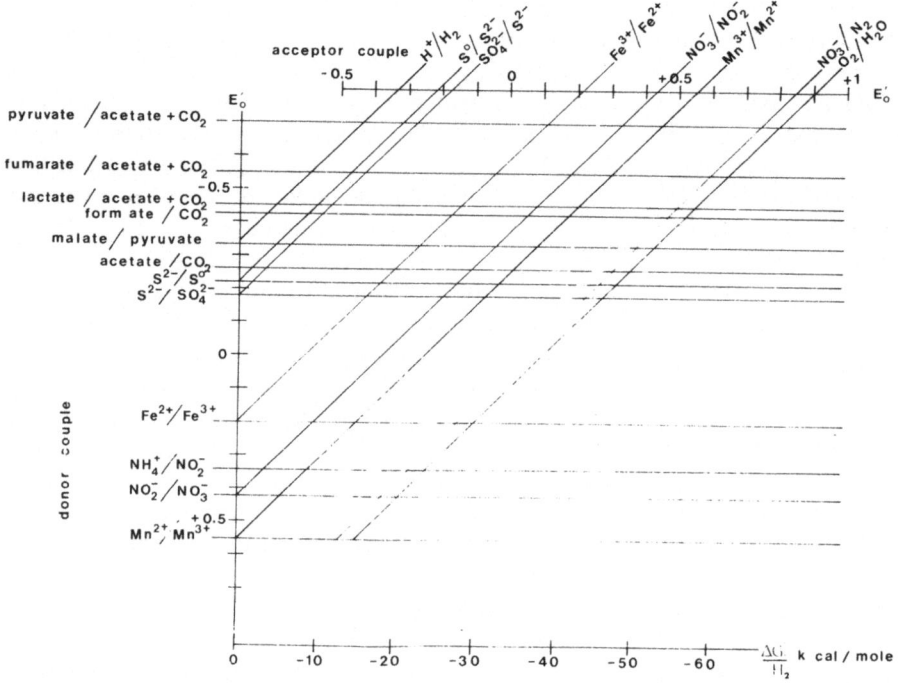

Fig. 2. Diagram describing the standard free energy change for each two-electron transfer. For each metabolic system, the standard free energy is the intersection of the horizontal line corresponding to the donor couple with the oblique line corresponding to the acceptor couple. (Thermodynamic data from Pfennig and Biebl [198], Pitts *et al.* [200] and Senez [220]).

It is apparent from this section that the sulfur species present in the different zones of a flooded rice soil are determined by the redox conditions and the variety of microorganisms which can exist under those conditions.

4. Transformations of sulfur in flooded rice fields

As discussed above a great variety of sulfur compounds, both organic and inorganic, and redox conditions can exist in a flooded rice field. Even in a permanently flooded field discrete oxidized and reduced zones exist; an oxidized layer occurs at the water-soil interface and the root rhizosphere is oxic due to the transmission of oxygen from the atmosphere through air passages in the leaves, stems, nodes and roots. The oxidized zone can be quite considerable in a planted soil (see for example Fig. 2 in [301]). Thus many of the transformations which occur in flooded soils are characteristic of upland, or well-drained, soils. For example, in the aerobic zone of a flooded soil, sulfate can be produced from the decomposition of organic matter. On the other hand the interaction which exists between the rice plant and the filamentous bacteria *Beggiatoa* [132] is peculiar to the flooded system. Flooding produces hydrogen sulfide and inhibits oxygen release from rice; *Beggiatoa* significantly increases oxygen release from rice seedlings and reduces the accumulation of hydrogen sulfide in soil.

The transformations which occur are mainly biological, although non-biological reactions such as the oxidation of iron sulfides can be involved, and we have grouped them into five categories; (a) immobilization, or assimilation of sulfur into organic compounds, (b) mineralization, or decomposition of organic sulfur compounds, (c) production of sulfides, (d) production of volatile sulfur compounds, and (e) oxidation of sulfur and inorganic sulfur compounds. Reactions (a), (b) and (d) can take place in both aerobic and anaerobic zones but reactions (c) and (e) should be confined to anaerobic and aerobic environments respectively.

Relatively little work has been carried out and published on sulfur transformations in flooded rice soils but the reactions which occur in the oxidized zones would be expected to be the same as those occurring in upland soils. However, the rates of the reactions may well be different. The transformations of sulfur in oxidized systems have been reviewed recently [37, 41, 84, 283] and as no significant developments have been made since those reviews were published they will not be discussed here.

4.1. Mineralization and immobilization

Little information is available on the mineralization and immobilization of sulfur in flooded rice soils. In one of the few studies on this topic, Sachdev and Chhabra [214] used the radioactive isotope to study the immobilization of inorganic sulfate in flooded soils. They found that 37.8 percent of the added sulfur could be recovered as organic sulfur; this was considerably greater than the incorporation under aerobic conditions. Very little of the sulfur (2.6 percent) was incorporated into the amino acid fraction.

The enzyme arylsulfatase which may be involved in the mineralization of a certain fraction of soil organic sulfur was studied in rice soils by Han and Yoshida [301] who found that the arylsulfatase activity of rhizosphere soil was greater than that of the non-rhizosphere soil.

4.2. Production of sulfide

The main process leading to the production of sulfide in flooded soils is dissimilatory sulfate reduction and, as discussed above, this reaction follows the reduction of ferric iron and immediately preceeds methanogenesis [282]. Dissimilatory sulfate reduction is the property of a few specialized anaerobic bacteria which use sulfate as the terminal electron acceptor in respiration [202, 213].

Three genera of sulfate reducing bacteria have been described. These are,

(i) *Desulfovibrio*: these organisms are motile, heterotrophic, small, obligate anaerobes which are generally spiral-shaped [206].

(ii) *Desulfotomaculum*: the cells are larger, motile, heterotrophic rods that form heat resistant spores [64].

(iii) *Desulfomonas*: the cells are non-motile rods [172].

Desulfovibrio desulfuricans seems to be mainly responsible for sulfate reduction in flooded soils [85] but *Desulfotomaculum* was found to be the dominant form in dryland soils [246]. Jacq [124] found two groups of sulfate-reducing bacteria in West African paddy fields, (i) 'lactate-utilizing' *Desulfovibrio* and *Desulfotomaculum* strains, and (ii) an 'acetate-utilizing' *Desulfotomaculum* strain living in symbiosis with *Clostridia* strains. This strain probably originated from cattle rumen because the fields under study have been used as pasture during the dry season (Pfennig, personal communication).

Sulfate reducing bacteria tolerate high concentrations of salt and hydrogen sulfide and they function best in the pH range 5.5–9.0 [228]. Connell and Patrick [71] found that under the conditions of their experiments, i.e. at controlled redox potentials, sulfate reduction was confined to the pH range 6.5–8.5. However, intense sulfate reduction has been found to occur in rice fields outside this pH range. Garcia *et al.* [86] found a negative correlation between pH and number of sulfate reducers in 27 West African alluvial soils with low initial pH (< 4.5). Also addition of lime significantly increased the production of all sulfide fractions, except acid-soluble, from ammonium sulfate applied to paddy soils [95] and sulfate reduction appeared to be faster in a calcareous soil than in an alluvial soil or a sandy soil [165].

The clay content of a soil also appears to affect sulfate reduction [86] and this may be attributed to a decrease in oxygen supply with increased clay content.

Sulfate-reducing bacteria use a variety of fermentation products (such as lactate) or molecular hydrogen to reduce sulfate [228].

$$2 CH_3 CHOH COOH + SO_4^{2-} \rightarrow 2 CH_3 COOH + 2H_2O + 2CO_2 + S^{2-} \qquad (8)$$

$$4H_2 + SO_4^{2-} \rightarrow S^{2-} + 4H_2O \qquad (9)$$

The end product is sulfide ion or hydrogen sulfide gas which the cells excrete, in contrast to the assimilating sulfate reducers such as plants, which assimilate the products of sulfate reduction.

Recently, Widdel and Pfennig [285] isolated *Desulfotomaculum acetoxidans* which metabolizes acetate while reducing sulfate to sulfide, thus repudiating the long-held belief that sulfate-reducing bacteria could not use acetate as an electron donor and that acetate accumulated during the oxidation of lactate and other aliphatic acids. In addition, Badziong *et al.* [20, 21] showed that *Desulfovibrio vulgaris* (Marburg) could grow using acetate and carbon dioxide as carbon sources and hydrogen and sulfate as energy sources, while Laanbroek and Pfennig [149] have observed the oxidation of short chain fatty acids, including acetate, in fresh water and marine sediments during sulfate reduction.

Since virtually all strains of sulfate-reducing bacteria described in rice fields before 1977 were isolated using lactate as the electron donor, sulfide accumulation in these fields was mainly attributed to lactate-oxidising strains.

Sulfate reduction increases with period of submergence and the rate of increase is related to the amount of organic matter in the soil [183]. Sulfide production is also increased by the addition of organic matter to soil [77, 294] and influenced by the type of organic matter added [45]. However, the C/N ratio of the straw added to a flooded soil had no effect [77].

In a study of the microdistribution of sulfate reducers in Japanese paddy soils, Wakao and Furusaka [273, 274, 275] and Wakao et al. [277] found that the distribution pattern was linked to the dispersion of organic debris in soil. A similar localization of sulfate reducers around decomposing organic matter was found in the 1–2 cm surface layer, at the soil–water interface, in African paddy soils rich in organic matter [86], (Jacq unpublished). In this zone, fresh organic matter is provided by decaying algae and cyanobacteria, mainly during the cropping period.

In planted rice fields organic substrates may be provided to sulfate-reducing bacteria from seeds or root exudates, litter or gums. The activity of sulfate-reducing bacteria in the spermosphere and rhizosphere was extensively studied by Jacq [120, 121, 122, 124] and Garcia et al. [86]. Highly significant spermosphere and rhizosphere effects were established during field surveys and microplot experiments on acid paddy soils originating from reclaimed mangroves or fluvial and estuarine deposits. Germinating seeds or rice roots were coated with large quantities of black ferrous sulfide and the rice plant was susceptible to the sulfide produced by sulfate-reducing bacteria, not only during germination (spermosphere effect) but also during three stages of growth, transplanting, tillering and flowering (rhizosphere effects) [120, 122].

Localization of sulfide in the spermosphere has not been reported elsewhere but localization in the rhizosphere has also been found in Japanese rice soils [144, 145]. These authors found that the positive effect of the rice rhizosphere on sulfate reduction was greater under flooded conditions. They observed that sulfate-reducing bacteria were stimulated immediately after transplanting rice into flooded soils and that the rice roots, at later stages, were coated by ferrous sulfide.

It appears that most of the sulfate-reducing bacteria are found in the outer rhizosphere [31] where they number in excess of 10^7 bacteria g^{-1} [124, 145]. This stimulation seems to be due to the exudation of substrates such as organic acids [53, 121, 244], amino acids [53, 121, 152, 158, 215, 216] and carbohydrates [53, 121, 158].

The number of sulfate-reducing bacteria is generally higher in flooded than in dryland rice fields, and they are not confined to the lower, most reduced horizons but are more abundant in the plow layer [113, 246, 282]. A survey of initial populations at planting showed that these bacteria varied from 60 to 7×10^5

cells g^{-1} dry soil in the plow layer depending on the physical and chemical properties of the soil [86]. This substantial inoculum probably explains the rapid initiation of sulfate reduction when soils are rewetted. Even though the bulk of sulfate reducers is found in the plow layer it is apparent from the discussion above that they would not be evenly distributed in that layer; their distribution being affected by the amount and distribution of organic matter, spermosphere and rhizosphere effects and the presence of micro-aggregates which control the supply of oxygen [275].

Takai and Tezuka [246] suggested that the spore-forming sulfate-reducing bacteria (i.e. *Desulfotomaculum*) were more resistant to the air-drying and aeration of soil which occurs between successive crops. In Japan they found that the total number of spore-forming sulfate reducers did not decrease in one year while in West Africa, Jacq [124] found that the non-sporing forms could decrease sharply from $10^6 - 10^8$ cells g^{-1} at harvest to between $10^1 - 10^2$ cells g^{-1} at the end of the dry season eight months later.

The resistance of *Desulfotomaculum* strains to air-drying does not explain the repartition of sulfate-reducing bacteria in the rice soils of the central Ivory Coast and those of Senegal. *Desulfotomaculum* strains represented about half of the sulfate-reducing bacteria in the irrigated rice fields of the central Ivory Coast which are cultivated 8 months of the year but very few were found in the coastal paddy soils of Senegal which are cultivated for 4 months only [124]. This difference may be explained by the recent finding [68] that the adenosine triphosphate balance during growth is three times more favourable for *Desulfovibrio* than for *Desulfotomaculum* in sulfate rich media but is identical in sulfate limited media, and the knowledge that the Senegal rice fields are richer in sulfate than those of the continental basins of the Ivory Coast. The Senegal coastal areas are saline or irrigated with brackish water.

While most of the sulfide formed in paddy soils appears to come from the dissimilatory reduction of sulfate it can also be produced by other processes, viz. the reduction of elemental sulfur, the decomposition of organic compounds and the reduction of sulfate, sulfite and thiosulfate by other organisms.

A new genera of bacteria capable of reducing elemental sulfur and oxidizing acetate has been isolated by Pfennig and Biebl [198]. This bacterium *Desulfuromonas acetoxidans* occurs in African rice fields in large numbers (10^9 cells g^{-1} [124], Traoré, Jacq and Mouraret, unpublished). The ecological significance of such sulfur reducing bacteria is not known but they may be involved in the biological transformation of sulfur containing fertilizers such as sulfur coated urea in paddy soils [123]. However, no correlation was observed between the numbers of these bacteria in rice soils and the subsequent accumulation of sulfide.

Some sulfide is certainly produced during the metabolism of protein in paddy soils [45, 261] and many heterotrophic microorganisms can convert organic sulfur to sulfide under anaerobic conditions [133].

As stated above some organisms other than those classified as sulfate reducers can convert sulfate to sulfide. Wu *et al.* [291] isolated from rice soils a denitrifying organism *Pseudomonas putida* which, in the absence of nitrate, would use sulfate as an electron acceptor and reduce it to hydrogen sulfide. This reaction seemed to be responsible for the so-called 'suffocating' disease of rice in poorly drained paddy soils in Taiwan.

Spirillum 5175 isolated from an anaerobic culture of *Desulfuromonas* growing on acetate and elemental sulfur was found to reduce sulfite, thiosulfate and elemental sulfur but not sulfate [289]. This organism has been found to grow syntrophically with *Chlorobium*, a photosynthetic sulfur bacteria. As *Desulforomonas, Chlorobium*, elemental sulfur and acetate have been found in rice fields it seems likely that *Spirillum* bacteria will also be found in these soils (Jacq unpublished).

Sulfite, polythionate, thiosulfate and elemental sulfur are reduced more readily than sulfate, and many organisms other than the sulfate reducing bacteria can produce sulfide from these compounds [228].

4.3. Production of volatile sulfur compounds

Hydrogen sulfide can be produced in waterlogged soils as a result of sulfate reduction or protein decomposition [45, 262]. However, the concentration of water-soluble hydrogen sulfide seems to be extremely low under most conditions due to its reaction with ferrous iron and other cations to form insoluble sulfide [13, 72, 202]. Even when sulfide accumulated in the rhizosphere to the extent of 30 ppm only about 3 ppm could be found in the pore-water of the clays and less than 1 ppm sulfide was found in the water surrounding the soil (Jacq, unpublished). When soils are high in sulfate and available organic matter, and deficient in the cations which precipitate sulfide significant amounts of water-soluble sulfide can be found [45, 235]. This situation can be prevented by addition of iron oxide (Fe_2O_3) to the soil [72, 294], and in the case of acid sulfate soils by addition of lime [13].

Apart from a recent publication by Jørgensen *et al.* [129] on the emission of hydrogen sulfide from coastal environments there does not appear to be any evidence for the emission of appreciable hydrogen sulfide from flooded soils in the field [57]. There is also very little evidence for its emission, in significant amounts, from flooded soils in laboratory experiments. No trace of hydrogen sulfide could be detected in the atmosphere above soil when inorganic and organic sulfur compounds, plant materials, animal manures and sewage sludges were incubated in waterlogged soils [26, 27, 28, 56]. Harter and McLean [101] were unable to detect hydrogen sulfide emission from a waterlogged soil which produced large quantities of sulfide.

Some publications, however, indicate that hydrogen sulfide is emitted during incubation studies with soil under anaerobic conditions [45]. The emission in these cases may be caused by the experimental conditions and may have no relevance to the field situation. Any hydrogen sulfide generated in a flooded soil in the field and not trapped by reaction with cations, e.g. Fe^{++} has to pass through the oxidised layer of soil and then through the oxidized water layer. The chances of emission before reaction with iron or oxidation to sulfur seem to be small. Nevertheless measurements of hydrogen sulfide emission to the atmosphere need to be made in the field environment under a range of conditions before this supposition can be proven.

A range of volatile organic sulfur compounds are emitted from flooded soils but again most of the measurements have been made under controlled conditions in the laboratory [57, 133]. The volatiles isolated in significant amounts include carbon disulfide, carbonyl sulfide, methyl mercaptan, dimethyl sulfide and dimethyl disulfide, but the total amounts released from soils or decomposing organic material under waterlogged conditions appear to be very small [57].

Takai and Asami [243] found that more methyl mercaptan was produced at the higher temperatures in mid summer which accompany the lowering of the redox potential and the increase in production of hydrogen sulfide. Addition of green or stable manure to a flooded soil increased the production of methyl mercaptan [11].

4.4. Oxidation of sulfur and inorganic sulfur compounds

Sulfide produced in reduced sites of waterlogged soils as a result of the decomposition of sulfur-containing organic compounds or by dissimilatory sulfur- and sulfate-reduction may be precipitated as insoluble sulfides or converted to sulfur and sulfate by chemical or biological oxidation. The rate of sulfide reoxidation at the marine sediment-water interface has been found to exceed significantly the rate of sulfide generation [22, 40, 128] but little information concerning this equilibrium is available for the flooded rice soil. In the following section the main groups of sulfur-oxidizing bacteria are described, and their possible contribution to sulfide reoxidation in the surface oxidized layer, the reduced plow layer and the rhizosphere of rice is discussed.

Sulfur oxidizing bacteria

(i) *The thiobacilli.* Although several bacterial genera can make use of the oxidation of reduced sulfur compounds as a source of electrons and energy for growth, complete chemolithotrophy has only been clearly demonstrated in the *Thiobacillus* genus. The biology and general metabolism of this group of bacteria have been reviewed [140, 271] and 7 groups described by application of numerical analysis to 93 strains using 38 tests and 106 characteristics [110].

Group 1, corresponding to *Thiobacillus thioparus*, oxidizes sulfide, sulfur, thiosulfate and thiocyanate under aerobic conditions. Growth occurs in the pH range 7.8–4.5 and is strictly autotrophic [213]. This bacterium has been isolated from paddy soils (Baldensperger, unpublished data), but it is unlikely that this strict autotrophic and aerobic microorganism is of importance in flooded rice soils because organic compounds present in the oxidized layers would be expected to promote the growth of heterotrophic strains.

The *Thiobacillus thiooxidans* species Group 2 resembles *T. thioparus* with respect to aerobiosis, autotrophy and sulfur compounds oxidized, but, unlike the latter, grows best around pH 2 [213] with the production of strong acid. In a medium containing sulfur, pH values below 1.0 may be generated [142]. Although *Thiobacillus thiooxidans* has been found to occur in flooded rice soils at populations up to 10^6 cells/gm [177] their contribution to sulfur oxidation in flooded soils of near neutral pH is unlikely to be significant.

Group 3 bacteria, represented by *Thiobacillus denitrificans*, are also autotrophic but can oxidise reduced sulfur compounds in the absence of oxygen at the expense of nitrate which is simultaneously reduced to molecular nitrogen [14]. *Thiobacillus denitrificans* has been isolated from flooded rice fields [25], and its contribution to the reoxidation of sulfide may be gauged from the observation that sulfide accumulation in rhizospheric soil samples was related to the ratio of *Thiobacillus denitrificans* and *Desulfovibrio* organisms [126].

In sulfur and thiosulfate-containing media *Thiobacillus ferrooxidans* (Group 4) resembles *T. thiooxidans* with respect to final pH but, unlike the latter, does not oxidize sulfide and may use ferrous iron as an energy source. Kelly and Tuovinen [141] have shown that the specialized iron oxidizing bacterium *Ferrobacillus ferrooxidans* can also grow on sulfur or thiosulfate and therefore ferrobacilli cannot be distinguished from thiobacilli. *Thiobacillus thiooxidans* plays a vital function in the oxidation of pyrite [253], a constituent of certain estuarine and coastal soils being reclaimed for rice growing. The contribution of *Thiobacillus ferrooxidans* to neutral paddy fields is not known but it could play an important role in the oxidation of pyrite during the reclamation of estuarine and coastal soils for rice growing.

Three further groups of *Thiobacilli* can be recognized on the basis of their sources of energy and carbon. Group 5 are obligate chemolithotrophs but are facultative heterotrophs. These organisms, such as *Thiobacillus neapolitanus*, oxidize sulfur with the production of sulfuric acid, which becomes inhibitory when the pH falls to 2.8. Growth is increased when glucose is used as a carbon source. Facultative heterotrophy might therefore be an advantageous characteristic of *T. neapolitanus* in the natural environment.

Organisms which are facultative chemolithotrophs (also called mixotrophic thiobacilli [210]) and facultative autotrophs (Group 6) have been extensively studied and several species such as *Thiobacillus novellus*, [227] *Thiobacillus intermedius* [156] and *Thiobacillus delicatus* [171] have been isolated. These bacteria are neutrophilic and can grown in a glucose-salt medium if reduced sulfur is provided for assimilation. Such versatile thiobacilli have been shown to compete successfully in mixed cultures containing specialized chemolithotrophic thiobacilli and obligate heterotrophs [90]. Although they have been isolated from the same habitats as obligate chemolithotrophs [236] their role in waterlogged rice soils is unknown. Thermophilic [287] and acidophilic [93] strains of mixotrophic thiobacilli have also been described.

Group 7 includes the facultative chemolithotrophs but obligate heterotrophs such as *Thiobacillus perometabolis* [157] and *Thiobacillus rubellus* [171]. They have been isolated from acid mine water [180] and oxidize sulfide, sulfur, thiosulfate and sulfite with sulfuric acid as the oxidation product. No reports on their isolation from flooded soils have been sighted.

(ii) *Other non-filamentous chemolithotrophic sulfur-oxidizing bacteria.* A wide variety of genera of aerobic or microaerophilic facultative autotrophic sulfur-oxidizing bacteria have been described, with spherical (*Sulfolobus* sp., *Thiovulum* sp.), cylindrical (*Macromonas* sp.) or spirillal (*Thiospira* sp. and *Thiomicrospira* sp.) shape. The physiology and ecology of these genera have been reviewed [127] but their possible role in flooded rice fields remains unknown.

(iii) *Gliding sulfur-oxidizing bacteria.* The cells of these bacteria are arranged in chains within trichomes which show a gliding motion when in contact with a substrate. One representative of this group is *Beggiatoa* which, unlike many of the organisms discussed above, has been isolated from rice soils and has been shown to play a vital role in this ecosystem [132, 201]. The catalase-like activity surrounding the root tips of rice is favourable for the growth of this microorganism and it significantly reduces hydrogen sulfide concentrations in flooded soil samples. *Beggiatoa* has been a center of controversy since it was first described as autotrophic in 1887. In a recent paper Strohl and Larkin [230] reported the separation of strains isolated from freshwater sediments into 5 groups. All strains deposited sulfur in the presence of hydrogen sulfide and grew heterotrophically. The available literature suggests that *Beggiatoa* are of prime importance in the reoxidation of sulfide produced by sulfur-reducing bacteria in the root zone of rice [132, 201].

(iv) *Phototrophic sulfur bacteria.* The colourless sulfur bacteria are typical gradient microorganisms which can thrive only in the redox discontinuity layer where both reduced sulfur compounds and electron acceptors are present at the same time. On the other hand the phototrophic bacteria, such as *Chromatium* and *Chlorobium*, do not depend on this unstable ecological niche if their energy source, light, penetrates the sulfide containing environment [197]. The biology and ecology of purple and green sulfur bacteria has been reviewed [15, 134, 150, 161, 197] and their sulfide-oxidizing activity at the soil-water interface of the flooded rice field and in the rice rhizosphere has been demonstrated [97, 147]. In the photic layer which is estimated to reach to about 2.5 mm below the surface, both cyanobacteria and flexibacteria in addition to phototrophic sulfur bacteria have the ability of anoxygenic sulfide oxidation [130].

(v) *Heterotrophic organisms.* The role of heterotrophic fungi, actinomycetes and bacteria in the oxidation of sulfur in soil has often been overlooked. These organisms oxidise sulfur by reactions incidental to their normal metabolism [228].

Vitolins and Swaby [272] found that heterotrophic yeasts and several genera of heterotrophic and facultative autotrophic bacteria were far more numerous than the strict autotrophs in soils and may play an important role in the oxidation of sulfur in many soils.

5. Interactions of the carbon, nitrogen and sulfur cycles in flooded soils

As described above organic compounds added to flooded rice fields as plant residues, animal manure, floodwater or as the result of photosynthetic activity are oxidized by organolithotrophic, chemolithotrophic and fermentative micro-organisms according to a sequential order of availability of oxidants. After the depletion of oxygen, nitrate and ferric iron, the metabolic pathways of organic compounds depend on the availability of inorganic sulfur compounds.

Some of the information concerning the oxidation of organic substrates by sulfur- and sulfate-reducing bacteria is summarized in Table 6. It is apparent that both groups are involved in the oxidation of low molecular weight compounds. Acetate is oxidized during sulfate reduction by *Desulfotomaculum acetoxidans* [285], equation (10),

$$CH_3COO^- + SO_4^{2-} \rightarrow 2HCO_3^- + SH^- \tag{10}$$

and during sulfur reduction by *Desulfuromonas acetoxidans* [198], equation (11),

$$CH_3COO^- + 4S^0 + 4H_2O \rightarrow 2HCO_3^- + 4SH^- + 5H^+ \tag{11}$$

with the final production of bicarbonate and sulfide in mixed [61, 62, 63] and in pure cultures (Pfennig, personal communication). Long-chain even-numbered fatty acids and long-chain odd-numbered fatty acids are oxidized to acetate, and propionate plus acetate respectively and subsequently to bicarbonate.

Hydrogen is commonly produced by anaerobic bacteria and it is used as a substrate by methanogenic bacteria (Table 6). However, hydrogen is not usually produced in the presence of sulfate. When sulfur compounds are limiting, a mutualistic association exists between methanogenic bacteria and hydrogen producing bacteria [290].

The spore forming *Desulfotomaculum* is unable to produce hydrogen as this sulfate reducer lacks the C_3 cytochrome which is necessary for electron transfer to hydrogenase (Le Gall, personal communication). However, certain strains of *Desulfovibrio* growing on sulfate limited media [102] or on lactate-sulfate media (Le Gall, personal communication) can produce hydrogen. Thus interspecific hydrogen exchange is theoretically possible, but doesn't appear to occur because methanogenic bacteria are inhibited by the sulfide produced by sulfate-reducing bacteria. This has been demonstrated *in vitro* [60] and in soil [181, 288].

In the waterlogged rice soil any interactions between methanogenic and sulfur-reducing bacteria are likely to be the result of competition for organic substrates (as both groups are involved in the oxidation of similar compounds) and the inhibitory effect of sulfide [65, 303].

There is some evidence that fatty acids cause a disease in rice [107, 291] and the effects of butyric acid on rice have been demonstrated. Methanogenic and sulfate-reducing bacteria may therefore play a vital role in detoxification processes in paddy soils by their mutualistic oxidation of fatty acids.

The possible pathways for the decomposition of organic molecules in the

Table 6. Carbon substrates oxidized by sulfate-reducing, sulfur-reducing and methanogenic bacteria

Genus	Morphology	Substrates used	End products
Sulfate- and sulfur-reducing bacteria			
Desulfovibrio 'sapovorans'	vibrios	long-chain fatty acids	propionate + acetate
Desulfolobus	spherical cells	propionate	acetate + HCO_3^-
Desulfovibrio	vibrios	lactate	acetate + HCO_3^-
Desulfotomaculum	vibrios, spore-forming	acetate	HCO_3^-
Desulfuromonas	vibrios	acetate	HCO_3^-
Methanogenic bacteria			
All genera		H_2, HCO_3^-	CH_4
Methanobacterium	rods (variable)	formate	$CH_4 + HCO_3^-$
Methanococcus	cocci	formate	$CH_4 + HCO_3^-$
Methanospirillum	vibrios	formate	$CH_4 + HCO_3^-$
Methanosarcina	cocci in regular cubical packages	acetate, methanol	$CH_4 + HCO_3^-$

presence and absence of oxidized sulfur compounds are summarized in Fig. 3. The present information indicates that methanogenic and sulfate-reducing bacteria are important in the carbon flow in the reduced layer of flooded soils, and that the sulfate reducers have a more important role which may benefit from the methanogenic bacteria when sulfate concentrations are low [176].

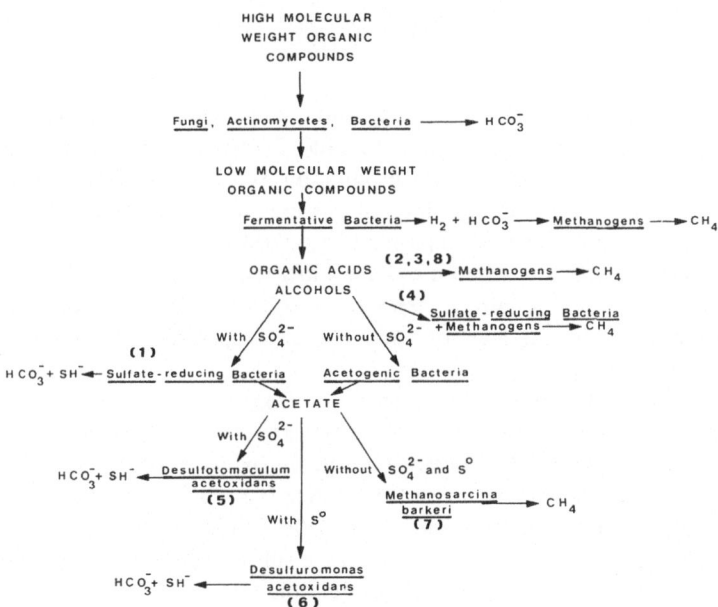

Fig. 3. Carbon flow in flooded rice soils as affected by the concentrations of inorganic sulfur. References: 1, Pfennig, personal communication, 2 [221], 3, [304], 4 [60], 5 [285], 6 [198], 7 [159], 8 [222].

The sulfur cycle is also linked to the nitrogen cycle through the reactions of organisms such as the denitrifiers *Thiobacillus denitrificans* and *Pseudomonas putida* and nitrogen fixing bacteria such as *Chromatium* and *Thiospirillum*.

Thiobacillus denitrificans oxidizes sulfur compounds anaerobically in the presence of nitrate [14] while *Pseudomonas putida*, a denitrifier isolated from the rice rhizosphere, reduces sulfate to hydrogen sulfide when nitrate is absent [291].

The photosynthetic sulfur bacteria can be found in reasonably high concentrations ($10^2 - 10^3$ cells g^{-1} soil) in certain rice soils (e.g. in Japan [187] and the tropics [146]) and their potential contribution to nitrogen fixation in lowland rice culture has been assessed [94]. Using propanil to inhibit the nitrogenase activity of blue-green algae it was concluded that photosynthetic bacteria (predominantly cells resembling the genera *Chromatium* and *Thiospirillum*) could contribute more to nitrogen fixation than the rhizosphere microflora and contribute as much as blue-green algae.

6. Agronomic implications of the sulfur cycle

Many of the soils used for rice production suffer from a deficiency of sulfur which restricts growth, an excess of soluble sulfide which is toxic or accumulations of pyrite which result in highly acid soils. These and other problems of rice production which can be overcome to a large extent by manipulation of the sulfur cycle are discussed below.

6.1. Sulfur deficiency in rice

Deficiencies of sulfur for the growth of plants have been reported with increasing frequency throughout the world and this has focussed greater attention on the importance of this nutrient in crop production and quality. Whilst most of the responses to sulfur have been obtained in pastures it has also been found to limit the production of cereals in several parts of the world [32].

Even though cereals are believed to have a low requirement for sulfur, numerous experiments have shown a direct response to sulfur. Quite often the response to sulfur is determined by the level of other nutrients especially nitrogen, phosphorus and potassium [32, 225].

The importance of sulfur for the nutrition of flooded rice has only recently been appreciated [42, 43, 44, 117, 278, 279, 280, 281, 302] even though responses to sulfur in field experiments had been found in Burma as early as 1938 [219]. The recognition of the importance of sulfur for rice is probably due to the increasing use of high analysis fertilizers containing little sulfur, the introduction of high yielding varieties with a greater requirement for sulfur [42, 302] and greater awareness of the role of this nutrient in crop production.

Sulfur deficiency symptoms in flooded rice are similar to those observed in upland crops. The first sign of deficiency is the yellowing of the new growth and this extends to a general yellowing of the whole plant. The leaves on sulfur deficient plants appear narrow and the stems are thin [44, 234]. The most striking effect of sulfur deficiency in rice is the reduction in tiller numbers. Sulfur deficiency not only reduces panicle number but also reduces the number of filled grains per panicle and thus the final yield of grain [280]. Sulfur deficiency usually delays the development of the rice crop [44, 278] which results in delays in harvesting which can result in further reductions in yield of grain through rain or wind damage or disease.

6.2. Responses to sulfur fertilization

Responses of flooded rice to application of sulfur have been obtained in the field in Bangladesh (Myers, private communication), Burma [219], India [2, 199, 218], Indonesia [43, 44, 103, 117, 160], Taiwan [151, 297] and Brazil [280, 281].

Sulfur fertilization markedly increased tiller production [44, 280] and the effect on tiller production can be observed as early as 17 days from transplanting [44].

Sulfur deficiency severely limits the production of rice grain. For example, in the lower Amazon Basin the yields in the absence of applied sulfur were of the order of 1.3 tonnes ha^{-1} compared with ~5.6 tonnes for the healthy crop [280]. Blair et al. [43] also obtained large responses in grain yield to applied sulfur; responses ranged up to 278 percent, with the average grain response at 28 sites being 18.6 percent.

As fertilization with sulfur has such a large effect on tiller production it is apparent that the timing of fertilizer application to sulfur deficient rice is very important. The results of Wang et al. [280] show that sulfur should be applied as early as possible in the growth cycle and certainly before the active tillering stage. Applications made after this stage were still effective but the benefits were less [280].

In addition to the responses to applied sulfur in the field a number of responses have been obtained in pot experiments (e.g. Bangladesh [3, 137, 138], India [1], Indonesia [115, 118, 153], Japan [8], Philippines [155], Senegal [123] and Sri Lanka [232]). However, it is very difficult to apply these results to the field because of differences in growth rate, nutrient supply, supply of sulfur from outside sources etc. [302].

6.3. Uptake of sulfur

Very few definitive studies have been made of the mechanisms of sulfur uptake by flooded rice but the available evidence suggests that sulfate is the main chemical species absorbed. Studies with upland crops have shown that plants can absorb sulfur as inorganic sulfate or in the form of small organic compounds, such as the sulfur containing amino acids cystine and methionine [30, 166]. Flooded rice should also be able to assimilate these forms of sulfur but the amounts actually present in flooded soils at any one time appear to be very low. It also appears that plants do not have the capacity to assimilate the larger organic forms of sulfur found in soil otherwise no sulfur deficient plants would be found. Rice plants have the capacity to absorb soluble sulfide but this chemical species is toxic [184, 185, 186, 247] and, under normal conditions, it appears that rice assimilates very little, if any, sulfide [185, 186]. It therefore seems most likely that flooded rice absorbs sulfate that has been formed from sulfide or sulfur in the oxidized layer at the soil—water interface or in the rice rhizosphere by Beggiatoa [132, 201] or by Thiobacillus [96, 126]. No information on the relative role of these groups of bacteria in this process is available. There seems to be little doubt that organisms and mechanisms are available in the rice rhizosphere for the conversion of sulfide to sulfate.

Engler and Patrick [75] made a study of the capacity of rice roots to oxidize a number of metallic sulfides labelled with ^{35}S in flooded soil. They found that the sulfur from sodium, manganese, iron, zinc and copper sulfides was apparently partially oxidized in the soil adjacent to the growing roots of the rice plant. The degree of oxidation, and subsequent uptake, was directly related to the solubility of the sulfides. In the case of insoluble sulfides the spatial distribution of sulfides and roots will obviously affect the uptake of sulfur. The more active the plant and the greater the proliferation of its root system the greater the amount of sulfide oxidized and absorbed.

Han [96] studied the uptake of added sulfide and sulfate by flooded rice using sodium sulfide and potassium sulfate labelled with the ^{35}S isotope. He found that rice absorbed sulfur from both sources and that there was little difference in the amounts absorbed from the two sources. As might be expected the added sulfate was rapidly converted to sulfide in the flooded soil and thereafter little or no sulfate could be detected in the soil. Any sulfate formed by oxidation of sulfide at the soil-water interface or in the rhizosphere (by *Thiobacillus* [96] or *Beggiatoa* [201]) must have been rapidly absorbed by the plant.

6.4. Sulfur requirement of rice

The amounts of sulfur removed in a rice crop depends on many factors including the yield of the crop, the portion of the crop removed and the treatment of the residues. Consequently many different figures have been quoted for the removal of sulfur in a rice crop. Tisdale and Nelson [254] suggested a removal of approximately 3 kg of sulfur from a field producing $\sim$ 4.5 tonnes of grain ha^{-1}, Beaton and Fox [33] quoted a figure of 20 kg S ha^{-1} for the production of grain and stubble while Wang [278] and Jacq [123] showed that 8–18 kg S ha^{-1} is required to produce between 4 and 9 tonnes of grain ha^{-1}. Wang [278], in a more detailed study, showed that the amount of sulfur removed varied with variety and yield. Nitrogen fertilization had a large effect on the rate of removal of sulfur as did water management [278] and boron supply [155].

Blair *et al.* [42] found that the amount of sulfur removed in grain and straw varied from 0.34–1.6 and 0.35–1.6 kg S ton^{-1} respectively. They point out that the loss of sulfur is higher in systems where the straw is cut close to the ground and burned after threshing than in the traditional system where only the panicles are removed and the straw incorporated into the soil.

In soils with a limited supply of available sulfur the sulfur removed must be replaced. Wang *et al.* [281] found that in the lower Amazon Basin, where the natural supplies of sulfur are very limited, 10 to 20 kg S ha^{-1} should be applied to all varzea soils to maintain high levels of rice production. Blair *et al.* [44] found that the requirement varied with site and that yield increased with application up to 60 kg S ha^{-1} at one site.

This is obviously more sulfur than that removed in the rice crops and it suggests that only part of the sulfur applied is recovered by the crop. Some may be lost by leaching, immobilization or converted to insoluble sulfides [278].

Little work has been done on sources of sulfur for rice but it appears that super-phosphate, ammonium sulfate, gypsum and elemental sulfur are equally effective [44, 281]. Sulfur coated urea also appears to be an effective source of sulfur for rice [123, 151].

It seems that fertilizer sulfur should be applied to the surface of a paddy soil so that it remains in or can be converted to the sulfate form [42].

6.5. Sulfide toxicity and physiological diseases of rice

Although large quantities of sulfide can be produced in waterlogged soils, the concentration of water-soluble hydrogen sulfide is usually extremely low [109, 167, 175] due to its reaction with ferrous iron and possibly other cations to form insoluble sulfides [202]. However, in certain situations sufficient sulfide remains in solution to cause disorders in the rice plant [169, 184, 185, 244, 247]. This may happen when excessive amounts of sulfur are present in or are added to soils low in active iron or other cations, in sandy soils naturally low in cations, in other soils as a result of cation depletion due to many years of continuous rice cultivation [254] and even in iron-excess soils [5, 131, 200].

Some workers have attempted to relate the concentration of sulfide in soil to the severity of the disease. For example, Jacq [122], using the method of Chaudhry and Cornfield [66] to determine total sulfide (gaseous and soluble hydrogen sulfide, soluble sulfide and precipitated metal sulfides), found that $\sim$4, 8–15 and 20–40 ppm S was toxic to germinating seeds, seedlings at tillering and flowering plants, respectively.

Others have attempted to relate hydrogen sulfide concentration in solution to the toxicity but according to Ayotade [13] there is little reliable data for flooded soils because of the absence of a precise and sensitive method for its determination. Mitsui et al. [170] reported that free hydrogen sulfide concentrations as low as 0.07 ppm were toxic to rice seedlings in culture solutions and toxic situations have been found in rice paddies at the 0.1 ppm level [200]. More recent work has demonstrated that the concentration of hydrogen sulfide at which toxicity develops varies with the cultivar and is related to the ability of the plant to excrete oxygen [131].

Hydrogen sulfide has been reported as the causal factor in 12 of the 27 physiological disorders of rice [233, 252]. Injury to rice seeds during germination has been reported during in vitro experiments [233] and in West African acid soils [86, 120, 121]. A few days after sowing germinating seeds are covered by a black coating of iron sulfide and further growth is prevented in certain of these soils where high initial populations of sulfate-reducing bacteria increase rapidly after flooding (up to 10^6 bacteria g^{-1} soil within a few days).

Sulfide also affects plants at more advanced stages of growth and the main symptoms found [121, 122] are: (i) roots stunted or partly destroyed and covered with black iron sulfide; (ii) wilting, commencing with the older basal leaves, which turn yellow and then brown from the margins to the base; (iii) reduction in growth and death of 30–100% of the crop according to the severity of the disease; (iv) an increase in the number of empty panicles.

The diagnosis of sulfide toxicity is often obscured by associated or subsequent causal factors and it is often difficult to determine which agent was responsible for the initiation of the damage to the plant. Some of the diseases and associated causal factors are given in Table 7.

Joshi *et al.* [131] concluded that straighthead disease is primarily caused by free hydrogen sulfide, and that symptomless mild sulfide disease associated with yield reduction and late-season decline is caused by soluble sulfides. They also suggest that straighthead is an acute manifestation of a more general susceptibility of rice to mild sulfide disease.

Late season diseases are typical of degraded soils from which cations have been leached due to many years of continuous rice cultivation and to the sandy soils in Japan and Korea which are normally low in cations. This disease is termed 'Akiochi' in Japan [168, 254]. Hollis *et al.* [105, 106] showed that the highest concentrations of hydrogen sulfide in United States rice soils occurred at the flowering stage which generally corresponds with the most reduced status of the soil.

Okajima and Takagi [184, 185, 186] investigated the mechanism of injury caused by hydrogen sulfide and found that the efficiency of nutrient absorption was reduced, the translocation of inorganic nutrients in the plant was inhibited and the formation and translocation of carbohydrates was reduced. Subsequent workers have shown that (i) enzymes, such as catalase, peroxidase, ascorbic acid oxidase and polyphenol oxidase, are inactivated [4, 111], (ii) aerobic respiration of roots is inhibited [292], (iii) oxidative power of the roots is decreased [4, 191, 249, 251], (iv) the regulation of iron uptake is hindered [191], and (v) weakened plants may be susceptible to fungal diseases (*Helminthosporium*) and nematodes [106].

Sulfide-induced diseases may be alleviated by controlling the ecological conditions favourable to sulfide producing microorganisms and by suitable agricultural practices which include water and fertilizer management and choice of cultivar. Among the methods advocated for the control of sulfide induced diseases are; (i) mid-summer drainage to decrease reduced compounds [89], (ii) use of pregerminated seeds in the acid soils of West Africa [121], (iii) limit incorporation of organic matter which is used as a substrate for sulfate-reducing bacteria [203], (iv) addition of oxidants such as nitrate [72, 74, 284, 293, 295], manganese dioxide [74, 85], ferric phosphate [74] and calcium chlorate [284] to delay reduction of sulfate, (v) avoid the use of sulfate and sulfur containing fertilizers [7, 191] and (vi) to select non-susceptible cultivars [131, 158, 260]. Selection of cultivars may well prove to be the most economic method of controlling this problem and its success seems to be related to the greater release of oxygen to the rhizosphere by the non-susceptible varieties [131] and the consequent lowering of the sulfide concentration in the vicinity of the rice root.

Soluble sulfides in soil can also be beneficial to rice because of their toxicity to nematodes [80, 125, 211]. Jacq and Fortuner [80, 125] found that soluble sulfide

Table 7. Diseases of the rice plant associated with production of sulfide [106, 233, 252]

Physiological disease	Role of sulfides	Associated causal factors	Country
Bruzone [259, 260, 262, 263, 265, 266, 306]	Certain	Organic acids, fungus	Hungary
Root-rot[a] [17, 264]	Certain		Hungary, Japan
Akiochi [18, 104, 105, 106, 191, 245, 248, 251, 296]	Certain	Organic acids, fungus, excess Fe^{++}	Japan, Korea, USA
Bronzing or browning[b] [112, 188, 250, 305]	Certain	Excess Fe^{++}	Ceylon, India, Hungary
Acid sulfate soil diseases [86, 106, 120, 121]	Certain	Excess Fe^{++}	Vietnam, West Africa
Straighthead and mild sulfide diseases [5, 12, 105, 106, 131]	Certain		USA
Straighthead [11, 119, 243]	Possible	Thiol compounds	Japan
Akagare type II[c] [19, 158, 239]	Possible	Zinc deficiency, organic acids	Japan
Branca [106]	Possible		Portugal
Kuttipachal [106]	Possible		India

[a] Root-rot may be identical to bruzone or a form of 'akiochi'
[b] Bronzing may be induced by ferrous iron
[c] Zinc deficiency is the main causal agent

added at concentrations normally found in African soils, or produced microbiologically, would limit nematode populations.

6.6. Reclamation of mangrove soils for rice production

Many studies have been undertaken to extend the areas that could be used for rice-growing by reclaiming large areas of mangrove soils [34, 98, 99, 100, 179] and this topic was discussed during the International Symposium on Acid Sulfate Soils at Wageningen in 1972. The reclamation treatments reviewed [46, 204] include empoldering, prolonged submergence and leaching with rain or seawater and addition of lime.

It was clear from early studies that water management systems which allowed drying of the surface soil would not be successful for reclamation. Systems which permitted drying resulted in aerobic oxidation of the sulfides and polysulfides present in mangrove soils before reclamation and generation of considerable quantities of acid [34]. It may be assumed that acidophilic strains of the thiobacilli groups (viz. *Thiobacillus thiooxidans*) were responsible.

Continuous flooding and percolation (improved by the addition of 4 tonnes gypsum ha^{-1}) allowed the anaerobic oxidation of reduced sulfur and removal of oxidation products without acid production (e.g. by *Thiobacillus denitrificans*) [179]. Sulfur oxidation and numbers of anaerobic thiobacilli were found to be significantly higher in gypsum treated soils than in control soils [23], and as a result of this increased activity and water percolation the sulfur content decreased.

Depending on the cost of providing deep-soil drainage and the availability of gypsum, continuous percolation may prove to be a valuable technique for reclaiming mangrove soils for rice production.

The transformations and microorganisms involved in these reclamation treatments were discussed above.

7. Conclusions

It is apparent from the material presented that the sulfur cycle in a rice field is a very complicated system. A great variety of reactions takes place, mostly mediated by a multitude of microorganisms and affected by a wide range of environmental variables.

A great deal is known about the microbiology and the major sulfur cycle processes occurring, but the influence of soil characteristics and environmental factors on many of the organisms and transformations is not known. Virtually no information is available concerning the mineralization and immobilization of sulfur in the flooded system, the source of the mineralized sulfur or the fate of sulfur immobilized.

In the flooded system more detail is required on the distribution and content of

303

sulfur in soil, the nature of the organic fraction, the factors affecting the balance between mineralization and immobilization, and the input and output of sulfur if best use is to be made of the fertilizer sulfur now being added to many soils. Also very little is known on the mechanism of sulfur uptake by rice in the flooded system and of the role played by organisms such as *Beggiatoa* in this process.

Many challenging problems remain to be solved in this fascinating ecosystem. Even though rice is one of the world's most important food crops very few research workers are involved in a study of the sulfur cycle and thus further progress is likely to be slow.

References

1. Acharya, N. 1973 A radiotracer investigation of the effect of sulphate and phosphate application on paddy. Indian J. Agric. Chem. 6, 23.
2. Aiyar, S.P. 1945 A chlorosis of paddy (*Oryza sativa* L.) due to sulphate deficiency. Current Sci. 14, 10.
3. Alam, S.M. and Karim, M. 1972 Effect of sulphur on S uptake and dry matter yield of rice plant with respect to soil sulphur. Pakistan J. Sci. Res. 24, 222.
4. Allam, A.I. and Hollis, J.P. 1972 Sulfide inhibition of oxidases in rice roots. Phytopathology 62, 634–639.
5. Allam, A.I., Pitts, G. and Hollis, J.P. 1972 Sulfide determination in submerged soils with an ion-selective electrode. Soil Sci. 114, 456–467.
6. Andal, R., Bhuvaneswari, K. and Subba Rao, N.S. 1956 Root exudate of paddy. Nature (London) 178, 1063–1064.
7. Aomine, S. 1962 A review of research on redox potentials of paddy soils in Japan. Soil Sci. 94, 6–13.
8. Araki, M. 1955 Studies on sulfur deficiency symptoms of the rice plant. Brief records. J. Sci. Soil Manure Japan 1, 80–85.
9. Armstrong, W. 1967 The use of polarography in the assay of oxygen diffusion from roots in anaerobic media. Physiol. Plant 20, 533–540.
10. Armstrong, W. 1969 Rhizosphere oxidation in rice; an analysis of intervarietal difference in oxygen flux from the roots. Physiol. Plant 22, 296–303.
11. Asami, T. and Takai, Y. 1963 Formation of methyl mercaptan in paddy soils II Soil Sci. Plant Nut. 9, 23–27.
12. Atkins, J.G., Beachell, H.M. and Crane, L.E. 1956 Reaction of rice varieties to straighthead. Texas Agric. Exp. Stn. Prog. Rep. 1865. 2p.
13. Ayotade, K.A. 1977 Kinetics and reactions of hydrogen sulphide in solution of flooded rice soils. Plant and Soil 46, 381–389.
14. Baalsrud, K. and Baalsrud, K.S. 1954 Studies on *Thiobacillus denitrificans*. Arch. Mikrobiol. 20, 34–62.
15. Baas-Becking, L.G.M. and Ferguson Wood, E.J. 1955 Biological processes in the estuarine environment. I. and II, Ecology of the sulphur cycle. Proc. Koninkl. Nederl. Akademie Van Westenschappen, Amsterdam, 58, ser. B, 160–181.
16. Baas-Becking, L.G.M., Kaplan, I.R. and Moore, D. 1960 Limits of the natural environment in term of pH and Eh. J. Geol. 68, 243–284.
17. Baba, I. 1955 Varietal differences of the rice plant in relation to the resisting capacity to root-rot disease induced by hydrogen sulphide, and a convenient method to test them. Proc. Crop Sci. Soc. Jap. 23, 167–168.

18. Baba, I. 1958 Nutritional studies on the occurrence of *Helminthosporium* leaf spot and 'Akiochi' of the rice plant. Bull. Nat. Inst. Agric. Ser. D. (Plant Physiol. Genet. Crops Gen.) Jap. 7, 1—157.
19. Baba, I., Inada, K. and Tajima, K. 1965 Mineral nutrition and the occurrence of physiological diseases. In: Proceedings of a symposium on the mineral nutrition of the rice plant. Published for the International Rice Research Institute by Johns Hopkins Press, Baltimore, Maryland. 494 pp.
20. Badziong, W., Thauer, R.K. and Zeikus, J.G. 1978 Isolation and character-ization of *Desulfovibrio* growing on hydrogen plus sulfate as the sole energy source. Arch. Microbiol. 116, 41—49.
21. Badziong, W., Ditter, B. and Thauer, R.K. 1979 Acetate and carbon dioxide assimilation by *Desulfovibrio vulgaris* (Marburg) growing on hydrogen and sulfate as sole energy sources. Arch. Microbiol. 123, 301—305.
22. Bågander, L.E. 1980 Bacterial cycling of sulfur in a Baltic sediment: an *in situ* study in closed systems. Geomicrobiol. J. 2, 141—159.
23. Baldensperger, J.F. 1976 Use of respirometry to evaluate sulphur oxidation in soils. Soil Biol. Biochem. 8, 423—427.
24. Baldensperger, J.F. 1981 Short-term variations of microbiological and physicochemical parameters in submersion water over a rice field. Ann. Microbiol. Inst. Pasteur Paris. 132 B, 101—122.
25. Baldensperger, J.F. and Garcia, J-L. 1975 Reduction of oxidized inorganic nitrogen compounds by a new strain of *Thiobacillus denitrificans*. Arch. Microbiol. 103, 31—36.
26. Banwart, W.L. and Bremner, J.M. 1975 Formation of volatile sulfur com-pounds by microbial decomposition of sulfur-containing amino acids in soils. Soil Biol. Biochem. 7, 359—364.
27. Banwart, W.L. and Bremner, J.M. 1976 Volatilization of sulfur from unamended and sulfate-treated soils. Soil Biol. Biochem. 8, 19—22.
28. Banwart, W.L. and Bremner, J.M. 1976 Evolution of volatile sulfur com-pounds from soils treated with sulfur-containing organic materials. Soil Biol. Biochem. 8, 439—443.
29. Barber, D.A., Ebert, M. and Evans, N.T.S. 1962 The movement of $^{18}O_2$ through barley and rice plants. J. Exp. Bot. 13, 397—403.
30. Bardsley, C.E. 1960 Absorption of sulfur from organic and inorganic sources by bush beans. Agron. J. 52, 485—486.
31. Bauzon, D. and Diem, H.G., reported by Dommergues, Y. 1978 Microbiol activity of microenvironments in paddy soils. In: Environmental Biogeo-chemistry and Geomicrobiology. Krumbein, W.E. (ed.), Ann Arbor Science, Michigan, 1, 245—253.
32. Beaton, J.D. 1966 Sulfur requirements of cereals, tree fruits, vegetables and other crops. Soil Sci. 101, 267—282.
33. Beaton, J.D. and Fox, R.L. 1971 Production, marketing and use of sulfur products. In: Fertilizer Technology and Use, pp. 335—379. Olson, R.A., Army, T.J., Hanway, J.J. and Kilmer, V.J. (eds.), Soil Sci. Soc. Am. Inc. Madison, Wisconsin.
34. Beye, G. 1973 Acidification of mangrove soils after empoldering in lower Casamance. Effects of the type of reclamation used. In: Proc. Int. Symp. Acid Sulphate Soils 2, pp. 359—372. Dost, H. (ed.), Int. Inst. Land Reclamation and Improvement. Wageningen, The Netherlands.
35. Beye, G., Touré, M. and Arial, G. 1975 Acid sulfate soils of West Africa: problems of their management for agricultural use. International Rice Research Conference, IRRI, Los Baños, Philippines, 10 pp.

36. Bhan, C. and Tripathi, B.R. 1973 The forms and contents of sulphur in some soils of U.P. J. Indian Soc. Soil Sci. 21, 499–504.
37. Biederbeck, V.O. 1978 Soil organic sulfur and fertility. In: Soil Organic Matter, pp. 274–310. Schnitzer, M. and Khan, S.U. (eds.), Elsevier, Amsterdam.
38. Billen, G. 1975 Nitrification in the Scheldt estuary (Belgium and The Netherlands). Estuar. Coasts Mar. Sci. 3, 79–89.
39. Billen, G. 1976 The dependence of the various kinds of microbial metabolism on the redox state of the medium. Comm. SCOR/UNESCO Workshop on the Biogeochemistry of Estuarine Sediments. Melreux, 29 Nov–3 Dec, 254–261.
40. Blackburn, T.H., Kleiber, P. and Fenchel, T. 1975 Photosynthetic sulfide oxidation in marine sediments. Oikos 26, 101–108.
41. Blair, G.J. 1971 The sulphur cycle. J. Aust. Inst. Agric. Sci. 37, 113–121.
42. Blair, G.J., Mamaril, C.P. and Momuat, E.O. 1978 The sulphur nutrition of rice. Contr. Centr. Res. Inst. Agric. Bogor. No. 42, 13 pp.
43. Blair, G.J., Mamaril, C.P., Pangerang Umar, A., Momuat, E.O. and Momuat, C. 1979 Sulfur nutrition of rice. I. A survey of soils of South Sulawesi, Indonesia. Agron. J. 71, 473–477.
44. Blair, G.J., Momuat, E.O. and Mamaril, C.P. 1979 Sulfur nutrition of rice. II. Effect of source and rate of S on growth and yield under flooded conditions. Agron. J. 71, 477–480.
45. Bloomfield, C. 1969 Sulfate reduction in waterlogged soils. J. Soil Sci. 20, 207–221.
46. Bloomfield, C. and Coulter, J.K. 1973 Genesis and management of acid sulfate soils. Adv. Agron. 25, 265–326.
47. Bockris, J.O.M. and Reddy, A.K.N. 1970 Modern Electrochemistry. Plenum Press, N.Y. 1432 pp.
48. Boulègue, J. 1976 Equilibres dans le système $H_2S–S_8$ colloide–H_2O. C.R. Acad. Sci. Paris 283, sér. D. 591–594.
49. Boulègue, J. 1979 Evolution des composés soufrés dans la sédimentation récente. Coll. Int. C.N.R.S. 293, Biogéochimie de la matière organique à l'interface eau-sédiment marin, 93–101.
50. Boulègue, J. and Michard, G. 1978 Constantes de formation des ions polysulfurés S_6^{2-}, S_5^{2-} et S_4^{2-} en phase aqueuse, J. Fr. Hydrol. 9, 27–34.
51. Boulègue, J. and Michard, G. 1979 Sulfur speciations and redox processes in reducing environments. In: Chemical modeling in aqueous systems. Jenne, E.A. (ed.), ACS Symposium series, 93, 25–50.
52. Boulègue, J., Ciabrini, J-P., Fouillac, C., Michard, G. and Ouzounian, G. 1979 Field titrations of dissolved sulfur species in anoxic environments. Geochemistry of Puzzichello waters (Corsica-France). Chemical Geology 25, 19–29.
53. Boureau, M. 1977 Application de la chromatographie en phase gazeuse à l'étude de l'exsudation racinaire du riz. Cah ORSTOM Sér. Biol. 12, 75–81.
54. Brammer, H. 1978 Rice soils of Bangladesh, In: IRRI, Soils and Rice, pp. 35–55. Los Baños, Philippines.
55. Breck, W.G. 1972 Redox potential by equilibrium. J. Mar. Res. 30, 121–139.
56. Bremner, J.M. 1977 Role of organic matter in volatilization of sulfur and nitrogen from soils. In: Proc. Symp. Soil Organic Matter Studies. Braunschweig, Federal Republic of Germany, 1976. International Atomic Energy Agency, Vienna. 2, 229–240.
57. Bremner, J.M. and Steele, C.G. 1978 Role of microorganisms in the atmospheric sulfur cycle. Adv. Microbial Ecol. 2, 155–201.

306

58. Bristow, J.M. 1975 The structure and function of roots in aquatic vascular plants. In: The Development and Function of Roots, pp. 221—236. Torrey, J.G. and Clardson, D.T. (eds.), Academic Press, New York.
59. Bromfield, A.R. 1974 The deposition of sulphur in the rainwater of Northern Nigeria. Tellus 26, 408—411.
60. Bryant, M.P., Campbell, L.L., Reddy, C.A. and Crabill, M.R. 1977 Growth of *Desulfovibrio* in lactate or ethanol media low in sulfate in association with H_2 utilizing methanogenic bacteria. Appl. Environ. Microbiol. 33, 1162—1169.
61. Cahet, G. 1966 Substrats énergétiques naturels des bactéries sulfatoréductrices. C.R. Acad. Sci. Paris 263, sér. D, 691—692.
62. Cahet, G. 1970 Aspects chemotrophiques en sédiments lagunaires. Cas du soufre. Vie et Milieu 21, 1—33.
63. Cahet, G. 1975 Transfert d'énergie en milieu sédimentaire. Cas des sulfatoréducteurs. II. Relations syntrophiques avec diverses microflores. Vie et Milieu 25, 49—66.
64. Campbell, L.L. and Postgate, J.R. 1965 Classification of the spore-forming sulfate-reducing bacteria. Bact. Rev. 29, 359—363.
65. Cappenberg, Th.E. 1974 Interrelations between sulfate-reducing and methane-producing bacteria in bottom deposits of a fresh-water lake. I. Field observations. Antonie van Leeuwenhoek 40, 285—295.
66. Chaudhry, I.A. and Cornfield, A.H. 1966 Determination of sulphide in waterlogged soils. Plant and Soil 3, 474—479.
67. Chen, K.Y. and Morris, J.C. 1972 Kinetics of oxidation of aqueous sulfides by O_2. Environ. Sci. Technol. 6, 529—537.
68. Chi Li Liu and Peck, H.D. Jr. 1981 Comparative bioenergetics of sulfate reduction in *Desulfovibrio* and *Desulfotomaculum* spp. J. Bacteriol. 145, 966—973.
69. Choudhri, M.B. 1978 Rice soils of Pakistan. In: IRRI, Soils and Rice, pp. 147—161. Los Baños, Philippines.
70. Chow, W.T. and Ng, S.K. 1969 A preliminary study on acid sulphate soils in West Malaysia. Malaysian Agric. J. 47, 253—267.
71. Connell, W.E. and Patrick, W.H. Jr. 1968 Sulfate reduction in soil: effects of redox potential and pH. Science (Washington D.C.) 159, 86—87.
72. Connell, W.E. and Patrick, W.H. Jr. 1969 Reduction of sulfate to sulfide in waterlogged soil. Soil Sci. Soc. Amer. Proc. 33, 711—715.
73. Elgabaly, M.M. 1978 Rice soils of Egypt and other Near East countries. In: IRRI, Soils and Rice, pp. 135—146. Los Baños, Philippines.
74. Engler, R.M. and Patrick, W.H. Jr. 1973 Sulfate reduction and sulfide oxidation in flooded soil as affected by chemical oxidants. Soil Sci. Soc. Amer. Proc. 37, 685—688.
75. Engler, R.M. and Patrick, W.H. Jr. 1975 Stability of sulfides of manganese, iron, zinc, copper and mercury in flooded and nonflooded soil. Soil Sci. 119, 217—221.
76. Farwell, S.O., Sherrard, A.E., Pack, M.R. and Adams, D.F. 1979 Sulfur compounds volatilized from soils at different moisture content. Soil Biol. Biochem. 11, 411—415.
77. Fawzi Abed, M.A.H. 1976 Sulfate reduction in poorly-drained soils as influenced by organic matter and soil texture. Beitrage zur Tropischen Landwirtschaft und Veterinärmedizin No. 1, 89—93.
78. Fenchel, T.M. and Riedl, R.J. 1970 The sulfide system: a new biotic community underneath the oxidized layer of marine sand bottoms. Mar. Biol. 7, 255—268.

79. Flach, K.W. and Slusher, D.F. 1978 Soils used for rice culture in the United States. In: IRRI, Soils and Rice, pp. 199–214. Los Baños, Philippines.

80. Fortuner, R. and Jacq, V.A. 1976 *In vitro* study of toxicity of soluble sulphides to three nematodes parasitic on rice in Senegal. Nematologica, 22, 343–351.

81. Freney, J.R. 1961 Some observations on the nature of organic sulphur compounds in soil. Aust. J. Agric. Res. 12, 424–432.

82. Freney, J.R., Melville, G.E. and Williams, C.H. 1975 Soil organic matter fractions as sources of plant-available sulphur. Soil Biol. Biochem. 7, 217–221.

83. Freney, J.R., Stevenson, F.J. and Beavers, R.H. 1972 Sulfur-containing amino acids in soil hydrolysates. Soil Sci. 114, 468–476.

84. Freney, J.R. and Swaby, R.J. 1975 Sulphur transformations in soils. In: Sulphur in Australasian Agriculture, pp. 31–39. McLachlan, K.D. (ed.), Sydney University Press, Sydney.

85. Furusaka, C. 1968 Studies on the activity of sulfate reducers in paddy soils. Bull. Inst. Agric. Res. Tohoku Univ. 19, 101–184 (in Japanese).

86. Garcia, J-L., Raimbault, M., Jacq, V., Rinaudo, G. and Roger, P.A. 1974 Activités microbiennes dans les sols des rizières du Sénégal: Relations avec les caractéristiques physico-chimiques et influence de la rhizosphère. Rev. Ecol. Biol. Sol, 11, 169–185.

87. Gibbs, R.J. 1972 Water chemistry of the Amazon River. Geochim. Cosmochim. Acta 36, 1061–1066.

88. Goldhaber, M.B. and Kaplan, I.R. 1974 The sulfur cycle. In: The Sea, pp. 569–655. Goldberg, E.D. (ed.), John Wiley & Sons, New York.

89. Gotoh, S. and Yamashita, K. 1966 Oxidation-reduction potential of a paddy-soil *in situ* with special reference to the production of ferrous iron, manganous manganese and sulfide, Soil Sci. Plant Nutr. 12, 24–32.

90. Gottschal, J.C., De Vries, S. and Kuenen, J.G. 1979 Competition between the facultatively chemolithotrophic *Thiobacillus* A2, and obligately chemolithotrophic *Thiobacillus* and a heterotrophic *Spirillum* for inorganic and organic substrates. Arch. Microbiol. 121, 241–249.

91. Gourmelon, C., Boulègue, J. and Michard, G. 1977 Oxydation partielle de l'hydrogène sulfuré en phase aqueuse. C.R. Acad. Sci. Paris. 284(c), 269–272.

92. Greenwood, D.J. and Goodman, D. 1964 Oxygen diffusion and aerobic respiration in soil spheres. J. Sci. Food Agric. 15, 579–588.

93. Guay, R. and Silver, M. 1975 *Thiobacillus acidophilus* sp. nov.; isolation and some physiological characteristics. Can. J. Microbiol. 21, 281–288.

94. Habte, M. and Alexander, M. 1980 Nitrogen fixation by photosynthetic bacteria in lowland rice culture. Appl. Environ. Microbiol. 39, 342–347.

95. Haldar, B.R. and Barthakur, H.P. 1976 Sulphide production in some typical rice soils of Assam. J. Indian. Soc. Soil Sci. 24, 387–395.

96. Han, K.W. 1973 Sulfur availability to rice plants and its transformations in rice soil and rhizosphere. M.Sc. Thesis, University of Philippines, Los Baños.

97. Haque, M.Z., Kobayashi, M., Fujii, K. and Takahashi, E. 1969 Seasonal changes of photosynthetic bacteria and their products. Soil Sci. Plant Nutr. 15, 51–55.

98. Hart, M.G.R. 1959 Sulphur oxidation in tidal mangrove soils of Sierra Leone. Plant and Soil 11, 215–236.

99. Hart, M.G.R. 1962 Observations on the source of acid in empoldered mangrove soils. I. Formation of elemental sulphur. Plant and Soil 17, 87–98.

100. Hart, M.G.R. 1963 Observations on the source of acid in empoldered

mangrove soils. II. Oxidation of soil polysulphides. Plant and Soil 19, 106–114.

101. Harter, R.D. and McLean, E.O. 1965 The effect of moisture level and incubation time on the chemical equilibria of a Toledo clay loam soil. Agron. J. 57, 583–588.

102. Hatchikian, E.C., Chaigneau, M. and Le Gall, J. 1975 Analysis of gas production by growing cultures of three species of sulfate-reducing bacteria. Proc. Symp. Microbial Production and Utilization of Gases, pp. 109–118. Akademie der Wissenschaften zu Göttingen.

103. Hauser, G.F. and Sadikin, R. 1956 The productivity of the soils of East Central Java based on yields of Savah rice. Contrib. Gen. Agric. Res. Stn. Bogor 144, 1–98.

104. Hollis, J.P. 1967 Toxicant diseases of rice. Louisiana Agric. Exp. Sta. Bull. 614, 24 pp.

105. Hollis, J.P., Allam, A.I. and Pitts, G. 1972 Sulfide diseases of rice. Abs. Phytopathology, 62, 764–765.

106. Hollis, J.P., Allam, A.I., Pitts, G., Joshi, M.M. and Ibrahim, I.K.A. 1975 Sulfide diseases of rice on iron-excess soils. Acta Phytopathol. Acad. Sci. Hung. 10, 329–341.

107. Hollis, J.P. and Rodriguez-Kabana, R. 1967 Fatty acids in Louisiana rice fields. Phytopathology 57, 841–847.

108. Houghton, C. and Rose, F.A. 1976 Liberation of sulfate from sulfate esters by soils. Appl. Environ. Microbiol. 31, 969–976.

109. Hutchinson, G.E. 1957 A Treatise on Limnology. Vol. 1. Wiley, New York.

110. Hutchinson, M., Johnstone, K.I. and White, D. 1969 Taxonomy of the genus *Thiobacillus*: the outcome of numerical taxonomy applied to the group as a whole. J. Gen. Microbiol. 57, 397–410.

111. Inada, K. 1965 Bronzing disease of rice plant in Ceylon. I. Effect of field treatments on bronzing occurrence and changes in leaf respiration induced by the disease. Nippon Sakumotsu Gakkai Kiji 33, 309–314.

112. Inada, K. 1965 Bronzing disease of rice plant in Ceylon. II. Cause of the occurrence of bronzing. Nippon Sakumotsu Gakkai Kiji 33, 315–323.

113. Ishizawa, S. and Toyoda, H. 1964 Microflora of Japanese soils. Nogyo Gijutsu Kenkyusho Kokoku 14B: 204–284. [in Japanese].

114. Ishizuka, Y. and Tanaka, A. 1959 Inorganic nutrition of rice plants, IV. Effect of calcium, magnesium and sulfur levels in culture solution on yields and chemical composition of the plant. J. Sci. Soil and Manure, Japan 30, 414–416 [in Japanese].

115. Ismunadji, M. and Miyake, M. 1978 Sulfur and amino acid content of brown rice. Mimes Central Research Institute of Agriculture, Bogor, Indonesia, 8 pp.

116. Ismunadji, M. and Zulkarnaini, I. 1977 Sulphur deficiency in lowland rice in Indonesia. In: Proceedings of the International Seminar on Soil Environment and Fertility Management in Intensive Agriculture, Tokyo, pp. 647–652.

117. Ismunadji, M. and Zulkarnaini, I. 1978 Sulphur deficiency of lowland rice in Indonesia. Sulphur in Agriculture 2, 17–19 and 22.

118. Ismunadji, M., Zulkarnaini, I. and Miyake, M. 1975 Sulphur deficiency in lowland rice in Java. Contr. Centr. Res. Inst. Agric. Bogor 14, 17 pp.

119. Iwamoto, R. 1969 Straighthead of rice plants affected by functional abnormality of thiol-compound metabolism. Mem. Tokyo Univ. Agric. 13, 62–80.

120. Jacq, V.A. 1973 Biological sulphate-reduction in the spermosphere and the rhizosphere of rice in some acid sulphate soils of Senegal. In: Proc. Int. Symp.

Acid Sulphate Soils, 2, pp. 82–98. Dost, H. (ed.), Int. Inst. Land Reclamation and Improvement, Wageningen, The Netherlands.

121. Jacq, V.A. 1975 La sulfato-réduction en relation avec l'excrétion racinaire. Soc. Bot. Fr. Coll. Rhizosphère 122, 169–181.

122. Jacq, V. 1977 Sensibilité du riz aux sulfures d'origine microbienne. Cah. O.R.S.T.O.M. Sér. Biol. 12, 97–99.

123. Jacq, V.A. 1978 Utilisation du 'Sulfur Coated Urea' en rizière et production de sulfures toxiques. Cah. O.R.S.T.O.M. Sér. Biol. 13, 133–136.

124. Jacq, V.A. 1980 Biological sulphur cycle in paddy fields: populations of microorganisms (sulphate- and sulphur-reducing bacteria, sulphooxidizers) in the spermosphere and rhizosphere of rice, and resulting accumulation of toxic sulphides. IInd Int. Symp. Microbial Ecology. Univ. Warwick, Coventry (GB).

125. Jacq, V.A., and Fortuner, R. 1979 Biological control of rice nematodes using sulphate reducing bacteria. Rev. Nematol. 2, 41–50.

126. Jacq, V.A. and Roger, P.A. 1978 Evaluation des risques de sulfato-réduction en rizière au moyen d'un critère microbiologique mesurable in situ. Cah. O.R.S.T.O.M. Sér. Biol. 13, 137–142.

127. Johnson, C.L. 1977 Chemoautotrophic bacteria. In: Handbook of microbiology, 2nd ed., pp. 285–300. Laskin, A.I. and Lechevalier, H.A. (eds.), C.R.C. Press Inc., Cleveland.

128. Jørgensen, B.B. and Fenchel, T. 1974 The sulfur cycle of a marine sediment model. Mar. Biol. 24, 189–201.

129. Jørgensen, B.B., Hansen, M.H. and Ingvarson, K. 1978 Sulfate reduction in coastal sediments and the release of H_2S to the atmosphere. In: Environmental Biogeochemistry and Geomicrobiology, 1, 245–253. Krumbein, W.E. (ed.), Ann Arbor Science, Michigan.

130. Jørgensen, B.B., Revsbech, N.P., Blackburn, T.H. and Cohen, Y. 1979 Diurnal cycle of oxygen and sulfide microgradients and microbial photosynthesis in cyanobacterial mat sediment. Appl. Environ. Microbiol. 38, 46–58.

131. Joshi, M.M., Ibrahim, I.K.A. and Hollis, J.P. 1975 Hydrogen sulfide: effects on the physiology of rice plants and relation to straighthead disease. Phytopathology 65, 1165–1170.

132. Joshi, M.M., and Hollis, J.P. 1977 Interaction of Beggiatoa and rice plant: detoxification of hydrogen sulfide in the rice rhizosphere. Science (Washington D.C.) 195, 179–180.

133. Kadota, H. and Ishida, Y. 1972 Production of volatile sulfur compounds by microorganisms. Ann. Rev. Microbiol. 26, 127–138.

134. Kaiser, P. 1966 Ecologie des bactéries photosynthétiques. Rev. Ecol. Biol. Sol 3, 409–472.

135. Kapoor, R.K., Khemani, L.T. and Ramana Murty, Bh. V. 1972 Chemical composition of rain water and rain characteristics at Delhi. II. Tellus, 24, 575–579.

136. Kapoor, R.K. and Paul, S.K. 1980 A study of the chemical components of aerosols and snow in the Kashmir region. Tellus 32, 33–41.

137. Karim, M., Alam, S.M. and Rahman, M. 1970 Annual Tech. Report. Atomic Energy Centre. Dacca, Bangladesh.

138. Karim, M. and Majlish, M.A.K. 1958 A study on the formative effects of sulphur on rice plant. Pakistan J. Sci. Res. 10, 52.

139. Kawalec, A. 1973 World distribution of acid sulphate soils: references and map. In: Proc. Int. Symp. Acid Sulphate Soils, 1, pp. 292–295. Dost, H. (ed.), Int. Inst. Land Reclamation and Improvement Wageningen, The Netherlands.

140. Kelly, D.P. 1971 Autotrophy. Concepts of lithotrophic bacteria and their organic metabolism. Ann. Rev. Microbiol. 25, 177–210.

141. Kelly, D.P. and Tuovinen, O.H. 1972 Recommendation that the names *Ferrobacillus ferrooxidans* Leathen and Braly and *Ferrobacillus sulfooxidans* Kinsel be recognized as synonyms of *Thiobacillus ferrooxidans* Temple and Colmer. Int. J. Syst. Bacteriol. 22, 170–172.

142. Kempner, E.S. 1966 Acid production by *Thiobacillus thiooxidans*. J. Bact. 92, 1842–1843.

143. Khemani, L.T. and Ramana Murty. Bh. V. 1968 Chemical composition of rain water and rain characteristics at Delhi. Tellus 20, 284–291.

144. Kimura, M., Wada, H. and Takai, Y. 1977 Rhizosphere of rice plant. III. Microbiological features of the rhizosphere (2), Nippon Dojohiryo Gaku Zasshi 48, 111–114. [in Japanese].

145. Kimura, M., Wada, H., and Takai, Y. 1979 The studies on the rhizosphere of paddy rice. VI. The effects of anaerobiosis on microbes. Soil Sci. Plant Nutr. 25, 145–153.

146. Kobayashi, M., Takahashi, E. and Kawaguchi, K. 1966 Distribution of nitrogen-fixing microorganisms in paddy soils of Southeast Asia. Soil Sci. 104, 113–118.

147. Kobayashi, M. and Haque, M.Z. 1971 Contribution to nitrogen fixation and soil fertility by photosynthetic bacteria. Plant and Soil Spec. Vol. 43, 443–456.

148. Kumada, K. 1949 Investigation on the rhizosphere of rice seedling. (1) On the microscopic structure of the rhizosphere and oxidative power of root. J. Sci. Soil Manure 19, 119–124 [in Japanese].

149. Laanbroek, H.J. and Pfennig, N. 1981 Oxidation of short-chain fatty acids by sulphate-reducing bacteria in freshwater and in marine sediments. Arch. Microbiol. 128, 330–335.

150. Larsen, H. 1953 Green sulfur bacteria. Kon. Norshe Vidensk. Selskabs Skrifter, NR1, 1–187.

151. Lee, C.C. 1973 An observation on the fertilizer effect of locally produced sulfur-coated urea used for paddy. Tech. Bull. Taiwan Fertilizer Co. No. 48.

152. Leelavathy, K.M. 1970 Amino-acids and sugars in the exudates from germinating seeds of rice. Proc. Indian Acad. Sci., Ser. B. 72, 81–90.

153. Leijder, R.A. and Aldjabri, M. 1972 Sulphur deficiency under conditions of wet rice cultivation with specific reference to a vertisol near Ngawi, East Java. Newsletter, Soil Study Group, Bogor 1/2: 21.

154. Livingstone, D.A. 1963 Chemical composition of rivers and lakes. Geological Survey Professional Paper 440-G. United States Government Printing Office, Washington, 64 pp.

155. Lockard, R.G., Ballaux, J.G. and Liongson, E.A. 1972 Response of rice plants grown in three potted Luzon soils to additions of boron, sulfur and zinc. Agron. J. 64, 444–447.

156. London, J. 1963 *Thiobacillus intermedius*, nov. sp. a novel type of facultative autotroph. Arch. Mikrob. 46, 329–337.

157. London, J. and Rittenberg, S.C. 1967 *Thiobacillus perometabolis*, nov. sp., a non-autotrophic *Thiobacillus*. Arch. Mikrob. 59, 218–225.

158. MacRae, I.C. and Castro, T.F. 1966 Carbohydrates and amino acids in the root exudates of rice seedlings. Phyton. 23, 95–100.

159. Mah, R.A., Smith, M.R. and Baresi, L. 1978 Studies on an acetate-fermenting strain of *Methanosarcina*. Appl. Environ. Microbiol. 35, 1174–1184.

160. Mamaril, C.P., Pangerang Umar, A., Manwan, I. and Momuat, C.J.S. 1976 Sulphur response of lowland rice in South Sulawesi, Indonesia. Contr. Centr. Res. Inst. Agric. Bogor, 22, 12 pp.

161. Matheron, P. 1976 Contribution à l'étude écologique, systématique et physiologique des *Chromatiaceae* et des *Chlorobiaceae* isolées des sédiments marins. Thèse Univ. Aix Marseille II 193 pp.
162. Matheron, R. and Baulaigue, R. 1976 Sur l'écologie des *Chromatiaceae* et des *Chlorobiaceae* marines. Ann. Microbiol. Inst. Pasteur Paris 127, A, 515–520.
163. Matsuo, H., Pecrot, A.J. and Riquier, J. 1978 Rice soils of Europe. In: IRRI, Soils and Rice, pp. 193–198. Los Baños, Philippines.
164. Matsuzaka, Y. 1978 Rice soils of Japan. In: IRRI, Soils and Rice, pp. 163–177. Los Baños, Philippines.
165. Metwally, A.I., El-Damaty, A. and Yani, Y.G. 1978 Chemical changes accompanying waterlogging. 1. Effect of sulphate and organic matter. Acta Agronomica Acad. Scientiarum. Hung. 27, 133–139.
166. Miller, L.P. 1947 Utilization of dl-methionine as a source of sulfur by growing plants. Boyce Thompson Inst. Contrib. 14, 443–456.
167. Misra, R.D. 1938 Edaphic factors in the distribution of aquatic plants in English lakes. J. Ecol. 26, 411–451.
168. Mitsui, S. 1956 Inorganic Nutrition, Fertilization and Soil Amelioration of Lowland Rice (3rd Ed.). Yokendo, Toyko, Japan.
169. Mitsui, S., Aso, S. and Kumazawa, K. 1951 Dynamic studies on the nutrient uptake by crop plants I. The nutrient uptake of rice roots as influenced by hydrogen sulfide. J. Sci. Soil Manure Japan 22, 46–52.
170. Mitsui, S., Aso, S., Kumazawa, K. and Ishiwara, T. 1954 The nutrient uptake of rice plants as influenced by hydrogen sulfide and butyric acid abundantly evolving under waterlogged soil condition. Trans. 5th Int. Congr. Soil Sci. 2, 364–368.
171. Mizoguchi, T., Sato, T. and Okabe, T. 1976 New sulfur-oxidizing bacteria capable of growing heterotrophically, *Thiobacillus rubellus* nov. sp. and *Thiobacillus delicatus* nov. sp., J. Ferm. Technol. 54, 181–191.
172. Moore, W.E.C., Johnson, J.L., and Holdeman, L.V. 1976 Emendation of Bacteroidaceae and *Butyrivibrio* and descriptions of *Desulfomonas* gen. nov. and ten species in the genera *Desulfomonas, Butyvibrio, Eubacterium, Clostridium* and *Ruminococcus*. Int. J. Syst. Bact. 26, 238–253.
173. Moorman, F.R. 1978 Morphology and classification of soils on which rice is grown. In: IRRI, Soils and Rice, pp. 255–272. Los Baños, Philippines.
174. Morris, J.C. and Stumm, W. 1967 Redox equilibria and measurements of potentials in the aquatic environment. In: 'Equilibrium Concepts in Natural Water Systems'. Adv. Chem. Series 67, 270–285. Washington D.C.
175. Mortimer, C.H. 1941 The exchange of dissolved substances between mud and water in lakes. J. Ecol. 29, 280–329.
176. Mountfort, D.O., Asher, R.A., Mays, E.L. and Tiedje, J.M. 1980 Carbon and electron flow in mud and sand flat intertidal sediments at Delaware inlet, Nelson, New Zealand. Appl. Environ. Microbiol. 39, 686–694.
177. Mouraret, M. and Baldensperger, J.F. 1977 Use of membrane filters for the enumeration of autotrophic *Thiobacilli*. Microbial Ecology 3, 345–359.
178. Murthy, R.S. 1978 Rice soils of India. In: IRRI, Soils and Rice, pp. 3–17. Los Baños, Philippines.
179. Mutsaars, M. and Van Der Velden, J. 1973 Le dessalement des terres salées du fleuve Sénégal. Bilan des trois années d'expérimentations (1970–1973). Rapp. FAO, 74 pp.
180. Myers, P.S. and Millar, W.N. 1975 Non-autotrophic *Thiobacillus* in acid mine water. Appl. Environ. Microbiol. 30, 884–886.
181. Nikaido, M. 1977 On the relation between methane production and sulfate

312

reduction in bottom muds containing sea water sulfate. Geochem. J. 88, 199–206.

182. Nissenbaum, A. and Kaplan, I.R. 1972 Chemical and isotopic evidence for the *in situ* origin of marine humic substances. Limnol. Oceanogr. 17, 570–582.

183. Ogota, G. and Bower, C.H. 1965 Significance of biological sulfate reduction to soil salinity. Soil Sci. Soc. Amer. Proc. 29, 23–25.

184. Okajima, H. and Takagi, S. 1953 Physiological behavior of hydrogen sulfide in the rice plant. Part 1. Effect of hydrogen sulfide on the absorption of nutrients. Rep. Inst. Agr. Res. Tohoku Univ. D5, 21–31.

185. Okajima, H. and Takagi, S. 1955 Physiological behavior of hydrogen sulfide in the rice plant. Part 2. Effect of hydrogen sulfide on the content of nutrients in the rice plant. Rep. Inst. Agr. Res. Tohoku Univ. D6, 89–99.

186. Okajima, H. and Takagi, S. 1956 Physiological behavior of hydrogen sulfide in the rice plant. Part 4. Effect of hydrogen sulfide on the distribution of radioactive P^{32} in the rice plant. Rep. Inst. Agr. Res. Tohoku Univ. D7, 107–113.

187. Okuda, A., Yamaguchi, M. and Kamata, S. 1957 Nitrogen-fixing micro-organisms in paddy soils. III. Distribution of non-sulfur purple bacteria in paddy soils. Soil Sci. Plant Nutr. 2, 131–133.

188. Ota, Y. 1968 Occurrence of the physiological disorder of rice called 'bronzing'. Bull. Natn. Inst. Agric. Sci. Tokyo D18, 97–104.

189. Panabokke, C.R. 1978 Rice soils of Sri Lanka. In: IRRI, Soils and Rice, pp. 19–33. Los Baños, Philippines.

190. Paramananthan, S. 1978 Rice soils of Malaysia. In: IRRI, Soils and Rice, pp. 87–98. Los Baños, Philippines.

191. Park, Y.D. and Tanaka, A. 1968 Studies of the rice plant on an 'akiochi' soil in Korea. Soil Sci. Plant Nutr. 14, 27–34.

192. Patnaik, S. 1978 Natural sources of nutrients in rice soils. In: IRRI. Soils and Rice, pp. 501–519. Los Baños, Philippines.

193. Patrick, W.H. Jr. and Delaune, R.D. 1972 Characterization of the oxidized and reduced zones in flooded soil. Soil Sci. Soc. Amer. Proc. 36, 573–576.

194. Patrick, W.H. Jr. and Mikkelsen, D.S. 1971 Plant nutrient behavior in flooded soil. In: Fertilizer Technology and Use. pp. 187–215. Olson, R.A., Army, T.J., Hanway, J.J. and Kilmer, V.J. (eds.), Soil Sci Soc. Am. Madison, Wisconsin.

195. Patrick, W.H. Jr. and Reddy, C.N. 1978 Chemical changes in rice soils. In: IRRI, Soils and Rice, pp. 361–379. Los Baños, Philippines.

196. Pearsall, W.H. and Mortimer, C.H. 1939 Oxidation-reduction potentials in waterlogged soils, natural waters and muds. J. Ecol. 27, 483–501.

197. Pfennig, N. 1975 The phototrophic bacteria and their role in the sulfur cycle. Plant and Soil (Special Vol.) 43, 1–16.

198. Pfennig, N. and Biebl, H. 1976 *Desulfuromonas acetoxidans* gen. nov. and sp. nov., a new anaerobic, sulfur-reducing, acetate-oxidizing bacterium. Arch. Microbiol. 110, 3–12.

199. Pillai, P.B. and Singh, H.G. 1974 Effect of sulphur in preventing the occurrence of chlorosis in paddy seedlings. Agric. Res. J. Kerala 12, 49–55.

200. Pitts, G., Allam, A.I. and Hollis, J.P. 1972 Aqueous iron-sulfur systems in rice field soils of Louisiana, Plant and Soil 36, 251–260.

201. Pitts, G., Allam, A.I. and Hollis, J.P. 1972 *Beggiatoa*: occurrence in the rice rhizosphere. Science (Washington D.C.) 178, 990–992.

202. Ponnamperuma, F.N. 1972 The chemistry of submerged soils. Adv. Agron. 24, 29–96.

203. Ponnamperuma, F.N. 1977 Physico chemical properties of submerged soils in relation to fertility. IRRI Research Paper Series No. 5. 32 pp.
204. Ponnamperuma, F.N. and Beye, G. 1973 Amelioration of three acid sulphate soils for lowland rice. Proc. Int. Symp. Acid sulphate soils, 2, 391–406. Dost, H. (ed.), Int. Inst. Land Reclamation and Improvement, Wageningen, The Netherlands.
205. Pont, D. 1977 Recherches sur l'évolution saisonnière du peuplement de Copépodes, Cladocères et Ostracodes des rizières de Camargue. Thèse Univ. Sci. Tech. Languedoc. Montpellier, France. 24 pp.
206. Postgate, J.R. and Campbell, L.L. 1966 Classification of *Desulfovibrio* species, the non-sporulating sulfate-reducing bacteria. Bact. Rev. 30, 732–739.
207. Probert, M.E. 1976 The composition of rainwater at two sites near Townsville, Qld. Aust. J. Soil Res. 14, 397–402.
208. Putnam, H.D. and Schmidt, E.L. 1959 Studies on the free amino acid fractions of soils. Soil Sci. 87, 22–27.
209. Raymundo, M.E. 1978 Rice soils of the Philippines. In: IRRI, Soils and Rice, pp. 115–133. Los Baños, Philippines.
210. Rittenberg, S.C. 1969 The roles of exogenous organic matter in the physiology of chemolithotrophic bacteria. Adv. Microbiol. Physiol. 3, 159–196.
211. Rodriguez-Kabana, R., Jordan, J.W. and Hollis, J.P. 1965 Nematodes: biological control in rice fields: role of hydrogen sulfide. Science (Washington D.C.) 148, 524–526.
212. Rojanasoonthon, S. 1978 Rice soils of Thailand. In: IRRI, Soils and Rice, pp. 73–85. Los Baños, Philippines.
213. Roy, A.B. and Trudinger, P.A. 1970 The Biochemistry of Inorganic Compounds of Sulphur. Cambridge University Press, Cambridge, U.K.
214. Sachdev, M.S. and Chhabra, P. 1974 Transformations of ^{35} S-labelled sulfate in aerobic and flooded soil conditions. Plant and Soil 41, 335–341.
215. Sadhu, M.K. and Das, T.M. 1968 Amino acids liberated by growing rice seedlings. Bull. Botan. Soc. Bengal 22, 219–220.
216. Sadhu, M.K. and Das, T.M. 1971 Root exudates of rice seedlings. The influence of one variety on another. Plant and Soil 34, 541–546.
217. Saito, M. and Watanabe, I. 1978 Organic matter production in rice field flood water. Soil Sci. Plant Nut. 24, 427–444.
218. Saran, A.B. 1949 Some observations on an obscure disease of paddy, *Oryza sativa*. Current Sci. 18, 378–379.
219. Sen, A.T. 1938 Further experiments on the occurrence of depressed yellow patch of paddy in the Mandalay farm. Burma Dept. Agric. Rep. 1937–38, 35 pp.
220. Senez, J.C. 1973 Eléments de bioénergétique. Ediscience, Paris, 79 pp.
221. Shlomi, E.R., Lamkhorst, A. and Prins, R.A. 1978 Methanogenic fermentation of benzoate in an enrichment culture. Microbial Ecology 4, 249–261.
222. Smith, M.R. and Mah, R.A. 1978 Growth and methanogeneis by *Methanosarcina* strain 227 on acetate and methanol. Appl. Environ. Microbiol. 36, 870–879.
223. Soepraptohardjo, M. and Suhardjo, H., Rice soils of Indonesia. In: IRRI, Soils and Rice, pp. 99–113. Los Baños, Philippines.
224. Soo, S.W. 1972 Semi-detailed survey of the Kedah/Perlis coastal plain. West Malaysia Soil Survey Report 1 Soil and Analytical Services Branch. Div. Agric. Kuala Lumpur.
225. Spencer, K. and Freney, J.R. 1980 Assessing the sulfur status of field-grown wheat by plant analysis. Agron. J. 72, 469–472.

314

226. Stangel, P. 1979 Nitrogen requirement and adequacy of supply for rice production. In: IRRI Nitrogen and Rice, pp. 45–69. Los Baños, Philippines.
227. Starkey, R.L. 1935 Isolation of some bacteria which oxidize thiosulfate. Soil Sci. 39, 197–215.
228. Starkey, R.L. 1966 Oxidation and reduction of sulfur compounds in soils. Soil Sci. 101, 297–306.
229. Stevenson, F.J. 1956 Isolation and identification of some amino compounds in soils. Soil Sci. Soc. Am. Proc. 20, 201–208.
230. Strohl, W.R. and Larkin, J.M. 1978 Enumeration, isolation and characterization of *Beggiatoa* from freshwater sediments. Appl. Environ. Microbiol. 36, 755–770.
231. Stumm, W. 1966 Redox potential as an environmental parameter: conceptual significance and operational limitation. Proc. Int. Water Pollution Res. Conf. (3rd., Munich) 1, 283–306.
232. Subbiah, B.V. and Venkateswarlu, J. 1965 Availability and transformation of sulfur in rice soils. In: Radioisotopes and Radiation in Soil-Plant Nutrition Studies IAEA, Vienna, pp. 563–572.
233. Subramoney, N. 1965 Injury to paddy seedlings by production of H_2S under field conditions. J. Indian Soc. Soil Sci. 13, 95–98.
234. Suzuki, A. 1977 Effect of sulphur nutrition on some aspects of amino acid metabolism and the diagnosis of sulphur deficiency in crop plants. Bull Nat. Inst. Agric. Sci. Japan. No. 29, 49–106.
235. Suzuki, S. and Shiga, H. 1956 Studies on physical and chemical characteristics of Akiochi paddy soils. Part 2. Relation between production of free hydrogen sulfide and Akiochi degree. Bull. Chugoku Agr. Exp. Sta. 3, 69–80.
236. Swaby, R.J. and Vitolins, M.I. 1968 Sulphur oxidation in Australian soils. Trans. 9th Int. Congr. Soil Sci. 4, 673–681.
237. Tabatabai, M.A. and Bremner, J.M. 1972 Forms of sulfur and carbon, nitrogen and sulfur relationships in Iowa soils. Soil Sci. 114, 380–386.
238. Tadano, T. and Tanaka, A. 1970 Studies on the iron nutrition of rice plants. (3). Iron absorption affected by potassium status of the plant. J. Sci. Soil Manure 41, 142–148 [in Japanese].
239. Takahashi, Y. 1970 Nutrition of the rice plant in relation to the occurrence of 'akagare' disease. Bull. Nat. Inst. Agric. Sci. D 21, 1–59.
240. Takahasi, J. 1964 Natural supply of nutrients in relation to plant requirements. In: IRRI, The Mineral Nutrition of the Rice Plant, pp. 271–294. John Hopkins Press, Baltimore.
241. Takai, Y. 1969 The mechanism of reduction in paddy soil. Japan Agric. Res. Quarterly 4, 20–23.
242. Takai, Y. 1978 Reduction mechanism of paddy soils. In: Suidendojogaku, pp. 23–55. Kawaguchi, K. (ed.), Kodansha, Tokyo [in Japanese].
243. Takai, Y. and Asami, T. 1962 Formation of methyl mercaptan in paddy soils. I. Soil Sci. Plant Nutr. 8, 40–44.
244. Takai, Y. and Kamura, T. 1969 The mechanism of reduction in waterlogged paddy soil. Folia Microbiol. 11, 304–313.
245. Takai, Y., Koyama, T., and Kamura, T. 1969 Effects of rice plant roots and percolating water in the reduction process of flooded paddy soil in pot. V. Microbial metabolism in reduction process of paddy soils. Nippon Dojohiryo Gaku Zasshi, 40, 15–19 [in Japanese].
246. Takai, Y. and Tezuka, C. 1971 Sulfate-reducing bacteria in paddy and upland soils. Nippon Dojohiryo Gaku Zasshi 42, 145–151 [in Japanese].
247. Takijima, Y. 1963 Studies on behavior of the growth inhibiting substances in

paddy soils with special reference to the occurrence of root damage in the peaty paddy field. Bull. Nat. Inst. Agri. Sci. B. 13, 117–252.

248. Takijima, Y. 1964 Studies on the mechanism of root damage of rice plant in the peat paddy fields. 1. Root damage and growth inhibitory substances found in the peaty and peat soil. Soil Sci. Plant Nutr. 10, 231–238.

249. Takijima, Y. 1965 Studies on the mechanism of root damage of rice plant in the peat paddy fields. 2. Status of roots in the rhizosphere and the occurrence of root damage. Soil Sci. Plant Nutr. 11, 204–211.

250. Takijima, Y., Wijayaratna, H.M.S. and Seneviratne, C.J. 1970 Nutrient deficiency and physiological disease of lowland rice in Ceylon. IV. Remedy for bronzing disease of rice. Soil Sci. Plant Nutr. 16, 17–23.

251. Tanaka, A., Mulleriyawa, R.P., and Yasu, T. 1968 Possibility of hydrogen sulfide induced iron toxicity of the rice plant. Soil Sci. Plant Nutr. 14, 1–6.

252. Tanaka, A. and Yoshida, S. 1970 Nutritional disorders of the rice plant in Asia. International Rice Research Institute. Tech. Bull. 10, 51 pp.

253. Temple, K.L. and Delchamps, E.W. 1953 Autotrophic bacteria and the formation of acid in bituminous coal mines. Appl. Microbiol. 1, 255–258.

254. Tisdale, S.L. and Nelson, W.L. 1966 Soil Fertility and Fertilizers. Macmillan, New York.

255. Tomlison, T.E. 1957 Changes in sulfide containing mangrove soil on drying and their effect upon the suitability of the soil for the growth of rice. Emp. J. Exp. Agric. 25, 108–118.

256. Tomlison, T.E. 1957 Relationship between mangrove vegetation, soil texture and reaction of surface soil, after empoldering saline swamps in Sierra Leone. Trop. Agric. (Trin.) 34, 41–50.

257. UNESCO (United Nations Educational, Scientific, and Cultural Organization). 1974 FAO-UNESCO Soil Map of the World, 1:500,000. Vol 1, Legend, Paris, 59 pp.

258. USDA (United States Department of Agriculture). 1975 Soil Conservation Service, Soil Survey Staff. Soil taxonomy: a basic system of soil classification for making and interpreting soil surveys. USDA Agric. Handbook 436. US Government Printing Office Washington DC, 260 pp.

259. Vámos, R. 1958 Hydrogen sulphide, the cause of bruzone (akiochi) disease of rice. Soil Plant Food 4, 37–40.

260. Vámos, R. 1959 'Brusone' disease of rice in Hungary. Plant and Soil 11, 65–77.

261. Vámos, R. 1964 The release of hydrogen sulfide from mud. J. Soil Sci. 15, 103–109.

262. Vámos, R. 1965 The biological effects of sulphate reduction in waterlogged soils. Agrokemia es Talajtan. 14, 115–116.

263. Vámos, R. 1966 The effect of H_2S on the IAA content of the rice plant and on the development of its adventitious roots. Acta Biol. 12, 67–72.

264. Vámos, R. 1968 The factors of the root-rot of the rice plant. Il Riso, 189–199.

265. Vámos, R. and Kovacs, E. 1962 A study on the Eh_7 conditions of the rhizosphere in rice varieties resistant and susceptible to 'bruzone'. Acta Agron. Acad. Sci. Hung. 11, 369–382.

266. Vámos, R. and Köves, E. 1972 Role of the light in the prevention of the poisoning action of hydrogen sulphide in the rice plant. J. Appl. Ecol. 9. 519–525.

267. Van Breemen, N. and Pons, L.J. 1978 Acid sulfate soils and rice. In: IRRI, Soils and Rice, pp. 739–761. Los Baños, Philippines.

316

268. Van Raalte, M.H. 1941 On the oxygen supply of rice roots. Ann. Bot. Garden Buitenzorg 50, 99—114.
269. Van Raalte, M.H. 1944 On the oxidation of the environment by the roots of rice (*Oryza sativa* L.). Ann. Bot. Garden Buitenzorg 54, 15—34.
270. Venkateswarlu, J., Subbiah, B.V. and Tamhane, R.V. 1969 Vertical distribution of forms of sulfur in selected rice soils of India. Indian J. Agric. Sci. 39, 426—431.
271. Vishniac, W. and Santer, M. 1957 The *Thiobacilli*. Bacteriol. Rev. 21, 195—213.
272. Vitolins, M.I. and Swaby, R.J. 1969 Activity of sulphur-oxidising microorganisms in some Australian soils. Aust. J. Soil Res. 7, 171—183.
273. Wakao, N. and Furusaka, C. 1972 A new agar plate method for the quantitative study of sulfate-reducing bacteria in soil. Soil Sci. Plant Nutr. 18, 39—44.
274. Wakao, N. and Furusaka, C. 1973 Distribution of sulfate-reducing bacteria in paddy-field soil. Soil Sci. Plant Nutr. 19, 47—52.
275. Wakao, N. and Furusaka, C. 1976 Presence of micro-aggregates containing sulfate-reducing bacteria in a paddy-field soil. Soil Biol. Biochem. 8, 157—159.
276. Wakao, N. and Furusaka, C. 1976 Influence of organic matter on the distribution of sulfate-reducing bacteria in a paddy-field soil. Soil Sci. Plant Nutr. 22, 203—205.
277. Wakao, N., Hattori, T. and Furusaka, C. 1973 Study on the distribution patterns of sulfate-reducing bacteria in a paddy-field soil by Iδ-index. Soil Sci. Plant Nutr. 19, 201—203.
278. Wang, C.H. 1978 Sulphur fertilization of rice. Sulphur in Agriculture 2, 13—16.
279. Wang, C.H. 1979 Sulphur fertilization of rice — Diagnostic techniques. Sulphur in Agriculture 3, 12—15 and 18.
280. Wang, C.H., Liem, T.H. and Mikkelsen, D.S. 1976 Sulfur deficiency — a limiting factor in rice production in the lower Amazon basin. I. Development of sulfur deficiency as a limiting factor for rice production. IRI Research Institute Bulletin No. 47, 46 pp.
281. Wang, C.H., Liem, T.H. and Mikkelsen, D.S. 1976 Sulfur deficiency — a limiting factor in rice production in the lower Amazon Basin. II. Sulfur requirement for rice production. IRI Research Institute Bulletin No. 48, 30 pp.
282. Watanabe, I. and Furusaka, C. 1980 Microbial ecology of flooded rice soils. In: Advances in Microbial Ecology, 4, pp. 125—168. Alexander, M. (ed.), Plenum Pub. Corp, New York.
283. Weir, R.G. 1975 The oxidation of elemental sulphur and sulphides in soil. In: Sulphur in Australasian Agriculture, pp. 40—49. McLachlan, K.D. (ed.), Sydney University Press, Sydney.
284. Wen Lin Yuan and Ponnamperuma, F.N. 1966 Chemical retardation of the reduction of flooded soils and the growth of rice. Plant and Soil 25, 347—360.
285. Widdel, F. and Pfennig, N. 1977 A new anaerobic, sporing, acetate-oxidising, sulfate-reducing bacterium, *Desulfotomaculum* (emend.) *acetoxidans*. Arch. Microbiol. 112, 119—122.
286. Williams, C.H. 1974 The chemical nature of sulphur in some New South Wales soils. In: Handbook on Sulphur in Australian Agriculture, pp. 16—23. McLachlan, K.D. (ed.), CSIRO Australia, Melbourne.
287. Williams, R.A. and Hoare, D.S. 1972 Physiology of a new facultatively autotrophic thermophilic *Thiobacillus*. J. Gen. Microbiol. 70, 555—566.

288. Winfrey, M.R. and Zeikus, J.G. 1977 Effect of sulfate on carbon and electron flow during microbial methanogenesis in freshwater sediments. Appl. Environ. Microbiol. 33, 275–281.
289. Wolfe, R.S. and Pfennig, N. 1977 Reduction of sulfur by Spirillum 5175 and syntrophism with *Chlorobium*. Appl. Environ. Microbiol. 33, 427–433.
290. Wolin, M.J. 1975 Interactions between H_2-producing and methane-producing species. Proc. Symp. Microbial Production and Utilization of Gases. Akademie der Wissenschaften zu Göttingen, 141–150.
291. Wu. M.M.H., Wu, C.S., Chiang, M.H. and Chou, S.F. 1972 Microbial investigations on the suffocation disease of rice in Taiwan. Plant and Soil 37, 329–344.
292. Yamada, N. and Ota, Y. 1958 Study on the respiration of crop plants (8) Effect of hydrogen sulphide and lower fatty acids on root respiration of rice. Proc. Crop Sci. Soc. Japan 27, 155–160.
293. Yamane, I. and Koseki, K. 1976 Effect of some oxidants upon sulfate reduction and methane formation under submerged conditions. J. Sci. Soil Manure Japan 47, 58–62.
294. Yamane, I. and Sato, I. 1961 Metabolism in muck paddy soil. Part 3. Role of soil organic matter in the evolution of free hydrogen sulfide in water-logged soils. Rep. Inst. Agr. Res. Tohoku University Ser. D12, 73–86.
295. Yamane, I. and Sato, K. 1968 Initial drop of oxidation-reduction potential in submerged air-dried soils. Soil Sci. Plant Nutr. 14, 68–72.
296. Yamane, I. and Sato, K. 1970 Plant and soil in a lowland rice field added with forage residues. Rep. Inst. Agr. Res. Tohoku University 21, 79–101.
297. Yang, S.C. 1972 Effect of magypsum urea on rice. J. Taiwan Agric. Res. 21, 215–220.
298. Ye Goung, Khin Win and Win Htin 1978 Rice soils of Burma. In: IRRI, Soils and Rice, pp. 57–71. Los Baños, Philippines.
299. Yong Haw Shin 1978 Rice soils of Korea. In: IRRI, Soils and Rice, pp. 179–191. Los Baños, Philippines.
300. Yoshida, T. 1975 Microbial metabolism of flooded soils. In: Soil Biochemistry, 3, pp. 83–122. Paul, E.A. and McLaren, A.D. (eds.), Marcel Dekker, New York.
301. Yoshida, T. 1978 Microbial metabolism in rice soils. In: IRRI, Soils and Rice, pp. 445–463. Los Baños, Philippines.
302. Yoshida, S. and Chaudhry, M.R. 1979 Sulfur nutrition of rice. Soil Sci. Plant Nutr. 25, 121–134.
303. Zeikus, J.G. 1977 The biology of methanogenic bacteria. Bacteriol. Rev. 41, 514–541.
304. Zhilina, T.N. and Zavarzin, G.A. 1973 Trophic relationship between *Methanosarcina* and its associates. Mikrobiologiya 42, 235–241.
305. Zsoldos, F. 1959 Changes in free amino acids in rice seedlings due to the effect of factors rendering them susceptible of the browning disease. Acta Biol. 5, 71–76.
306. Zsoldos, F. 1962 Nitrogen metabolism and water regime of rice plant affected by 'brusone' disease. Plant and Soil 16, 269–283.

Taxonomic index

Subject index